System Dynamics and Control with Bond Graph Modeling

System Dynamics and Control
with Bond Graph Modeling

Javier A. Kypuros, Ph.D.
Department of Mechanical Engineering
University of Texas – Pan American
Edinburg, TX

March 7, 2013

CRC Press
Taylor & Francis Group
6000 Broken Sound Parkway NW, Suite 300
Boca Raton, FL 33487-2742

© 2013 by Taylor & Francis Group, LLC
CRC Press is an imprint of Taylor & Francis Group, an Informa business

No claim to original U.S. Government works

Printed on acid-free paper
Version Date: 20130204

International Standard Book Number-13: 978-1-4665-6075-8 (Hardback)

Library of Congress Cataloging-in-Publication Data

Kypuros, Javier A.
  System dynamics and control with bond graph modeling / author, Javier A. Kypuros.
  pages cm
  Includes bibliographical references and index.
  ISBN 978-1-4665-6075-8 (hardcover : alk. paper)
  1. Automatic control--Mathematics. 2. Automatic control--Graphic methods. 3. Bond graphs. I. Title.

TJ213.K97 2013
629.8072'8--dc23   2012050908

Visit the Taylor & Francis Web site at
http://www.taylorandfrancis.com

and the CRC Press Web site at
http://www.crcpress.com

To my family: Claudia, Nyssa, and Andrik

Contents

Preface

This text was written for those who teach/learn System Dynamics using bond graphs. It was designed from the onset to be undergraduate focused. As such the material is a synergy of bond graph concepts and a traditional System Dynamics curriculum. The intent was to present bond graphs as a more integrated tool within System Dynamics. Moreover, the intention was to develop a text that makes the bond graph methodology more accessible to undergraduate Engineering students. The text is purposefully designed to cater to a third or fourth year undergraduate Engineering student.

The prerequisites for this text include Linear Algebra, Ordinary Differential Equations, Engineering Mechanics, and Electrical Circuits. The reader may also benefit from exposure to Fluid Mechanics and Thermodynamics. The text includes ten chapters and can be divided into four parts – bond graph modeling, mathematical representations, analysis, and automatic control.

The first part, Chapters 1-3, focuses on synthesizing models of dynamic systems using the bond graph methodology. The first chapter is an introduction to model decomposition based on energy formalisms. This chapter introduces the reader to the needs and uses of System Dynamics. Further, the reader is shown how systems can be broken down into more basic components. The reader learns that energy and power are two unifying concepts that exist regardless of the energy domain. They are introduced to bonds, signals, and causality. In Chapter 2, the reader learns about basic bond graph elements and discovers how elements can be categorized based on their energy usage. Chapter 2 also explains the differences between linear and nonlinear systems. Since the text primarily targets linear systems, the chapter covers linearization. Finally, in Chapter 3 students learn how to synthesize bond graphs and derive differential equations.

Chapters 4-6 compose the second part which addresses state-space and transfer function representations of dynamic systems. Chapter 4 covers state-space representations. The readers learn how to convert the system of first-order differential equations derived using the bond graph model into a state-space representation. At this point, students are introduced to the use of MATLAB® for numerically simulating basic dynamic responses. (MATLAB is used through the second, third, and fourth parts of the text.) Using the state-space model, they simulate impulse, step, and ramp responses. Through Chapter 4, the focus is on time-domain representations. In Chapter 5, the students review Laplace Transforms in preparation for impedance methods and transfer function representations, which are introduced in Chapter 6. Chapter 6 covers some unique material on impedance bond graphs. The chapter is influenced by the unpublished work of Beaman and Paynter (Beaman and Paynter 1993). Students also learn how to simulate responses using transfer function representations derived from impedance bond graphs.

In Chapters 7-8, the text covers analysis, including time- and frequency-domain methods. Time-domain analysis is discussed in detail in Chapter 7. The chapter covers the characteristics of first- and second-order systems and explains how higher-order systems have responses that are the combinations of lower-order responses. In preparation for the introduction to classical control methods in Chapter 9, students learn about pole-zero analysis. Chapter 8 covers the frequency domain. Students discover methods used in the analysis of vibrating systems, AC circuits, and the like. They learn about concepts including phasor analysis, modal analysis, and bode plots.

In the final part, Chapters 9-10, the reader is introduced to automatic control methods that vary from traditional to more modern approaches. The focus in Chapter 9 is on classical methods of designing lead-lag and proportional-integral-derivative type compensators. The methods covered include the root locus method and bode plot analysis. Chapter 10, the final chapter, is intended to introduce the reader to more modern state-space approaches. The students learn how to assess controllability and observability using the state-space model. They discover how to design a compensator through pole placement and linear quadratic regulation. Additionally, they learn about the use and design of state observers.

Each chapter includes three types of exercises. The first are "review" problems or questions. These measure the reader's mastery or understanding of content. The second are simply "problems" like those commonly found in textbooks. These are designed so that the reader practices the concepts introduced in each chapter. They assess the reader's ability to implement the concepts to solve problems similar to examples in the chapter. The third and final type of exercises are "challenges." Challenges are semi-open-ended problems that require the student to transfer the knowledge learned in a manner more indicative of "real-world" problems. These challenges do not have one right answer; rather, the solution may vary based on the assumptions made by the student.

Though much of the text is unique material, it brings to bear concepts from Bond Graph Modeling, System Dynamics, and Automatic Controls. It has been influenced by several works including those of Paynter, Beaman, Ogata, Karnoop, Margolis, and Rosenburg (Beaman and Paynter 1993; Paynter 1961; Ogata 2002, 2004; Karnopp, Margolis, and Rosenberg 2000). This work has also been impacted by the course notes of Raul G. Longoria and Joseph Beaman from the Mechanical Engineering Department at the University of Texas at Austin. It is based primarily on my personal course notes for the System Dynamics course in the Mechanical Engineering Department at the University of Texas-Pan American where I have been teaching dynamic systems related courses for over a decade.

Additional material is available from the CRC Web site:

http://www.crcpress.com/product/isbn/9781466560758

MATLAB and Simulink are registered trademarks of The MathWorks, Inc. For product information, please contact:

The MathWorks, Inc.
3 Apple Hill Drive
Natick, MA 01760-2098 USA
Tel: (508)647-7000
Fax: (508)647-7001
E-mail: info@mathworks.com
Web: www.mathworks.com

Javier A. Kypuros

Author Biography

Javier A. Kypuros is a professor of mechanical engineering at the University of Texas-Pan American. He earned his PhD in mechanical engineering in 2001 at the University of Texas at Austin under the guidance of Dr. Raul Longoria, and his BSE in mechanical engineering from Princeton University in 1996. He has taught courses in the area of dynamic systems and control for over a decade, and has been awarded numerous grants from the National Science Foundation to develop and implement pedagogical innovations for engineering mechanics and system dynamics curricula.

Nomenclature

$1(t)$ unit step or Heaviside function

$[K]$ stiffness matrix

$[M]$ mass matrix

α angular acceleration (rad/s^2) or attenuation factor

β rotational damping constant (N-m-s/rad) or attenuation factor

δ relative displacement (m)

Γ hydraulic momentum (N-s/m^2)

γ phase margin (degrees)

κ rigidity (N-m/rad)

λ flux linkage (V-s) and eigenvalues

$\mathbf{u}(t)$ input vector

$\mathbf{x}(t)$ state vector

$\mathbf{y}(t)$ output vector

$\mathscr{C}$ state controllability matrix

$\mathscr{O}$ observability matrix

$\mathscr{P}$ power (W)

$\mathscr{L}$ the Laplace transform

ω angular velocity (rad/s)

ω_d	damping frequency (rad/s)
ω_n	natural frequency (rad/s)
$\overline{\mathscr{C}}$	output controllability matrix
$\bar{z}$	complex conjugate
ϕ	phase angle (rad)
τ	torque (N-m)
θ	angular displacement (rad)
$\tilde{\delta}(t)$	unit impulse or Dirac delta function
ζ	damping ratio
b	damping constant (N-s/m)
C	generalized compliance and capacitance (F)
C_f	hydraulic compliance (m^3/Pa)
D	PID step response delay time
e	effort and voltage (V)
E_m	electromotive force (V)
F	force (N)
f	flow
$f(t)$	a generic function of time
$G(s)$	transfer function
h	angular momentum (N-m-s)
I	generalized inertia
i	current (A)
I_f	kg/m^4
J	inertia (kg-m^2)

k	spring stiffness (N/m)
K_a	static acceleration error constant
K_{cr}	critical gain for PID tuning
K_c	overall compensator gain
K_D	PID derivative gain
K_I	PID integral gain
k_m	motor constant (V-s/rad or N-m/A)
K_P	PID proportional gain
K_p	static position error constant
K_v	static velocity error constant
L	inductance (H)
m	mass (kg)
M_p	maximum percent overshoot
P	pressure (Pa)
p	generalized momentum, translational momentum (N-s), or pole
Q	volumetric flow rate (m^3/s) or quality factor
q	generalized displacement and charge (C)
R	generalized resistance and electrical resistance (Ω)
R_f	Pa-s/m^3
T_{cr}	critical time for PID tuning
T_D	PID derivative time constant
t_d	delay time (s)
T_I	PID integral time constant
T_p	period of oscillation

t_p	peak time (s)
t_r	rise time (s)
t_s	settling time (s)
TR	transmissibility
V	volume (m^3)
v	velocity (m/s)
x	displacement (m)
Z	impedance
z	complex number or zero

Chapter 1

Introduction to System Dynamics

System Dynamics is the study of physical mechanisms and devices that exhibit dynamic behavior that can be characterized through the use of mathematical models for the purposes of analysis, design, and automation. It is a discipline that draws on a variety of subjects in order to examine whole mechanisms, devices, and physical phenomena from a "systems" perspective. A system is composed of more basic elements that function together to form the whole. The basic components of that system can be described by various mathematical models or constitutive relations. In order to understand the whole, we must understand how the more basic elements interface and interact. Thus far in your engineering studies, you have learned separately a variety of subjects that are brought to bear together and expanded upon in System Dynamics. Here are a series of questions to consider:

▷ What aspects of the system must you consider?

▷ What tools, models, or information will you need?

▷ How do you design or optimize the system to ensure reasonable performance?

▷ What metrics do you use to measure the system's performance?

▷ How do you automate or control a system?

▷ Where do you get started?

You probably have more questions than answers at this stage. As this course and text proceed, you will evolve an understanding of tools and skills employed in examining dynamic systems. Each chapter purposely builds upon prior chapters and looks ahead to preview connections with chapters to come. The book can be broken down into three major parts – Chapters 1-6, Chapters 7 and 8, and Chapters 9 and 10. The first part focuses on synthesis of system models and the derivation of differential and algebraic equations that represent the dynamic response. In the second part we explore methods for analyzing dynamic systems including metrics to quantify their characteristics. The last part is an introduction to classical and modern methods for designing automatic controls for linear, time-invariant systems.

1.1   Introduction

Engineering often entails the design, analysis, or control of dynamic systems – systems with physical states that vary with time. The term "*dynamic*" or "dynamical" is often used in reference to forces that generate motion but can be more generally used to refer to physical efforts that cause states to vary with time. The term "system" can refer to an interdependent group of components that coordinate to form a unified device, mechanism, or process. Note that a "system" can vary in complexity and can include subsystems. A system can be as simple as mass-spring-damper and LCR circuit or as complex as a vehicle and an airplane. "System Dynamics" draws on a variety of engineering specialties to form a unified approach to study dynamic systems. The rather generalized nature of the individual words and the resulting term purposely encompasses a large variety of physical systems. However, in this book we will focus on simple to moderately complex mechanical, electrical, hydraulic, and electromechanical systems.

The discipline of System Dynamics focuses on the synthesis of mathematical models to represent dynamic responses of physical systems for the purpose of analysis, design, and/or control. *Analysis* is the mathematical inspection of the dynamic characteristics of a system and its responses to inputs. *Design* is the methodical synthesis of a system or selection of its parameters to meet specified criteria. *Control* is the use of sensor and actuators to automate a process or system. The resulting mathematical models often take the form of sets of differential equations. Thus, many of the tools used for analyzing systems of differential equations including Linear Algebra and Laplace Transforms are also commonly used in System Dynamics. In this chapter, we will begin to explore how a physical system is decomposed to enable mathematical representation. We will briefly discuss the uses of System Dynamics and will introduce a unified graphical approach for synthesizing dynamic system models.

At the beginning of each chapter, we will introduce the "objectives" and "outcomes" for that chapter. As you review the material in each chapter, you should keep in mind the objectives and outcomes, and once completing a chapter, it is important that you assess how well you have met these criteria. "Objectives" are the scholastic skills or knowledge we seek or aim to obtain and master. "Outcomes" are potentially measurable results or products that if achieved lend to accomplishing the objectives.

The objectives and outcomes for this chapter follow.

▷ *Objectives.* In this chapter you will

1. come to a deeper understanding of the art of System Dynamics and the purpose it serves in the design, analysis, and control of physical systems, and

2. begin to conceptualize how a system is broken down into subsystems and components to enable synthesis of mathematical models that represent the dynamics.

▷ *Outcomes.* After completing this chapter, you will

1. be able to identify systems, subsystems, and components,

2. be able to identify potential applications of system dynamics in design and analysis of mechanisms, and

3. recognize and/or recall concepts used to represent dynamic responses in other Engineering courses you are or have previously taken.

Each chapter will end with a summary that highlights and reinforces the more important content. Furthermore, each chapter includes review questions, problems, and challenges. The review questions are short answer problems intended to focus your reading and help you recall the more important concepts. The "problems" are the typical exercises used to practice the fundamentals covered in each chapter. The challenges are semi-open-ended problems that are based on "real world" applications.

1.2   System Decomposition and Model Complexity

One of the underlying principles of the discipline of System Dynamics is the study of engineering problems at the system level. *Systems* are a combination of components or subsystems acting together to perform a task or objective (Ogata 2004). Moreover, the components or subsystems may operate in various energy domains (i.e., mechanical, electrical, hydraulic, etc.). As such, the principles, concepts, or theories used to describe the physical phenomena that govern the dynamics of such systems come from a variety of engineering disciplines. Amongst those are Dynamics, Electronics, Heat Transfer, and Fluid Mechanics. To derive a System Dynamics model, a system must be decomposed into subsystems and individual components – discrete physical entities.

Take for example the ball-and-beam electromechanical system depicted in Figure 1.1. The system is composed of a ball, beam, lever arm, crank, and motor. The ball rolls along a beam surface that is attached to a pivot at one end and is actuated at the other through a crank turned by a permanent magnet direct current (PMDC) motor. The problem incorporates aspects of translational 2D motion that can be modeled using Newtonian Mechanics; the ball and beam each experience generalized planar motion. A DC motor converts a voltage or current to generate a torque which rotates the crank. The physics dictating the motor dynamics involves Electromagnetics and Electrodynamics. Though at first this problem may be seemingly simple, once one breaks down the system into simpler discrete components, it becomes evident that the combination of translational and rotational mechanics with electrical dynamics makes this problem more than trivial. To derive a mathematical model of the system we will need to employ fundamentals from a variety of engineering subjects. Other engineering problems have physical interactions of various energy domains that are far more complicated.

System Dynamic models are mathematical depictions of physical dynamics and can vary in complexity depending on the intended use. This may be best explained through an example. Take for instance the vehicle suspension depicted in Figure 1.2 (a). The suspension can be modeled as depicted in either Figure 1.2 (b) or (c). Vehicle suspensions are often modeled to predict and analyze vibrations induced by road inputs – displacements of the road profile. Such models are used to optimize spring rates and damping constants to improve suspension performance over varying terrain. Additionally, these models can be used to improve ride comfort for the vehicle occupants and to

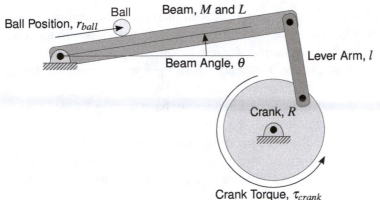

Figure 1.1: *Ball-and-beam system.*

determine stresses generated within the frame and chassis due to vibrations transferred through the tires and suspension.

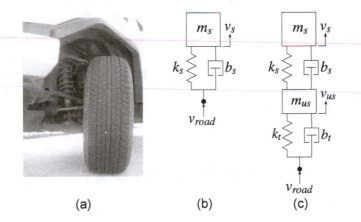

Figure 1.2: *(a) A vehicle suspension, (b) a simplified mass-spring-damper representation, and (c) a less simplified mass-spring-damper representation.*

The simple model that is depicted in Figure 1.2 (b) incorporates the compliance of the spring, the damping due to the shock absorber, and the mass of the front quarter of the vehicle. This model is adequate for predicting the vertical motion of the front quarter for relatively slow (low frequency) cyclic inputs. An example might be a series of regularly spaced speed bumps the vehicle travels over at slow to moderate speeds as it might do in a parking lot or residential area. At such speeds, the suspension absorbs the majority of the energy and accounts for most of the dynamics. Imagine, however, that the vehicle travels over a series of road markers embedded on the surface of a highway between lanes. At highway speeds, the suspension receives more frequent vertical displacements from the highway markers. Though of relatively small amplitude in comparison to the speed bumps, the frequency of the repeated displacements from the markers is much higher. At such frequencies, the vehicle suspension does not have sufficient time to respond and absorb the energy. Instead, the tires absorb much of the energy and play a more critical role in predicting the dynamic response and induced vibrations. Figure 1.2 (c) depicts a more elaborate mass-spring-damper model that includes the compliance of the tire (k_t), the damping of the tire (b_t) due to the compressed air within, and the combined unsprung mass (m_{us}) of the wheel, tire, brake rotor, and brake assembly. This model can be used to predict the dynamic response within the tire and the resulting effect on the suspension dynamics.

Thus, System Dynamics models can vary in complexity for a number of reasons. Amongst those are the interaction of multiple energy domains, the number of individual components involved, and the need for a more detailed representation. One of the skills essential to the discipline of System Dynamics is the ability to effectively dissect a system into subsystem and more basic components to better facilitate mathematical representations of the physics entailed. Throughout this book you will learn how to do just that. As you do, you will grow experience that will help you later as an engineer determine how to break a problem done into manageable parts, characterize those parts, and determine the level of complexity necessary to model the whole.

1.3   Mathematical Modeling of Dynamic Systems

System Dynamics makes use of mathematical representations of physical phenomena, concepts, and theories to formulate models that can be analytically or numerically solved. Mass-spring-damper schematics like those depicted in Figure 1.2 (b) and (c) are often used to derive mathematical models for vibrating structures. Equations are derived using concepts and theories from Newtonian Mechanics such as Newtons Second Law and Hooke's Law. Kirchhoff's Current and Voltage Laws are used to analyze dynamic currents and voltages within circuits. The Lumped Capacity Method is used in Heat Transfer to predict transient temperature response under special conditions. In Fluid Mechanics, differential equations such as the Conservation of Momentum equation are used to describe both steady and unsteady fluid flow.

Throughout your coursework you have been consistently exposed to models for describing dynamic responses. In this course, you will bring to bear this knowledge and expand on it to develop a unified approach or methodology for solving a large class of dynamic problems. Take, for example, the ball-and-beam problem previously introduced.

As we have already discussed, the ball–and–beam system depicted in Figure 1.1 is modeled using general planar motion including Newton's Second Law in translation and rotation,

$$\sum \mathbf{F} = m\mathbf{a} \text{ and } \sum \mathbf{M} = J\boldsymbol{\alpha}.$$

Additionally, the motor driving the crank is represented using Kirchhoff's Voltage Law and some basic Electromechanics,

$$e_{in}(t) - L\frac{di}{dt} - Ri - E_m = 0.$$

The electromotive force, E_m, can be related to the rotor angular velocity, ω_m, through a motor constant,

$$E_m = k_m \omega_m.$$

As depicted in Figure 1.3, the motor, like many, uses a gear train to multiply the torque. Basic planar kinetics can be used to show that

$$\frac{\tau_1}{\tau_2} = \frac{N_1}{N_2} = \frac{\omega_2}{\omega_1}.$$

As evidenced by this seemingly simple mechanism, a number of principles from a variety of courses often come into play when developing mathematical models of systems. A few such courses you may have or are currently taking include but are not limited to Dynamics, Fluid Mechanics, Electronic Circuits, and Heat Transfer.

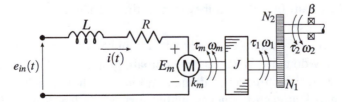

Figure 1.3: *A permanent magnet direct current (PMDC) motor schematic.*

In each model we devise existing *inputs*, *outputs*, and *states*. Inputs are variables that change the condition of the dynamic system and can include things such as external forces or torque, voltages sources, pressure sources, etc. They are sources that "drive" the system. The outputs are variables that are measured or observed to assess the dynamic condition of the system. In systems, these are often the variables that are measured using sensors. The states are variables that are used to mathematically model the dynamic behavior of the system. Figure 1.4 is a "black box" depiction of the system model and illustrates the relationships between the inputs, states, and outputs. We will discuss these relations in detail in later chapters starting with Chapter 2.

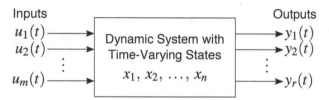

Figure 1.4: *"Black box" representation of dynamic system model.*

1.4 Analysis and Design of Dynamic Systems

Mathematical models are often used to analyze dynamic systems. Analysis has been identified as one of the primary reasons for modeling dynamic systems. We analyze systems to determine what makes them function or respond as they do so that we might be able to alter or optimize their responses. Dynamic systems are often characterized in the time or frequency domain. Time domain analysis entails evaluating how parameters alter the dynamic response as measured in time. Take for example the step response illustrated in Figure 1.5 (a). The spring rate and damping constant affect how the system follows the input. In particular they impact the overshoot and settling time of the step response.

Alternatively, systems exposed to cyclic inputs are often characterized in the frequency domain. Whereas time domain graphs plot how dynamic signals vary with time, frequency domain graphs plot how much of the signal response exists at each frequency within a range or *bandwidth* of frequencies. Frequency domain plots are often used to identify amplitude contribution present at each frequency.

Let us return to the suspension model we previously discussed. Imagine you wish to characterize a vehicle suspension. How might we characterize the response? What attributes typify a "good" response? As has already been mentioned, the spring stiffness and damping constant affect the dynamic response, in particular the maximum amplitude and the time the response takes to settle down – the settling time. One approach, as depicted in Figure 1.5 (a), is to measure the time response and to observe how changing the spring or damper constant affects the peak amplitude and the settling time. This is part of time domain analysis. If, however, you are more interested in how the amplitude varies with the frequency of a cyclic input like that illustrated in Figure 1.5, how would your analysis be different?

In Chapters 7 and 8, we will discover several forms of time and frequency domain analyses and will learn how each of these are used to optimize design of dynamic systems.

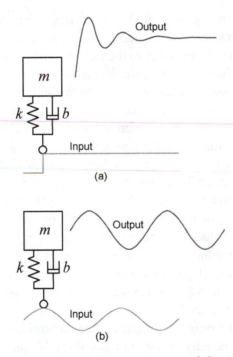

Figure 1.5: *Simple suspension (a) step response and (b) sinusoidal response.*

1.5 Control of Dynamic Systems

Analysis enables us to develop controls that alter the response of systems in order to achieve more desirable dynamics. Though varying system parameters can often garner the desired response, sometimes control systems are needed to meet design criteria. As illustrated in Figure 1.6, an automatic control system is a device or mechanism that measures the dynamic state you wish to steer (the output), compares it to the desired value you have specified (the reference), and enacts an effort (the input) to minimize the error between the two – the reference and measured output.

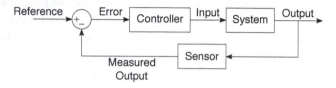

Figure 1.6: *Block diagram for a generic control system.*

Examples of such systems include vehicle cruise controls, heating and air conditioning temperature controls, and elevator controls. Control systems are employed in automated machinery in the manufacturing industry and human-machine interfaces such as modern surgical robots. Though most of the mathematical constructs and theories developed to facilitate design or synthesis of automatic controls have been around for less than a century, inventors have been developing mechanisms to automate processes for hundreds of years. Some of the first examples are the Greek and Roman water clocks, like that illustrated in Figure 1.7, which are attributed to Ctesibios (circa 270 B.C.) which incorporated a feedback system with complex gearing (Bennet 1996; Wikipedia 2009c). Heron of Alexandria is attributed with the development, in the first century A.D., of several automata or self-operating machines including the first recorded steam engine, a wind-powered organ, and a fire engine (Bennet 1996; Wikipedia 2009a). More recently in 1788, James Watt developed the first centrifugal governor also referred to as the fly-ball governor. The governor depicted in Figure 1.8 was used to control a throttle valve that regulated the flow of steam.*

Most of the commonly used mathematical control theorems were developed during the last century, including Bode Plots, Nyquist Stability Criteria, Lyapunov Stability Theory, Robust Control, etc. Many of the methods commonly used today were developed in response to needs resulting from World War II, the Cold War, and the Space Race (Wikipedia 2009b).

In System Dynamics, mathematical models are utilized to conduct analysis. The resulting analysis enables controller design. Controls generate efforts that alter the dynamic response in order to meet dynamic system design

---

*. The photograph in Figure 1.8 is attributed to Dr. Mirko Junge and can be found at Wikimedia Commons (Commons 2009).

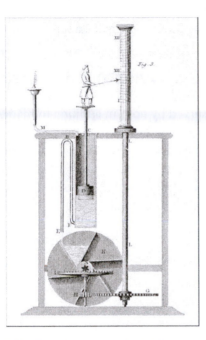

Figure 1.7: *Ctesibios water clock (Commons 2009).*

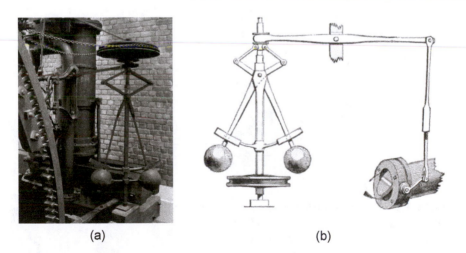

(a)   (b)

Figure 1.8: *(a) Illustration and (b) schematic of a fly-ball or centrifugal governor (Commons 2009).*

criteria. In the latter part of this book, we will learn some of the classi-cal approaches used for designing control systems including the Root Locus Method and the Bode Plot Frequency Response Method. Additionally, we will learn about more modern state-space approaches.

1.6 Diagrams of Dynamic Systems

As you may have observed by now, in addition to mathematical methods, we also have a number graphical approaches for representing dynamic systems. Some of these are depicted in Figure 1.9. By now, you have used free body diagrams to sum forces and moments and circuit schematics to analyze currents and voltages. Hydraulic circuits are also depicted using diagrams in many ways similar to circuit schematics.

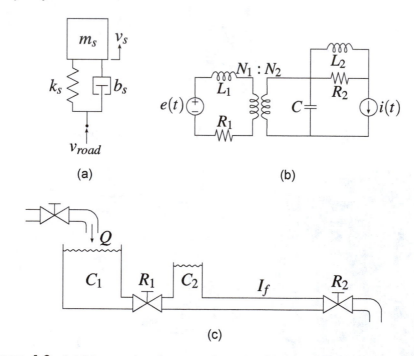

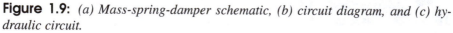

Figure 1.9: *(a) Mass-spring-damper schematic, (b) circuit diagram, and (c) hydraulic circuit.*

Schematics and diagrams provide us a means of cataloging information and parameters in a concise manner. Moreover, they provide an instrument by which we can dissect systems into more basic components. In System Dynamics, schematics and diagrams fulfill the old adage "a picture is worth a thousand words." In System Dynamics, however, a diagram is worth "a thousand" mathematical equations and parameters. They often illustrate more than equations alone can do. They can be used to garner an intuition for the physical response. As such, in this book, we will learn about a particular graphical approach developed to be more unified and generic than free body and circuit diagrams.

1.7 A Graph-Centered Approach to Modeling

Let us return to the ball-and-beam system previously depicted in Figure 1.1. What do the mechanical and electrical subsystems have in common? How do you model each? This points to a necessity to oftentimes analyze engineering problems from a more unified or generalized perspective. The unifying factor common amongst all the subsystems of the ball-and-beam is the fact that each has associated with it *energy* and *power*.

Bond Graphs are a graphical approach for diagramming the distribution and flow of power and energy within a dynamic system. They were originally developed in 1959 by the late Dr. Henry M. Paynter at MIT (Paynter 1961). More specifically, *bond graphing* is a unified graphical approach for modeling the storage, dissipation, and transformation of energy within a dynamic system. The structure of the bond graph has been designed to facilitate systematic derivation of differential equations governing the dynamic response of the system model. The bond graph approach leads directly to a computer simulation of the dynamic response because it delineates the causal relationships between elements or components of the model. That is, the bond graph accounts for the input/output relations between elements and subsystems of the model. The approach is generalized because it operates on the premise of energy and is not limited to a single energy domain.

In the chapter that follows, we will exploit this unifying factor to derive basic energy and power relations common to components in various energy domains.

1.8 Power and Energy Variables

In various energy domains, there are variables that when multiplied together give power. For example, force and velocity when multiplied give power (i.e., $\mathscr{P}(t) = F(t)v(t)$), and voltage and current also multiply to give power (i.e., $\mathscr{P}(t) = e(t)i(t)$). Power is generally defined as the multiplication of an *effort* (a force-like variable), $e(t)$, and a *flow* (a velocity-like variable), $f(t)$,

$$\mathscr{P}(t) = e(t)f(t). \tag{1.1}$$

By recalling which variables when multiplied give power, efforts and flows for various energy domains can be readily identified as illustrated in Table 1.1. Effort and flow can be related to the generalized energy variables *momentum*, $p(t)$, and *displacement*, $q(t)$. The generalized momentum is defined as the integral of the effort

$$p(t) \equiv \int e(t)\, dt\,, \tag{1.2}$$

which means then that the effort is the time rate of change of the momentum,

$$e(t) = \frac{dp}{dt} = \dot{p}. \tag{1.3}$$

This is readily evident when examining the mathematical formulation of Newton's Second Law. Newton's Second Law of motion as originally published in *Philosophiæ Naturalis Principia Mathematica* is as follows:

> The second law states that the net force on a particle is equal to the time rate of change of its linear momentum, $p = mv$,

$$\sum F = \frac{dp}{dt} = \frac{d}{dt}(mv) = m\dot{v} + \dot{m}v\,.$$

This simplifies to the more recognized form

$$\sum F = m\dot{v} = ma$$

if the mass, m, is constant as is the case for rigid bodies. The generalized displacement is defined as the integral of the flow,

$$q(t) \equiv \int f(t)\, dt, \tag{1.4}$$

which implies that the flow is the time rate of change of displacement,

$$f(t) = \frac{dq}{dt} = \dot{q}. \tag{1.5}$$

Translational velocity, v, for example, is the time rate of change of translational displacement, x,

$$v = \frac{dx}{dt}.$$

Momentum and displacement variables for various energy domains are given in Table 1.2.

Table 1.1: *Effort and flow variables.*

Domain	Effort	Flow	Power
Translational	F, force (N)	v, velocity (m/s)	$\mathscr{P} = Fv$
Rotational	τ, torque (N-m)	ω, angular velocity (rad/s)	$\mathscr{P} = \tau\omega$
Electrical	e, voltage (V)	i, current (A)	$\mathscr{P} = ei$
Hydraulic	P, pressure (Pa)	Q, flowrate (m³/s)	$\mathscr{P} = PQ$

Table 1.2: *Momentum and displacement variables.*

Domain	Momentum	Displacement
Translational	p, linear (N-s)	x, displacement (m)
Rotational	h, angular (N-m-s)	θ, angle (rad)
Electrical	λ, flux linkage (V-s)	q, charge (C)
Hydraulic	Γ, hydraulic (N-s/m²)	V, volume (m³)

As you may recall, energy is the time integral of power,

$$E(t) = \int \mathscr{P}(t)\,dt = \int e(t)f(t)\,dt. \tag{1.6}$$

By substituting Equations 1.3 or 1.5 into Equation 1.6, the potential energy equation,

$$E(t) = \int e(t)\frac{dq}{dt}\,dt = \int e(q)\,dq, \tag{1.7}$$

and kinetic energy equation,

$$E(t) = \int \frac{dp}{dt}f(t)\,dt = \int f(p)\,dp, \tag{1.8}$$

are derived in terms of the energy and power variables (i.e., e, f, p, and q). The relationships are summarized by the Tetrahedron of State (Paynter 1961; Karnopp, Margolis, and Rosenberg 2000) shown in Figure 1.10.

In this text, we categorize elements based on what they do with energy. In a system, energy can be *stored* (as potential or kinetic), *dissipated, supplied* (from and external source), *converted,* and *summed.* The following chapter describes the basic bond graph elements and how they are categorized based on energy and power. Generally speaking, a *constitutive relation* is the relation between two physical quantities, for instance the relation between a spring force and the spring deflection or the relation between the voltage drop across a resistor and the current flowing through it. As the tetrahedron of state illustrates, effort and flow variables can be related through integral,

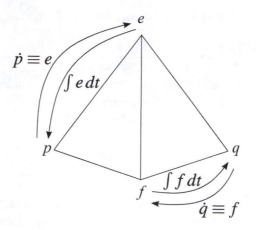

Figure 1.10: *Tetrahedron of State (redrawn from Paynter (1961) and Karnopp, Margolis, and Rosenberg (2000)).*

derivative, or algebraic relations. Newton's Second Law, for example, relates the sum of forces (effort) to the acceleration or the time rate of change of the velocity (flow).

1.9   Bonds, Ports, Signals, Inputs, and Outputs

In bond graphs, elements are connected by *bonds* through *power ports*. Each bond represents an effort-flow pair that when multiplied give the power entering or leaving the attached ports. In basic one-port elements, the power is usually assumed to be flowing from the system into the element. Take for example two elements or subsystems "A" and "B" as shown in Figure 1.11. Note that, in a bond graph, the half-arrow indicates the assumed direction of positive power flow and the bar at the beginning or end of the half-arrow indicates at which port *effort* is the input. Effort is an input to the port attached to the end of the bond that has the bar, and flow is an input at the opposite end of the bond. Thus, the input/output *causality*[†] of the bond graph in Figure 1.11 (a) is illustrated by the block diagram in Figure 1.11 (b), and the block diagram in Figure 1.11 (d) illustrates the input/output causality of the bond graph in Figure 1.11 (c). Note that for each case there is a distinct input

---

[†]. "Causality" is a term that refers to the cause-and-effect nature of mathematical functions. A simple function has an input and output, e.g.,

$$y = f(x),$$

where x is the input and y is the output.

and output, and only one variable (effort or flow) can be an input at any given port. Also make special note that any single port has two associated *signals* – effort and flow – and that one of these signals, as illustrated in Figure 1.11 (e) and (f), is the input to the port while the other is the output from the port. Notice that the signals which represent either an effort or a flow use a full arrow instead of a half arrow.

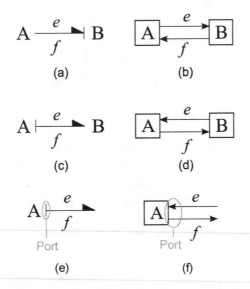

Figure 1.11: *(a) A bond graph with effort and power passing from element A to element B; (b) a block diagram with effort and power passing from A to B; (c) a bond graph with flow and power passing from A to B; (d) a block diagram with flow and power passing from A to B; (e) the port on a bond graph; and (f) the port on a block diagram.*

1.10   Word Bond Graphs

Word bond graphs are utilized to visualize how a system is broken down in terms of power usage and consumption. Such graphs identify the basic components of a system and their power connections. Words are used to specify the individual components pertinent to the dynamic system model. Bonds connect individual components which exchange power. Each bond is labeled with an effort-flow pair. By sketching a word bond graph, one practices decomposing a dynamic system into pertinent parts and begins to understand how the interaction of these components results in models used to analyze and simulate the dynamic system.

The process for sketching a word bond graph is rather straightforward. To sketch a word bond graph, follow these guidelines:

1. *Identify the basic components of the system.* Write in words the individual components that compose the overall system.

2. *Connect interacting components.* Connect components that exchange energy with bonds.

3. *Identify the effort-flow pairs.* For each bond connecting two components, determine the effort and flow that are exchanged between the attached components.

Figure 1.12 illustrates the word bond graph for a rear-drive vehicle. The basic components of the drivetrain are the engine, transmission, driveshaft, differential, and wheels. The wheels convert rotational energy into translation of the vehicle. The vehicle is acted upon by external forces due to wind and rolling resistance and gravity impedes the motion when climbing a hill. Notice that bonds connect the elements that exchange energy and effort-flow pairs are labeled on each bond.

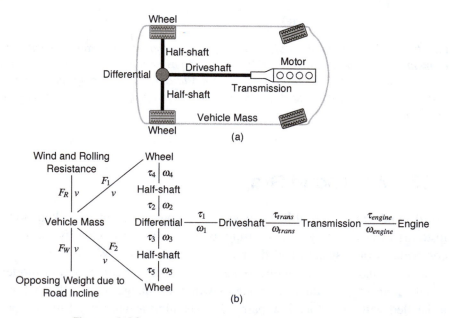

Figure 1.12: *Word bond graph for a rear-drive vehicle.*

1.11 Summary

▷ System decomposition consists of breaking down the system into basic components that can be readily characterized to enable modeling and mathematical representation.

▷ Model complexity depends on use of the system representation and the necessary accuracy of the predicted dynamic response.

▷ Mathematical models of dynamic systems commonly take the form of differential and algebraic equations. As such, mathematical methods such as Linear Algebra and Laplace Transforms are commonly used to analyze and design dynamic systems.

▷ Analysis is used to study dynamic systems and to characterize their responses. It can be used to determine how changes in system parameters vary the dynamic response.

▷ When the desired dynamic response cannot be achieved through parametric optimization, *automatic control systems* can be employed to compensate and alter the system response. Automatic controls are used to modify and/or automate dynamic responses.

▷ Schematics and diagrams are commonly used to model dynamic systems. Free body and circuit diagrams are just two examples.

▷ Bond graphing is a generalized graphical approach for diagramming the power distribution within a dynamic system.

▷ Bonds connect individual components and represent the transfer of power between those components.

▷ Signals are effort or flow variables that transfer between elements.

▷ Inputs are variables, external to the system, that drive the dynamics.

▷ States are dynamic variables that require differential equations to model their dynamic variation.

▷ Outputs are variables or states that are measured or observed.

▷ Word bond graphs are word-based sketches that illustrate the decomposition of the system into more basic components and help derive the constitutive interactions between the individual components.

1.12   Review

R1-1 Describe system decomposition in your own words.

R1-2 Explain the differences between inputs, outputs, and states.

R1-3 Give at least two examples of mathematical models of constitutive relations for basic electrical, mechanical, and/or hydraulic components.

R1-4 Provide three reasons for modeling dynamic systems.

R1-5 Provide examples of graphical models you have used in previous courses to represent physical systems or phenomena.

R1-6 Explain the relation between power and energy.

R1-7 How are effort and momentum generally related?

R1-8 How are flow and displacement generally related?

R1-9 Explain the relations and distinctions between bonds, ports, and signals. In particular, explain how a power bond differs from a signal bond. Also detail how ports are realized in block diagrams as opposed to bond graphs.

R1-10 Describe, in your own words, the process for sketching a word bond graph.

1.13   Problems

The following problems are intended to help you practice or review the basic skills introduced or discussed in this chapter, including system decomposition, mathematical models, and schematics.

P1-1 *Inputs and Outputs.* Consider the systems depicted in Figure 1.9; identify the inputs for each system. Also, determine potential outputs of interest. That is, what dynamic variables might one be interested in knowing or predicting?

P1-2 *Newton's Second Law for Rotation.* Recall the formulation of Newton's Second Law discussed in Section 1.8; derive the general form for Newton's Second Law for rotation and provide the formulation for rigid bodies in rotation.

P1-3 *Constitutive Relations.* Figure 1.13 illustrates a simple mass-spring system. The constitutive relation for the mass was previously described in Section 1.8. Determine the constitutive relation for the spring. A simple series circuit is shown in Figure 1.14. Provide the constitutive relations for the resistor and inductor.

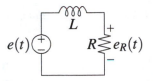

Figure 1.13: *Simple mass-spring system.*

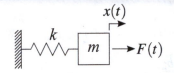
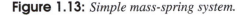

Figure 1.14: *A simple series circuit with voltage source, resistor, and capacitor.*

P1-4 *Mass-Spring-Damper Systems.* Decompose the mass-spring-damper systems depicted in Figure 1.15 into basic elements or subsystems. For each, sketch a word bond graph. Be sure to label the effort-flow pair for each bond. Note that $x_1(t)$, $x_2(t)$, and $x(t)$ are labels while $v(t)$ is a velocity source and $F(t)$ is an external force.

P1-5 *Rotational System.* Resolve the rotational systems in Figure 1.16 into their basic elements. Draw the word bond graph for each, and be sure to include the effort-flow labels on the bonds. An input torque τ_m is supplied for (a) and (b), while (c) has two input torques.

P1-6 *Electric Circuits.* For the circuits shown in Figure 1.17, identify and list the basic components. Provide word bond graphs for each. In (a) and (b), the voltage drop across the capacitor and the current through the inductor are labeled e_c and i_L, respectively.

P1-7 *Hydraulic Circuits.* Figure 1.18 depicts several hydraulic circuits. Identify the basic components that compose each and sketch their word bond graphs.

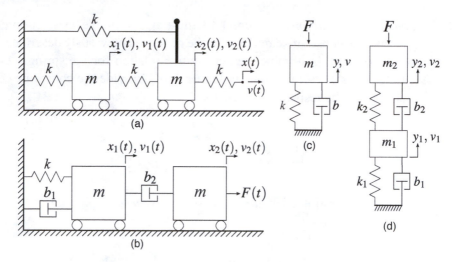

Figure 1.15: *Mass-spring-damper systems.*

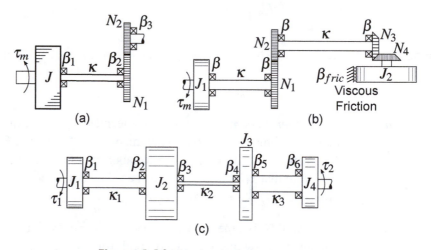

Figure 1.16: *Mechanical rotation systems.*

P1-8 *Permanent Magnet Direct Current (PMDC) Motor with Gear Reduction.* A schematic for an electromechanical system was illustrated in Figure 1.3. Use that schematic to discern the individual components of a geared PMDC motor. Note that $i(t)$ τ_m, ω_m, τ_1, ω_1, τ_2, and ω_2 are labels for local variables and do not indicate external sources. Draw a word bond graph.

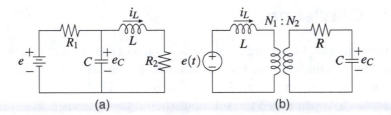

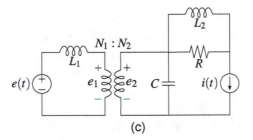

(c)

Figure 1.17: *Electric circuits.*

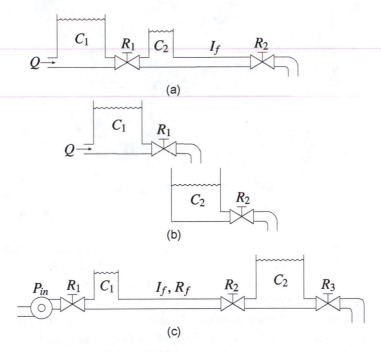

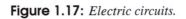

Figure 1.18: *Hydraulic circuits.*

1.14 Challenges

The problems that follow are intended for you to expand on the knowledge you garnered in this chapter. Use the basic skills you learned or practiced in this chapter to complete the challenges that follow. Challenges are intended to be somewhat open-ended and solutions should vary from person to person.

C1-1 *A Toy Electric Car.* A toy car is shown in the Figure 1.19. As the figure shows, the car has batteries, an electric drivetrain, gears, and more. Can you decompose the vehicle into more basic components? Identify the parts of the toy car and sketch a word bond graph.

Figure 1.19: *A toy electric buggy.*

C1-2 *A Basketball.* If released from rest at a height, a basketball bounces repeatedly, rebounding each time to a lesser height. That is, every time it bounces freely with no external input it will rebound with less energy to a lower maximum height than the previous rebound. The ball absorbs energy and releases some of the absorbed energy after impacting the ground. Depict a system of elements that would react in a similar manner. What are the basic components of that system and how are they similar to aspects of the basketball?

C1-3 *A Quarter-Car Suspension Model.* List all the primary components of the front corner suspension depicted in Figure 1.2 (c) and sketch the word bond graph. Be sure to include the wheel and tire combination. *Hint:* An inflated tire, like the basketball, will rebound repeatedly with decreasing energy if released from a height above ground.

C1-4 *An Electric Drill.* A battery-operated drill depicted in Figure 1.20 is used to drive a screw into the wall. The drill is made up of a battery, PMDC motor, gear reduction, and drill bit. As the screw is driven into the wall, friction impedes the twisting motion. Identify the basic components that make up the drill and draw the word bond graph. Hint: You can reference Problem P1-8 to help decipher some of the elements.

Figure 1.20: *A battery-powered drill.*

Chapter 2

Basic Bond Graph Elements

Recall the challenges from the previous chapter. Imagine that for each challenge you must identify the pertinent components of the vehicle and determine how they interface. You will need to devise mathematical representations that characterize the dynamics of the individual components. Ultimately, the individual representations must interface to result in an overall model of the system that can be used to analyze the dynamics and later design a control. Some things to consider are:

▷ What kind of "model" would be appropriate?

▷ What tools will be used to predict and analyze the dynamics?

▷ What components need to be modeled or considered?

▷ How do the individual components and subsystems interact with one another?

▷ How do you characterize individual components to facilitate an overall model?

▷ Are there any underlying concepts that will help categorize and represent the individual components?

2.1 Introduction

The previous chapter began to introduce the idea of decomposing dynamic systems into more basic elements. These basic elements can be categorized and mathematically represented based on how they use or convert energy.

The mathematical representations result in input-output relations. It was also discussed how systems operate in various power domains (i.e., mechanical, electrical, hydraulic, etc.) and how many operate in multiple power domains.

In this chapter, you will learn to identify and categorize basic components in dynamic systems based on the role each plays in terms of energy usage and conversion. Moreover, you will learn how this energy usage or conversion can result in mathematical input-output relations. These relations are based on energy formalisms and are designed to account for power flow and distribution in the dynamic system.

At first glance, this approach may seem foreign and unconventional. However, as you will begin to see in this chapter and come to appreciate more over the next few chapters, this approach provides a generalized method to analyze a great variety of systems and is particularly useful when deriving mathematical models for moderately complex systems that operate in a variety of power domains.

The objectives and outcomes for this chapter are detailed below.

▷ *Objectives:*

1. To be able to decompose dynamic systems into more basic elements that facilitate mathematical modeling,

2. To understand how energy usage and conversion are utilized to categorize basic elements, and

3. To be able to model the constitutive relations of basic dynamic system elements based energy.

▷ *Outcomes:* Upon completion of this chapter, you will

1. be able to categorize basic elements of dynamic systems,

2. be able to derive the mathematical input-output relations for each element,

3. begin to draw analogies between basic elements in different power domains, and

4. begin to understand the flow of "mathematical information" within dynamic system models.

2.2 Basic 1-Port Elements

As will be shown, the constitutive relationships for basic *1-port* elements relating the energy and power variables identify whether that element stores potential energy (C-element), stores kinetic energy (I-element), or dissipates energy (R-element).

2.2.1 R-Elements

R-elements are basic, 1-port elements that *dissipate energy*. They are characterized by a constitutive relationship that *directly relates effort to flow*,

$$e = \Phi_R(f) \text{ or } f = \Phi_R^{-1}(e), \tag{2.1}$$

where $\Phi_r(f)$ and $\Phi_R^{-1}(e)$ are in general nonlinear functions. As shown in Figure 2.1, examples of 1-port, R-elements are a mechanical damper, a rotational roller bearing, an electric resistor, and a hydraulic valve.

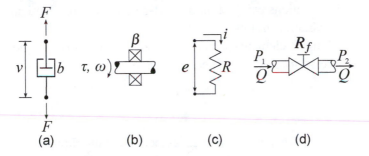

(a)   (b)   (c)   (d)

Figure 2.1: *One-port, R-element examples including (a) a mechanical damper, (b) a roller bearing, (c) an electric resistor, and (d) a hydraulic valve.*

Take for example a simple mechanical damper. The classic constitutive relation for a mechanical damper is that the force is proportional to the velocity differential, v, through the damping constant, b,

$$F_b = bv.$$

Note that in the above equation, the mechanical effort, F_b, is directly related to the mechanical flow, v. Though bearings are designed to minimize losses due to friction, they, like dashpots or dampers, dissipate energy and can be modeled in a similar fashion. The torque lost due to rotational damping in a bearing can be represented as

$$\tau = \beta\omega$$

where β is the rotational damping constant. What about a simple resistor? The voltage drop across a resistor, e, is proportional to the current flowing through the resistor, i, through the resistance, R:

$$e = iR.$$

Table 2.1: *Linear R-element resistance.*

Domain	Parameters	SI Units
Generalized	$R = e/f$	N/A
Translational	b, damping constant	N-s/m
Rotational	β, rotational damping constant	N-m-s/rad
Electrical	R, resistance	Ω (ohms)
Hydraulic	R_f, hydraulic resistance	Pa-s/m^3

Here the electrical effort, e, is directly related to the electrical flow, i. Similar constitutive relations can be found for R-elements in mechanical rotation systems and hydraulic circuits.

In general, for R-elements with a linear constitutive relation, the generalized element resistance, R, is defined as

$$R \equiv \frac{e}{f}. \tag{2.2}$$

Figure 2.2 shows the two causalities associated with Equation 2.3 (flow-in/effort-out (a) and (b)) and Equation 2.4 (effort-in/flow-out (c) and (d)). For linear R-elements, depending on the causality, Equation 2.1 simplifies to

$$e = Rf \tag{2.3}$$

or

$$f = \frac{e}{R}. \tag{2.4}$$

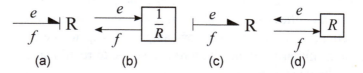

(a)   (b)   (c)   (d)

Figure 2.2: *Linear, 1-port, R-element representations including (a) a bond graph with effort in, (b) a block diagram with effort in, (c) a bond graph with flow in, and (d) a block diagram with flow in.*

Table 2.1 gives the parameters and SI units for the resistances in each energy domain. The table gives commonly used symbols for each parameter. Note that a linear hydraulic resistance occurs in pipes with laminar flow. The hydraulic resistance of a pipe, R_f, can be derived from the head loss equation (Fox and McDonald 1992). For laminar flow in a circular pipe

$$Q = \frac{\pi D^4}{128\mu L}\Delta P \Rightarrow R_f = \frac{\Delta P}{Q} = \frac{128\mu L}{\pi D^4}$$

where μ, L, and D are the kinematic viscosity, pipe length, and pipe diameter, respectively.

2.2.2   C-Elements

C-elements are basic elements that *store potential energy*. They are characterized by a constitutive relationship that *directly relates effort to generalized displacement*,

$$q = \Phi_C(e) \text{ or } e = \Phi_C^{-1}(q). \tag{2.5}$$

Note that in the above equation, generally either displacement is a nonlinear function of effort, or effort is a nonlinear inverse function of displacement. Figure 2.3 shows several commonly occurring 1-port C-elements.

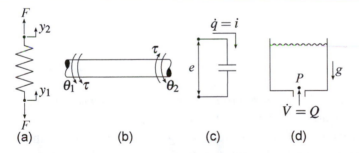

Figure 2.3: *One-port, C-element examples including (a) a spring, (b) a torsion shaft, (c) an electric capacitor, and (d) a hydraulic accumulator.*

Each of the elements in Figure 2.3 can be modeled by a linear constitutive relation linking the respective effort and displacement pairs. The spring force, for example, is

$$F = k(y_1 - y_1) = k\delta,$$

where k is the spring stiffness and Δy is the spring deflection or the relative displacement between the two ends of the spring. A torsion shaft has a similar relation between the torque and relative angular displacement,

$$\tau = \kappa(\theta_2 - \theta_1) = \kappa\Delta\theta.$$

The electrical analog of a spring is a capacitor. Whereas we are accustomed to characterizing a spring in terms of its stiffness, a capacitor is characterized in terms of its capacitance. The capacitor voltage is

$$e = \frac{q}{C}$$

where C is the capacitance. In all three cases, the effort is related directly to the respective displacement. If we draw further analogies between the constitutive relations of the spring and capacitor, capacitance is the analog of compliance or the inverse of stiffness (i.e., $1/k$).

Recall that bonds in a bond graph each have an associated effort and flow (not displacement). Additionally, each port attached to a bond has a distinct causality. For any 1-port, either the effort *or* the flow is the input, not both. On a 1-port C-element, if the effort is the input as depicted in Figure 2.4 (a) and (b), the displacement, q, is calculated from the effort, e, and the time derivative of the displacement is taken to generate the flow, f. If the element is linear, the generalized capacitance, C, is defined as

$$C \equiv \frac{q}{e}.$$

The flow-effort relation is

$$f = \frac{d}{dt}(Ce) = \frac{dq}{dt}, \tag{2.6}$$

and this formulation is referred to as *derivative causality* because a time derivative is used to relate flow to effort.

Alternatively, if, as depicted in Figure 2.4 (c) and (d), the flow is the input to the 1-port C-element, the flow, f, is integrated with respect to time to generate the displacement, q, and the displacement is used to calculate the the effort, e. If the element is linear, the displacement is divided by the capacitance, C, to calculate the effort,

$$e = \frac{q}{C}.$$

Here the effort-flow relation is

$$e = \frac{\int f dt}{C} = \frac{q}{C}, \tag{2.7}$$

and because an integral is used to relate effort to flow this formulation is referred to as *integral causality*.

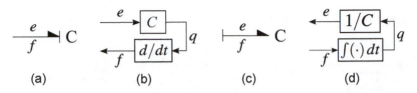

Figure 2.4: *Linear, 1-port, C-element representations including (a) a derivative causality bond graph, (b) a derivative causality block diagram, (c) an integral causality bond graph, and (d) an integral causality block diagram.*

Table 2.2: *Linear C-element compliance.*

Domain	Parameters	SI Units
Generalized	$C = q/e$	N/A
Translational	$1/k$, spring compliance	m/N
Rotational	$1/\kappa$, rotational compliance	rad/N-m
Electrical	C, capacitance	F (farad)
Hydraulic	C_f, hydraulic capacitance	m^3/Pa

C-element parameters for various energy domains are given in Table 2.2. Commonly used symbols and SI units are provided for each. Note a commonly occurring hydraulic C-element with linear constitutive relation is the accumulator depicted in Figure 2.3 (d). The hydraulic compliance can be readily derived by recalling the equation for hydrostatic pressure in a column of fluid

$$P = \rho g h = \rho g \frac{V}{A} \;\Rightarrow\; C_f = \frac{V}{P} = \frac{A}{\rho g}$$

where V is the volume of the fluid column with constant cross-sectional area A and fluid density ρ. Though the systems we will analyze are not hydrostatic, the associated volumetric rates are relatively slow-changing so that the hydrostatic assumption suffices.

2.2.3 I-Elements

I-elements are basic elements that *store kinetic energy*. They are characterized by a constitutive relationship that directly relates momentum to flow,

$$p = \Phi_I(f) \text{ or } f = \Phi_I^{-1}(p). \qquad (2.8)$$

Note that in the above equation, generally either momentum is a nonlinear function of flow, or flow is a nonlinear inverse function of momentum. Several common 1-port I-elements are depicted in Figure 2.5.

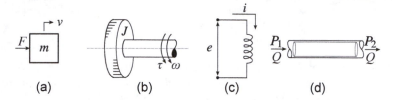

(a)	(b)	(c)	(d)

Figure 2.5: *One-port I-element examples including (a) a mass, (b) a rotational inertia, (c) an inductor, and (d) a slug of fluid.*

Each of the components depicted in Figure 2.5 can be modeled by a linear constitutive relation linking the respective momenta and flows. The momentum of a rigid body is

$$p = mv = \int F \, dt.$$

where v is the velocity, m is the mass of the rigid body, and F the force. The angular momentum of a rotational inertia, J, is

$$h = J\omega = \int \tau \, dt$$

where ω is the angular velocity of the rotational inertia J and τ the torque. For both the mass and rotational inertia, the respective momenta are directly proportional to the velocities. The equivalent of momentum in electric circuits is flux linkage (Chua 1971),

$$\lambda = Li = \int e \, dt,$$

where L is inductance, i current, and e the voltage. If the I-element is modeled by a linear constitutive relation, the generalized inertia, I, is

$$I \equiv \frac{p}{f}.$$

Like C-elements, 1-port I-elements have integral and derivative causalities. For rigid bodies, the mass is constant and the above equation simplifies to its more commonly used form. As previously stated, the force, $F(t)$, and moment, $p(t)$, are related through integration ($\int (\cdot) dt$) or differentiation (d/dt):

$$F = \dot{p} \quad \text{or} \quad p = \int F \, dt.$$

If the force, $F(t)$, is known it can be integrated, $\int(\cdot)dt$, to determine the momentum, $p(t)$, and then the momentum can be related to the velocity, $v(t)$. Conversely, if the velocity, $v(t)$, is known, it can be related to the momentum, $p(t)$, and the result can be differentiated, d/dt, to determine the force, $F(t)$.

For a linear I-element in integral causality (refer to Figure 2.6 (a) and (b)), the effort, e, is integrated to determine the momentum, p, and the momentum is divided by inertia, I, to calculate the flow,

$$f = \frac{\int e\,dt}{I} = \frac{p}{I}. \tag{2.9}$$

Otherwise, if the I-element has derivative causality (Figure 2.6 (c) and (d)), the flow, f, is multiplied by the inertia, I, to calculate the momentum, p, and the time derivative of the momentum is taken to generate the effort, e,

$$e = \frac{d}{dt}(If) = \frac{dp}{dt} = \dot{p}. \tag{2.10}$$

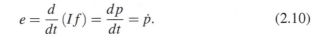

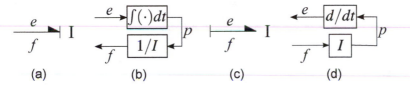

(a)   (b)   (c)   (d)

Figure 2.6: *Linear, 1-port, I-element representations including (a) an integral causality bond graph, (b) an integral causality block diagram, (c) a derivative causality bond graph, and (d) a derivative causality block diagram.*

Parameters, commonly used symbols, and units for linear I-elements in various energy domains are provided in Table 2.3. To derive the linear fluid inertia, we begin by applying Newton's Second Law to a control volume for a slug of fluid like that depicted in Figure 2.5 (d). Assuming incompressible flow and a constant control volume, the summation of forces on the slug of fluid in a pipe of length l results in the following:

$$\sum F_x = P_2 A - P_1 A = m\dot{v} + 0$$

$$\Delta P A = (\rho V)\frac{\dot{Q}}{A} = (\rho l)\frac{dQ}{dt}$$

where the volume is $V = lA$. The volumetric flow rate, $Q = dV/dt$, is directly proportional to the average fluid velocity, v, assuming a constant cross-sectional area, A (i.e., $Q = vA$ and $\dot{Q} = \dot{v}A$). To derive the hydraulic inertia,

Table 2.3: *Linear I-element inertance.*

Domain	Parameters	SI Units
Generalized	$I = p/f$	N/A
Translational	m, mass	kg
Rotational	J, rotational inertia	kg-m^2
Electrical	L, inductance	H (henrys)
Hydraulic	I_f, hydraulic inertia	kg/m^4

we solve the above relation for the pressure differential and use Equation 2.9,

$$\Delta P = \frac{\rho l}{A}\frac{dQ}{dt} \Rightarrow \Delta P\, dt = \frac{\rho l}{A} dQ \Rightarrow \int \Delta P\, dt = \frac{\rho l}{A} Q \Rightarrow I_f = \frac{\int \Delta P\, dt}{Q} = \frac{\rho l}{A}.$$

Hence, the hydraulic momentum is

$$\Gamma = \int \Delta P\, dt.$$

2.2.4   R-, C-, and I-Elements and the Tetrahedron of State

A modified tetrahedron of state (Paynter 1961; Karnopp, Margolis, and Rosenberg 2000) is depicted in Figure 2.7. It delineates which 1-port elements are associated with each edge including the edges that directly connect effort to flow (R-element), effort to displacement (C-element), and momentum to flow (I-element). The tetrahedron illustrates the causal relations of the R-, C-, and I-elements detailed in the previous sections. If we follow the edges of the tetrahedron, we can recall the relations between effort and flow. If, for example, we start with flow, f, at the bottom corner, we can integrate it to determine the displacement, q, at the right corner and then algebraically determine the effort as a function of the displacement, $e(q)$. This path represents integral causality for a C-element. In a similar fashion we can follow paths that represent the derivative causality for a C-element, the integral and derivative causalities for an I-element, and the causalities for an R-element.

Also, note Tables 2.4, 2.5, and 2.6 summarize the linear 1-port constitutive relations including the integral and derivative causalities for C- and I-elements.

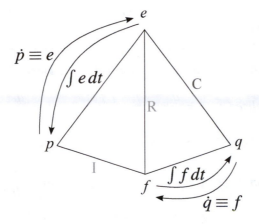

Figure 2.7: *Tetrahedron of state and the 1-port elements (redrawn from Paynter (1961) and Karnopp, Margolis, and Rosenberg (2000)).*

Table 2.4: *Constitutive relations for linear, 1-port, R-elements.*

Domain	Effort-In Relation	Flow-In Relation
Generalized	$f = \dfrac{e}{R}$	$e = Rf$
Translational	$v = \dfrac{F}{b}$	$F = bv$
Rotational	$\omega = \dfrac{\tau}{\beta}$	$\tau = \beta\omega$
Electrical	$i = \dfrac{e}{R}$	$e = Ri$
Hydraulic	$Q = \dfrac{P}{R_f}$	$P = R_f Q$

2.2.5   Effort and Flow Sources

Only two basic 1-port elements remain – the *effort source* and the *flow source*. These are used for idealized representations of elements such as voltage or current sources, vibration shakers, external forces, pressure sources, etc. As the names suggest, an effort source specifies an effort into the system, and a flow source specifies a flow into the system. Several examples are illustrated

Table 2.5: *Constitutive relations for linear, 1-port, C-elements.*

Domain	Linear	Integral	Derivative
Generalized	$e = \dfrac{q}{C}$	$e = \dfrac{\int f \, dt}{C}$	$f = \dfrac{d}{dt}(Ce)$
Translational	$F = kx$	$F = k \int v \, dt$	$v = \dfrac{d}{dt}\left(\dfrac{F}{k}\right)$
Rotational	$\tau = \kappa\theta$	$\tau = \kappa \int \omega \, dt$	$\omega = \dfrac{d}{dt}\left(\dfrac{\tau}{\kappa}\right)$
Electrical	$e = \dfrac{q}{C}$	$e = \dfrac{\int i \, dt}{C}$	$i = \dfrac{d}{dt}(Ce)$
Hydraulic	$P = \dfrac{V}{C_f}$	$P = \dfrac{\int Q \, dt}{C_f}$	$Q = \dfrac{d}{dt}(C_f P)$

Table 2.6: *Constitutive relations for linear, 1-port, I-elements.*

Domain	Linear	Integral	Derivative
Generalized	$f = \dfrac{p}{I}$	$f = \dfrac{\int e \, dt}{I}$	$e = \dfrac{d}{dt}(If)$
Translational	$v = \dfrac{p}{m}$	$v = \dfrac{\int F \, dt}{m}$	$F = \dfrac{d}{dt}(mv)$
Rotational	$\omega = \dfrac{h}{J}$	$\omega = \dfrac{\int \tau \, dt}{J}$	$\tau = \dfrac{d}{dt}(J\omega)$
Electrical	$i = \dfrac{\lambda}{L}$	$i = \dfrac{\int e \, dt}{L}$	$e = \dfrac{d}{dt}(Li)$
Hydraulic	$Q = \dfrac{\Gamma}{I_f}$	$Q = \dfrac{\int P \, dt}{I_f}$	$P = \dfrac{d}{dt}(I_f Q)$

in Figure 2.8. The ideal voltage and current sources obviously specify effort and flow, respectively, into a circuit. The external force is treated as an effort source, and the velocity input which could come from a shaker, for example, would be a flow source.

As illustrated in Figure 2.9, each source type has a single distinct causality. Effort sources specify effort out of the source element and into the system.

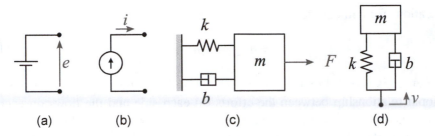

Figure 2.8: *Effort and flow source examples including (a) a voltage source, (b) a current source, (c) an external force input, and (d) an external velocity input.*

Conversely, flow sources specify flow out of the source element and into the system. Additionally, it is almost always assumed that positive power flow is from the effort or flow source to the attached element or junction.

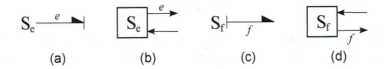

Figure 2.9: *Effort and flow source element representations including (a) an effort source bond graph, (b) an effort source block diagram, (c) a flow source bond graph, and (d) a flow source block diagram.*

2.3 Basic 2-Port Elements

Thus far, we have discussed energy storing elements, energy dissipating elements, and sources. There are elements that neither dissipate nor store energy but do transmit energy from one element or junction to another while often times interfacing between various energy domains. There are two basic types of *2-port* elements used to transmit energy. They are the *transformer* (–TF–) and the *gyrator* (–GY–). Ideally, in each, power is conserved (i.e., the power going in is equal to the power coming out).

2.3.1 Transformers (TF-Elements)

Ideal *transformers* are characterized by a constitutive relation that directly relates the effort at the first port, e_1, to the effort at the second port, e_2, and the flow at the second port, f_2, to the flow at the first port, f_1. The constitutive relations for a linear ideal transformer are

$$e_1 = ne_2 \text{ and } nf_1 = f_2 \tag{2.11}$$

where m is referred to as the *transformer modulus* and indicates the proportional relationship between the efforts on each side and the flows on each. Note, however, the constitutive relation could generally be nonlinear.

Examples of ideal transformers are depicted in Figure 2.10. In each example, the efforts on each side are proportional to one another as well as the flows. Take for example the gear pair (Figure 2.10 (b)). The output torque, τ_2, is proportional to the input torque, τ_1, through the gear ratio. Similarly, the angular velocity on the torque input side, ω_1, is proportional to the angular velocity on the torque output side, ω_2, through the gear ratio, also:

$$\frac{\tau_1}{\tau_2} = \frac{\omega_2}{\omega_1} = n$$

where n is the gear ratio, N_1/N_2. This can be confirmed through basic kinematics and by recognizing that, for an ideal gear pair, the power in one side is the power out the other.

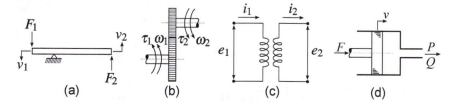

(a)   (b)   (c)   (d)

Figure 2.10: *TF-element examples including (a) a rigid lever, (b) a gear pair, (c) an electrical transformer, and (d) a hydraulic ram.*

Similar relations can be derived for the lever arm, electric transformer, and hydraulic piston. If the lever arm (Figure 2.10 (a)) is divided into two lengths – l_1 measured from the pivot out to F_1 and l_2 measured from the pivot to F_2 – the kinematic relation would be

$$\frac{F_1}{F_2} = \frac{v_2}{v_1} = \frac{l_2}{l_1}.$$

This results from the fact that every point on the lever has the same angular velocity,

$$\omega = \frac{v_1}{l_1} = \frac{v_2}{l_2}.$$

Given the transformer (Figure 2.10 (c)) winding ratio, N_1/N_2, the voltage and current relation is

$$\frac{e_1}{e_2} = \frac{i_2}{i_1} = \frac{N_1}{N_2}.$$

For the hydraulic ram (Figure 2.10 (d)), the force and pressure as well as the velocity and volumetric flow rate are related through the piston cross-sectional area, A,

$$\frac{F}{P} = \frac{Q}{v} = A.$$

Note that the first three examples transform energy within a single energy domain. The gear pair, for example, transforms rotational energy to rotational energy. The fourth example, the hydraulic piston, transforms mechanical energy to hydraulic.

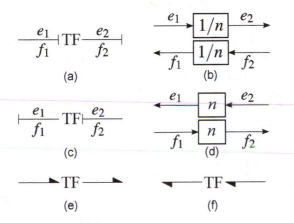

(a)

(b)

(c)

(d)

(e)

(f)

Figure 2.11: *TF-element representation including (a) an effort-in-effort-out bond graph, (b) an effort-in-effort-out block diagram, (c) a flow-in-flow-out bond graph, (d) a flow-in-flow-out block diagram, (e) a power-in-power-out bond graph, and (f) a power-out-power-in bond graph.*

Note that the transformer can have one of two possible causalities, illustrated in Figure 2.11 (a)-(d). The power flow direction (the half-arrow head) is left out of the bond graphs in (a) and (c) to solely emphasize the causality. Once the effort is specified into one port, the effort out the other port is immediately determined via the transformer modulus, n. Thus, a transformer can only have one port where the effort is specified as an input. Additionally, the positive power flow direction is assumed to go into one port and out the other, as depicted in Figure 2.11 (e) and (f). Thus, the half-arrows indicating positive power flow in Figure 2.11 (a) and (c) could appear in either direction so long as they follow the "flow-through" convention illustrated in (e) and (f).

Be careful to distinguish between causality and power direction. Remember, the half-arrow indicates the assumed direction of positive power flow, and the bar on each bond indicates the port where effort is an input. The bar *does not* indicate whether effort is positive or negative.

2.3.2   Gyrators (GY-Elements)

The second type of 2-port element is the gyrator. Like the transformer, the gyrator transmits energy. However, unlike the transformer, the direct relation is not between efforts or between flows. The direct relation is between the effort on one side and the flow on the other side. The constitutive relations for a linear ideal gyrator are

$$e_1 = r f_2 \text{ and } r f_1 = e_2 \tag{2.12}$$

where r is referred to as the *gyrator modulus* and indicates the proportional relation between effort on one side and flow on the other. Again, note that instead of a linear relation between the efforts and flows, there could in general be a nonlinear relation.

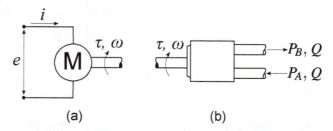

(a)   (b)

Figure 2.12: *GY-element examples including (a) an ideal electric motor and (b) an ideal centrifugal pump.*

Physical examples of ideal gyrators are far less common. Nonetheless, a commonly occurring gyrator in electromechanical systems is the ideal motor depicted in Figure 2.12 (a). In an ideal electric motor, the output torque, τ, is proportional to the input current, i, through the motor constant, k_m,

$$\tau = k_m i.$$

Alternatively, angular velocity, ω, can be specified as an input to generate the motor voltage, e, as an output,

$$e = k_m \omega.$$

Another less commonly occurring gyrator is an ideal centrifugal pump. Here the output pressure differential, $\Delta P = P_B - P_A$, is proportional to the input angular velocity, ω. Note for both the ideal motor and centrifugal pump the input/output relation is between an effort on one side and a flow on the other. Also note that both gyrators couple different energy domains at each port. The ideal motor couples mechanical rotation and electronics, and the centrifugal pump couples mechanical rotation and hydraulics.

Like the transformer, the gyrator can have one of two possible causalities shown in Figure 2.13. Unlike the transformer, however, either the efforts must be specified as inputs to both ports, or the flows must be specified as inputs to both ports. That is, once the effort into one port is specified as an input, the flow at the other port is immediately generated as an output. Conversely, once a flow is specified as an input to a port, the effort is immediately generated as an output at the opposite port.

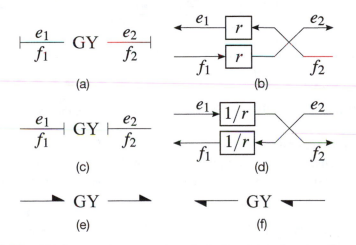

Figure 2.13: *GY-element representations including (a) a flow-in-effort-out bond graph, (b) a flow-in-effort-out block diagram, (c) an effort-in-flow-out bond graph, (d) an effort-in-flow-out block diagram, (e) a power-in-power-out bond graph, and (f) a power-out-power-in bond graph.*

2.4   Junction Elements

Thus far, elements to store, dissipate, and transmit energy have been introduced. Additionally, elements used to represent sources have also been presented. All that is left to comprise a basic bond graph are junctions. *Junctions* are used to interconnect the basic elements introduced in the preceding sections. They serve to represent the interaction of the constitutive relations associated with the 1- and 2-port elements. As with the basic 2-ports, junctions conserve power.

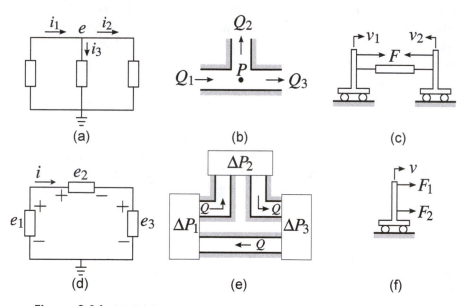

Figure 2.14: *(a)-(c) 0-junction examples and (d)-(f) 1-junction examples.*

Junctions naturally occur in dynamic systems. They are characterized by two conditions. Take for example the junctions illustrated in Figure 2.14. Let us discuss first examples (a)-(b). By examining these examples, two commonalities are revealed:

1. The *primary condition* in each example is that the components attached off the junction have a common *effort*. In the electric circuit, the generalized impedances connected in parallel have the same voltage drop across them, e. When pipe segments join together at a junction they share a common pressure, P. If you have an element like a damper or spring attached between two massless carts, the carts experience an equal and opposite force, F.

2. The *secondary condition* in each example is that there occurs a *summation of flows*. The node at the top of (a) connects three branches of the circuit, and, as dictated by Kirchhoff's Current Law, the currents sum to zero at a node (i.e., $i_1 - i_2 - i_3 = 0$). Similarly, in a hydraulic circuit, the flows at a junction sum to zero, $Q_1 - Q_2 - Q_3 = 0$. A velocity differential occurs between the two massless carts, $\Delta v = v_1 - v_2$.

In bond graphs, junctions that have common effort and sum flows (or calculate a flow differential) are referred to as *0-junctions*. A sample 0-junction is shown in Figure 2.15 (a). Junctions, in general, may have many ports attached (i.e., two or more ports). Because power is conserved at a junction the power going in must equal the power going out,

$$\mathscr{P}_{in} = \mathscr{P}_{out},$$

which can be rewritten as

$$\mathscr{P}_{in} - \mathscr{P}_{out} = 0.$$

For a junction with $j = 1, 2, 3, \ldots, n$ ports this can be generally written as

$$\sum_{j=1}^{n} \mathscr{P}_j = \sum_{j=1}^{n} e_j f_j = 0 \tag{2.13}$$

(i.e., the sum of the power *into* a junction is zero). For a common-effort junction, the primary condition dictates that

$$e_1 = e_2 = e_3 = \ldots = e. \tag{2.14}$$

The sum of the flows for a common effort (0-junction) results from substituting Equation 2.14 into Equation 2.13,

$$\sum_{j=1}^{n} \mathscr{P}_j = \sum_{j=1}^{n} e_j f_j = e \sum_{j=1}^{n} f_j = 0 \Rightarrow \sum_{j=1}^{n} f_j = 0. \tag{2.15}$$

Because the power flowing into a 0-junction is assumed to be positive, and the efforts on the attached bonds are all equal, the power direction indicates, relative to the junction, whether the *flow* on each attached bond is positive or negative. Therefore, the 0-junction example in Figure 2.15 (a) would have the following sum of flows:

$$f_1 - f_2 - f_3 = 0.$$

Moreover, because the efforts of each bond attached to a 0-junction are the same, only one bond attached can specify effort as an input. Returning to the example in Figure 2.15 (a), the causalities on the attached bonds indicate that f_2 and f_3 are inputs into the 0-junction used to calculate f_1 as an output. Therefore, the above equation should then be rewritten as

$$f_1 = f_2 + f_3$$

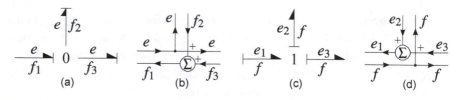

Figure 2.15: *(a) 0-junction bond graph, (b) 0-junction block diagram, (c) 1-junction bond graph, and (d) 1-junction block diagram.*

to emphasize the causality. Figure 2.15 (b) shows the corresponding block diagram.

Let us now turn our attention to the junction examples (d)-(f) in Figure 2.14. Upon close examination, two commonalities arise:

1. In each example, the *primary condition* is that the interconnected elements have *common flow*. Electrical components connected in series flow the same current, i. Similarly, in a hydraulic circuit if three elements, for example, are connected "in series" they all have the same flow rate, Q. Multiple force-generating elements attached to a massless cart will all move with the same velocity, v.

2. For each example, the *secondary condition* is that there exists a *summation of efforts*. Recall from Kirchhoff's Voltage Law that the voltage drops around a loop sum to zero, $e_1 - e_2 - e_3 = 0$. The same basic thing occurs in hydraulic circuits, $\Delta P_1 - \Delta P_2 - \Delta P_3 = 0$. When springs and dampers attach to the same point or to rigid body there occurs a summation of forces, $F_1 + F_2 = 0$.

In bond graphs, junctions that have common flow and sum efforts are referred to as *1-junctions*. For a common-flow junction

$$f_1 = f_2 = f_3 = \ldots = f. \tag{2.16}$$

By substituting the above into Equation 2.13, the sum of efforts for a 1-junction results in

$$\sum_{j=1}^{n} \mathscr{P}_j = \sum_{j=1}^{n} e_j f_j = f \sum_{j=1}^{n} e_j = 0 \;\Rightarrow\; \sum_{j=1}^{n} e_j = 0. \tag{2.17}$$

The power direction for bonds attached to a 1-junction indicate, relative to the junction, whether the *effort* on each attached bond is positive or negative. In addition, only one bond attached to a 1-junction can specify flow as an input.

Thus, the 1-junction example in Figure 2.15 has the following input/output relation that results from the sum of the efforts:

$$e_1 = e_2 + e_3.$$

To clarify, the power directions at a *0-junction* are used to indicate the sign convention for the summation of *flows* on the attached bonds, and the power directions at a *1-junction* are used to indicate the sign convention for the summation of *efforts* on the attached bonds. Table 2.7 summarizes the primary and secondary conditions and the sign conventions for 0- and 1-junctions.

Table 2.7: *Summary of junction conditions.*

Junction Type	Primary Condition	Secondary Condition	Bond Direction Indicates
0-Junction	common effort, $e_1 = e_2 = \ldots = e$	sum of flows, $\sum f_i = 0$	flow sign convention (sign of the flows)
1-Junction	common flow, $f_1 = f_2 = \ldots = e$	sum of efforts, $\sum e_i = 0$	effort sign convention (sign of the efforts)

Junctions are natural representations of some commonly used physical laws and mathematical formulations. Most have already been briefly mentioned. In mechanical translation, the 1-junction represents the summation of forces that occur at independently moving masses. Anything attached to the mass will move with the same velocity and will impose a force on the mass. Thus, a 1-junction in mechanical translation represents Newton's Second Law, or a summation of forces,

$$\sum F = \dot{p}.$$

In summing the forces, if we consider the inertial force, $\dot{p}$, in our summation, the forces indeed sum to zero:

$$\sum F - \dot{p} = 0.$$

As we shall see later when we discuss causality, the first formulation isolates the causes ($\sum F$) from the effect ($\dot{p}$). Similarly, in rotation, a 1-junction represents a summation of moments. In circuits, a 1-junction represents Kirchhoff's Voltage Law which states that the sum of the voltage drops around a

Table 2.8: *Summary of energy-domain-specific junction secondary conditions.*

Domain	1-Junction	0-Junction
Mechanical Translation	Newton's Second Law $\sum F = \dot{p}$	Relative velocity e.g., $\Delta v = v_2 - v_1$
Mechanical Rotation	Summation of the moments $\sum \tau = \dot{h}$	Relative angular velocity e.g., $\Delta \omega = \omega_2 - \omega_1$
Electric Circuits	Kirchhoff's Voltage Law $\sum e_j = 0$	Kirchhoff's Current Law $\sum i_j = 0$
Hydraulic Circuits	Pressure drops around a loop $\sum \Delta P = 0$	Sum of flow into a junction $\sum Q_j = 0$

closed loop is zero. Much like in electric circuits, the pressure drops around a loop in a hydraulic circuit must also sum to zero.

In a similar fashion, 0-junctions represent common engineering concepts. In translation, 0-junctions represent relative velocities or points where velocities are summed to determine a velocity differential,

$$\Delta v = v_2 - v_1 .$$

In this formation, the causes (v_1 and v_2) are isolated from the effect (Δv). The relative velocity, Δv, is technically still a velocity. Hence, if we consider it in our summation, the sum of the velocities is indeed zero,

$$v_2 - v_1 - \Delta v = 0 .$$

Relative or differential angular velocities occur in rotational systems. The 0-junction in electric circuits represents Kirchhoff's Current Law – the summation of currents that occur at a node. A hydraulic junction is the analog of an electric circuit node. The volumetric flow rates must sum to zero at a junction. Hence, the junction secondary conditions represent some commonly used concepts in Engineering. These formulations and laws are summarized in Table 2.8.

2.5 Simple Bond Graph Examples

All the necessary elements to synthesize a basic bond graph have been presented. In the next chapter, we will show how to synthesize bond graphs for moderately complicated systems and how to systematically derive, from the resulting bond graph, the corresponding state-space equations that describe the dynamic response of the system. However, to illustrate where we are ultimately headed, a few simple examples are covered in this section. These examples are meant only to introduce you to the application of bond graphs in system dynamics modeling. The chapter that follows will present in detail the actual steps for model synthesis and equation derivation.

Take as the first example the mass-spring-damper system depicted in Figure 2.16 (a). Upon investigation, it is apparent that the mass, spring, damper, and external force have one thing in common. The external force, F, applied at the left end of the spring is the same force the damper and mass experience; they all have the same force, namely, F. Thus, a common effort junction can be used to connect the effort source (F), the C-element (the spring), the R-element (the damper), and the I-element (the mass). Figure 2.16 (b) shows the bond graph and (c) illustrates the causality and mathematical relations represented by the bond graph.

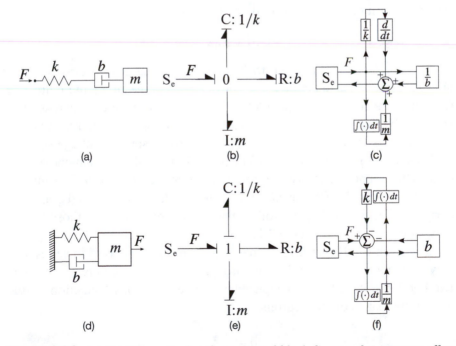

Figure 2.16: *(a)-(b) Schematic, bond graph, and block diagram for common effort mass-spring-damper system; (d)-(f) schematic, bond graph, and block diagram for a common flow mass-spring-damper system.*

The mass-spring-damper system in Figure 2.16 (d) differs from the one in (a). The force, mass, spring, and damper all move with the same velocity. Thus, they have common flow. The 1-junction in Figure 2.16 indicates that the four elements have the same velocity and that there occurs a summation of forces. Therefore, the time rate of change of deflection of the spring, x, is

$$\dot{x} = \frac{1}{m}p$$

and the time rate of change of the momentum, p, is

$$\dot{p} = F - kx - \frac{b}{m}p.$$

The resulting model is a set of two first-order differential equations. By recognizing that if the mass, m, is constant then

$$\dot{p} = \frac{d}{dt}(mv) = \frac{d}{dt}(m\dot{x}) = m\ddot{x},$$

the above equations can be combined to give the more familiar single second-order equation for a mass-spring-damper

$$m\ddot{x} + b\dot{x} + kx = F.$$

Electrical circuits analogous to the mass-spring-damper systems discussed above are illustrated in Figure 2.17. The elements of the circuit in Figure 2.17 (a) which are connected in parallel have the same voltage drop across them specified by the voltage input. Therefore, the corresponding bond graph has four elements attached off a 0-junction – an effort source (voltage source), an R-element (resistor), a C-element (capacitor), and an I-element (inductor). The 0-junction also represents the summation of currents at the top node due to Kirchhoff's Current Law. Note that the resulting bond graph and corresponding block diagram follow the exact same pattern as those for the mass-spring-damper in Figure 2.16 (a).

The circuit in Figure 2.17 (c) has four elements connected in series, and they have the same current flowing through them. The circuit is governed by Kirchhoff's Voltage Law. As depicted in Figure 2.17 (d), a 1-junction is used to indicate that a common current,

$$\dot{q} = \frac{1}{L}\lambda,$$

flows through the elements and to represent the sum of the voltage drops around the circuit loop,

$$\dot{\lambda} = e - \frac{1}{c}q - \frac{R}{L}\lambda.$$

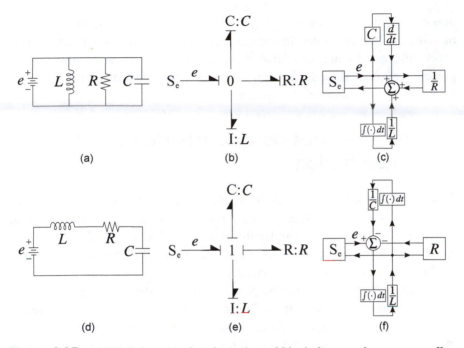

Figure 2.17: *(a)-(b) Schematic, bond graph, and block diagram for common effort electrical circuit; (d)-(f) schematic, bond graph, and block diagram for a common flow electrical circuit.*

By recalling that λ is the electrical momentum,

$$\lambda = Li = L\dot{q},$$

the above equations can be combined to get the more common second-order differential equation for a series LCR circuit,

$$L\ddot{q} + R\dot{q} + \frac{1}{C}q = e.$$

The block diagram in Figure 2.17 (e) shows the causality and mathematical relations represented by the bond graph in (d). Again, note that the bond graph and block diagram follow the same pattern as those for the mass-spring-damper system depicted in Figure 2.16 (d). Also note that the second-order differential equation for the circuit discussed above has the same format as the previously discussed second-order differential equation for the mass-spring-damper in Figure 2.16 (d).

As illustrated above, we will show in the following chapters that bond graphs can be used to derive a set of first-order differential equations that describe the dynamics of the system. As will be shown later, the advantage of using a set of first-order differential equations over the use of, say, an nth-order differential equation is that if the elements are linear or linearizable, powerful numerical methods from Linear Algebra can be used to facilitate analysis and simulation of the system.

2.6   Linear versus Nonlinear Systems and Linearization

Thus far we have assumed that the system elements are purely linear. That is, their constitutive relations can be represented by simple linear relations. However, we originally formulated the causal relations for R-, C-, and I-elements in terms of generally *nonlinear* functions. Under many conditions, nonlinear functions can be linearized.

Linear systems satisfy the properties of *superposition* and *homogeneity*. Nonlinear functions and systems do not. A system is said to be *additive*, or obey the principle of superposition, if given an input that is the summation of two signals $x_1(t)$ and $x_2(t)$, the overall response $y(t)$ is equal to the summation of the outputs resulting from the individual inputs,

$$y(x_1(t) + x_2(t)) = y_1(t) + y_2(t),$$

where $y_1(t)$ is the system response to $x_1(t)$ and $y_2(t)$ the response to $x_2(t)$. Linear systems are also *homogeneous*; that is, for any input $x(t)$ and real scalar a the output response to $ax(t)$ is the response to $x(t)$ scaled by a,

$$ay(x(t)) = y(ax(t)).$$

The above conditions for linearity can be summarized as follows:

$$y(a_1 x_1(t) + a_2 x_2(t)) = a_1 y_1(t) + a_2 y_2(t).$$

Take, for example, the simple mass-spring-damper system depicted in Figure 2.16 (d). You may recall from Dynamics or another course that the system can be represented by a second-order differential equation,

$$m\ddot{x} + b\dot{x} + kx = f(t).$$

It can be shown that the overall response of the system to a combination of two different sinusoids is equal to the sum of the responses to the individual sinusoids. In other words, if we took the separate responses of the system to two distinct inputs and summed the resulting output signals, we would get the same response that would result if the input were the combination of the two sinusoids. (We will revisit this in more detail later in the textbook.) Figure 2.18 illustrates the response $x(t)$ resulting from a forcing function $f(t) = 3\sin t + 5\cos 2t$, assuming the mass, damping constant, and spring rate are 2 kg, 0.01 kg/sec, and 5 N/m, respectively. The top subplot depicts the forcing function, $f(t) = f_1(t) + f_2(t)$, and its components, $f_1(t) = 3\sin t$ and $f_2(t) = 5\sin 2t$. The lower subplot depicts the responses to the individual components, $y_1(t)$ and $y_2(t)$, and the overall response, $y(t)$, to the combined input. Also illustrated in that subplot is the sum of the responses $y_1(t)$ and $y_2(t)$. Note that if one sums the responses, $y_1(t) + y_2(t)$, the result would be the same as the overall response $y(t)$.

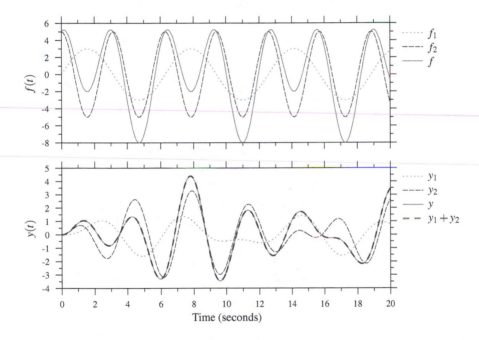

Figure 2.18: *An illustration of the additive and homogenous properties of linear systems.*

Imagine that the damper force is nonlinear (e.g., $F_{damper} = b\dot{x}|\dot{x}|$) so that the resulting second-order differential equation is

$$m\ddot{x} + b\dot{x}|\dot{x}| + kx = f(t).$$

This system is nonlinear because the damper force, in particular, does not obey the additive and homogenous properties. This is illustrated by the responses plotted in Figure 2.19. Note that the overall response, $y(t)$, to the combined input, $f(t)$, does not exactly match the sum of the individual response, $y_1(t) + y_2(t)$. There is a slight, yet noticeable, difference between the two. However, the fact that the overall and summed responses are similar points to a tool – linearization.

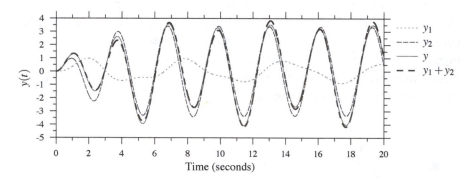

Figure 2.19: *An illustration of how nonlinear systems do not obey additive and homogenous properties.*

Many nonlinear systems behave approximately like linear systems when they function about an operating point, equilibrium for instance. This is illustrated in Figure 2.19 by the fact that, though different, $y(t)$ and $y_1(t) + y_2(t)$ are similar. For many nonlinear functions, reasonable linear approximations can be derived using the first few terms of the Taylor Series Expansion, about an operating point $\hat{x}$,

$$f(x) = f(\hat{x}) + \frac{f'(\hat{x})}{1!}(x - \hat{x}) + \frac{f''(\hat{x})}{2!}(x - \hat{x})^2 + \frac{f^{(3)}(\hat{x})}{3!}(x - \hat{x})^3 + \cdots$$

$$= \sum_{n=0}^{\infty} \frac{f^{(n)}(\hat{x})}{n!}(x - \hat{x})^n. \tag{2.18}$$

Note that the first few terms given above are purely linear and can be used to estimate the value of the nonlinear function near the operating point, $\hat{x}$. For a single variable nonlinear function, Equation 2.20 simplifies to

$$f(x) = f(\hat{x}) + \left.\frac{\partial f}{\partial x}\right|_{\hat{x}} (x - \hat{x}) + \text{H.O.T.} \tag{2.19}$$

(H.O.T. stands for higher-order terms in the series.) Notice that the first few terms are simply the equation for a line. Hence, a single variable nonlinear function is approximated by a line near the operating point. For a general multivariable nonlinear function the Taylor Series Expansion is

$$f(\mathbf{x}) = f(\hat{\mathbf{x}}) + \left.\frac{\partial f}{\partial x_1}\right|_{\hat{\mathbf{x}}}(x_1 - \hat{x}_1) + \left.\frac{\partial f}{\partial x_2}\right|_{\hat{\mathbf{x}}}(x_2 - \hat{x}_2) + \cdots$$

$$+ \left.\frac{\partial f}{\partial x_n}\right|_{\hat{\mathbf{x}}}(x_n - \hat{x}_n) + \text{H.O.T.},$$

(2.20)

where H.O.T. are the *higher order terms* of the series, $\mathbf{x} = [x_1\ x_2\ \ldots\ x_n]^T$ are the variables, and $\hat{\mathbf{x}} = [\hat{x}_1\ \hat{x}_2\ \ldots\ \hat{x}_n]^T$ is the operating point – the point about which the system tends to operate.

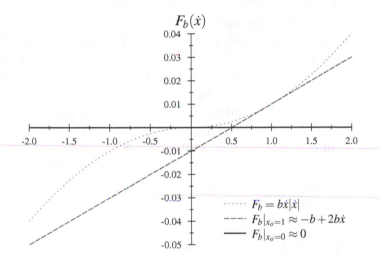

Figure 2.20: *The nonlinear damping force and its approximations at two operating points.*

Take, for instance, the nonlinear damper force, $F_b(\dot{x}) = b\dot{x}|\dot{x}|$. This force is similar to the force exerted due to drag on a rigid body moving through a fluid (e.g., a car moving through air),

$$F_d = \frac{1}{2}C_d A v|v|,$$

where C_d is the drag coefficient, A the cross-sectional area, and v the relative velocity. Using the first few terms of the Taylor Series, the linearized approximation of F_b is

$$F_b(\dot{x}) = b\dot{x}|\dot{x}| \approx b\dot{x}_o|\dot{x}_o| + \left[b|\dot{x}_o| + b\frac{\dot{x}_o^2}{|\dot{x}_o|}\right](\dot{x} - \dot{x}_o).$$

Near an operating point of $\dot{x}_o = 1$ the damping force can be approximated as

$$F_b(\dot{x})|_{\dot{x}_o=1} \approx b\,1|1| + \left[b|1| + b\frac{1^2}{|1|}\right](\dot{x}-1) = b + 2b(\dot{x}-1) = -b + 2b\dot{x}.$$

In the proximity of $\dot{x}_o = 0$ the damping force is approximately zero,

$$F_b(\dot{x})|_{\dot{x}_o=0} \approx b\,0|0| + \left[b|0| + b\frac{0^2}{|0|}\right](\dot{x}-0) = 0.$$

The damping force F_b and its approximations near $\dot{x}_o = 1$ and $\dot{x}_o = 0$ are illustrated in Figure 2.20. Note how each estimate provides an exact solution at the corresponding operating point, and how the estimate is decreasingly accurate as the function moves away from the operating point. The approximations are only reasonable in and around the respective points of operation. Hence, though linearization can be used to simplify nonlinear systems, it should only be used if the system dynamic states operate within a reasonable region of the operating point.

Many hydraulic and pneumatic elements exhibit nonlinear behavior over a large operating range. However, many operate over a limited range within which the nonlinearities can be reasonably approximated using linearization. Valves, for example, are often modeled by quadratic or cubic relations between the pressure drop and volumetric flow rate (e.g., $\Delta P = R_f Q^2$). Such formulations can be readily linearized.

2.7 Summary

▷ R-elements dissipate energy. They have a constitutive relation that directly relates effort to flow. They can exhibit one of two causalities – effort-in-flow-out or flow-in-effort-out. Power is generally assumed to flow from the system to the R-element.

▷ C-elements store potential energy. They have a constitutive relation that directly relates effort to displacement. They can exhibit one of two causalities – integral causality where flow is an input to the C-element and derivative causality where effort is an input to the C-element. Power is generally assumed to flow from the system to the C-element.

▷ I-elements store kinetic energy. They have a constitutive relation that directly relates momentum to flow. They can exhibit one of two causalities – integral causality where effort is an input to the I-element and derivative causality where flow is an input to the I-element. Power is generally assumed to flow from the system to the I-element.

▷ Effort sources provide an external effort as an input to the system.

▷ Flow sources provide an external flow as an input to the system.

▷ Transformers transmit and/or change the form of energy. They conserve power (i.e., the power in is equal to the power out). The efforts on either side are directly related, and the flows on either side are directly related. Only one bond attached to a transformer can specify effort as an input.

▷ Gyrators also transmit and change the energy form, and they also conserve power. The effort on one side is directly related to the flow on the other. Either both bonds must specify effort as an input, or both bonds must specify flow as an input.

▷ 0-junctions have common effort and sum flows. Only one bond can specify effort as input. The power direction specifies whether the flows on the attached bonds are positive or negative relative to the junction.

▷ 1-junctions have common flow and sum efforts. Only one bond can specify flow as input. The power direction specifies whether the efforts on the attached bonds are positive or negative relative to the junction.

▷ Linearization can be used to approximate nonlinear functions and systems. This can be accomplished by using the first few terms of the Taylor Series Expansion of the nonlinear terms.

2.8 Review

R2-1 Of the basic one-port elements, which are described by constitutive relations that algebraically relate effort and flow? Does this one-port store or dissipate energy?

R2-2 What type of energy do C-elements store? For C-elements, what two energy and power variables are directly related?

R2-3 I-elements store what form of energy? Constitutive relations for I-elements directly relate what two energy and power variables?

R2-4 What two types of basic sources exist? How do they differ?

R2-5 Explain the two basic ways in which energy can be converted. What types of elements are used to represent these forms of energy conversion? Explain the differences in input-output relations for the two.

R2-6 Describe in your own words the primary and secondary conditions for 1- and 0-junctions.

R2-7 Explain the differences between linear and nonlinear functions and systems.

2.9 Problems

For the systems introduced in Chapter 1, generate tables that identify each element, its parameter, its element type, and what it does with energy (e.g., Resistor 1, R_1, R-Element, and dissipates energy). A partial sample table for the first circuit in P2-3 follows. Do not include junctions and remember that sources supply energy and transformers and gyrators convert energy.

Element	Parameter	Element Type	Energy
voltage source	e	effort source	supplies energy
resistor 1	R_1	R-element	dissipates energy
⋮	⋮	⋮	⋮

P2-1 *Mass-Spring-Damper Systems (Figure 1.15).* Recall that for the mass-spring damper systems $x_1(t)$, $x_2(t)$, and $x(t)$ are labels while $v(t)$ is a velocity source and $F(t)$ is an external force.

P2-2 *Rotational Systems (Figure 1.16).* An input torque τ_m is supplied for (a) and (b) while (c) has two input torques.

P2-3 *Electric Circuits (Figure 1.17).* For the circuits, e or $e(t)$ is the input voltage source. The labels e_c and i_L are internal variables (not sources) that we will be solving for in later chapters.

P2-4 *Hydraulic Circuits (Figure 1.18).* Hydraulic circuits (a) and (b) have an input volumetric flow rate source that supplies Q. Hydraulic circuit (c) has an input pressure source.

P2-5 *Permanent Magnet Direct Current (PMDC) Motor with Gear Reduction (Figure 1.3).* A supply voltage $e_{in}(t)$ is applied to the input terminals of the PMDC motor illustrated in Figure 1.3. The internal variables

$i(t)$, E_m, τ_m, ω_m, τ_1, ω_1, τ_2, and ω_2 are labeled for convenience and will be used in later chapters to derive models of the system. They are not effort or flow sources.

P2-6 The inverted pendulum in Figure 2.21 can be modeled by the second-order, nonlinear differential equation

$$mL^2\ddot{\theta} = -mgL\sin\theta + \tau(t).$$

Linearize the differential equation.

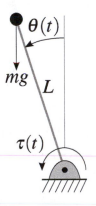

Figure 2.21: *An inverted pendulum.*

P2-7 In basic engineering mechanics courses, friction is approximated as a constant unidirectional force. In reality, friction, like drag, has directionality. It opposes the direction of motion. Moreover, friction models can be more complex than a constant force. Furthermore, it is quite nonlinear like the model in Figure 2.22. Some constitutive models attempt to capture some of the complexities of friction. One such model is (Armstrong and Wit 1995)

$$F_{fric} = \left[F_k + \left(F_s - F_k e^{-c_v|v|} \right) \right] \frac{v}{|v|} + bv$$

where $F_k = \mu_k N$ is kinetic Coulomb friction, $F_s = \mu_s N$ is the breakaway (or static) friction, c_v is a coefficient, b is the viscous friction coefficient, v is the relative velocity, and N is the normal force. Approximate the friction for a velocity v_o.

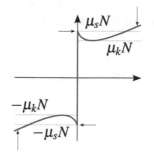

Figure 2.22: *Nonlinear friction model.*

P2-8 The flow, Q, through a valve can be modeled by a nonlinear function of the form

$$Q = K\sqrt{\Delta P}$$

where K is a coefficient and ΔP is the pressure drop across the valve. Linearize the function about a pressure drop of ΔP_o.

2.10   Challenges

As you did with the practice problems, generate tables for each of the real-world systems below. There is no one right answer. Identify the components you deem necessary to model the system. Provide a short justification explaining why you included the elements you did.

C2-1 *A Toy Electric Car.* The buggy previously illustrated in Chapter 1 is powered by batteries. A small motor drives a gear pair that twists the axle. This in turn rotates the rear wheels, which spin without slip along the ground or surface causing the toy car to propel forward.

C2-2 *A Quarter-Car Suspension Model.* The quarter-car suspension includes the components used to suspend one corner of a vehicle and isolate it from the road. It essentially filters the disturbances from the road so that the occupant does not feel the full impact, and it keeps the tire in contact with the road surface to improve traction.

C2-3 *An Electric Drill.* In a cordless drill, a battery supplies voltage to excite a motor. Most drills have a selectable "Hi-Lo" switch that varies the maximum torque the drill can generate. When used to insert a screw or drill a hole, the screw or drill bit rubs against a surface generating friction that impedes the rotation.

Chapter 3

Bond Graph Synthesis and Equation Derivation

Recall the challenges presented in previous chapters. The new task is to derive a mathematical model of each overall system. Each model will be used to predict the response and analyze the dynamic characteristics of the respective system. Each must account for the germane subsystems and their associated dynamic responses. Things you might consider are:

▷ What type of mathematical equations are needed?

▷ How are these equations systematically derived?

▷ How are the individual constitutive relations of the components connected to generate a mathematical model?

3.1   Introduction

It was made evident in the previous chapter that many dynamic systems can be decomposed into components that either store energy (C- and I-elements), dissipate energy (R-elements), or convert energy (transformers and gyrators). Additionally, elements are used to represent external inputs (efforts and flow sources) and common effort or common flow relations (0- and 1-junctions). A large class of dynamic systems can be broken down into such elements. It was also shown how each of the power domains discussed thus far – mechanical translation, mechanical rotation, electrical circuits, and hydraulic circuits – has analogous elements that serve each of the functions mentioned above. The question remains, though: How does one systematically organize these elements in such a way that appropriately represents the physical constraints and accounts for the energy flow in the system?

In this chapter, guidelines will be introduced for the four power domains that have been discussed thus far. The emphasis in this chapter will be first on synthesizing the bond graph representation and second on deriving the differential and algebraic equations used to mathematically represent the system's dynamics.

So, as a reader and novice practitioner of bond graph modeling, take a leap of faith and do not, at this time, concern yourself with the mathematics. We will address this later in the chapter. Simply focus on sketching an appropriately conceived bond graph. For now, concern yourself more with decomposing the dynamic systems presented and identifying, in terms of energy, the role each component plays. Furthermore, focus on the input/output relations and the flow of "mathematical signals" or information within the model. Once you begin to master this, we will delve later in the chapter into deriving mathematical equations.

The objectives and outcomes for this chapter are detailed below.

▷ *Objectives:*

1. To effectively use bond graphs to formulate models that facilitate deriving mathematical representations of dynamic systems,
2. To be able to systematically derive mathematical representations using bond graphs, and
3. To understand the flow of information within a system dynamics model and its relation to mathematical representations.

▷ *Outcomes.* Upon completion, you should

1. be able to synthesize bond graph models of mechanical, electrical, and hydraulic systems,
2. be able to annotate bond graphs to indicate appropriate power flow and causality, and
3. be able to derive mathematical models in the form of differential and algebraic equations using bond graph representations.

3.2 General Guidelines

At this point, there are some guidelines that have been introduced or implied in the previous chapter but are explicitly restated here for convenience. Generally, as depicted in Figure 3.1 (a), power is assumed to flow from the system to energy-storing elements. Thus, the bond graph half-arrows almost always point toward the R-, C-, and I-elements.

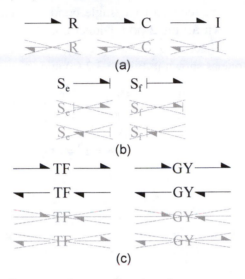

Figure 3.1: *Generally assumed power directions for energy-storing and dissipating elements*

Additionally, as shown in Figure 3.1 (b), effort and flow sources are usually assumed to supply power to the system. Effort sources specify an effort as an input to the system and flow sources specify a flow as an input to the system.

Two-port elements have a "power through" convention where the attached bonds point in the same direction. This is depicted in Figure 3.1 (c). On transformers and gyrators, one bond must specify power in and the other power out. For tree-like bond graphs without loops, the sign directions on the remaining bonds are not important and can be assigned somewhat arbitrarily.

Occasionally, as you synthesize a bond graph or simplify it, you end up with junction structures like those depicted in Figure 3.2. These structures can be readily simplified. If a 0- or 1-junction has only two bonds, then the efforts must be the same and so must the flows. This can be shown by mathematically analyzing the 2-port junction. Take for instance the 0-junction depicted in Figure 3.2 (a). The relations for the 0-junction are

$$e_1 = e_2 \text{ (common effort)}$$

and

$$f_1 - f_2 = 0 \text{ (sum of the flows).}$$

However, from the second equation we can show that $f_1 = f_2$, and thus not only do the efforts equal each other, but so do the flows. The junction itself is redundant and can be reduced to a single bond as depicted in the figure. The same can be shown for the 2-port 1-junction. When two 0-junctions or two 1-junctions are adjacent to one another, the attached junctions can be consolidated into one as shown in Figure 3.2 (c) and (d).

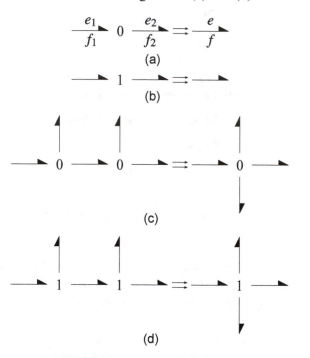

Figure 3.2: *(a) 2-port 0-junction, (b) 2-port 1-junction, (c) adjacent 0-junctions, and (d) adjacent 1-junctions*

3.3 Mechanical Translation

Mechanical systems that purely translate are categorized as mechanical translation systems. Examples include suspension and mass-spring-damper systems similar to those depicted in Figure 3.3. Though translation generally occurs in three dimensions, the problems discussed herein will be restricted to planar motion. However, the methods presented are generally applicable to three-dimensional motion.

Mass-spring-damper systems are assessed using basic Newtonian Mechanics. Before introducing the guidelines for synthesis of bond graphs for

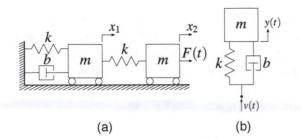

Figure 3.3: *(a) Mass-spring-damper system and (b) simple quarter-vehicle suspension model*

mechanical translation systems, let us review the basic mechanical translation elements. Masses store kinetic energy when in motion and are considered I-elements. The integrated states associated with masses are mass momenta and forces are often summed at a mass. Springs are compliant and store potential energy. Thus they are considered C-elements. Their associated integrated states are the spring deflections. The spring deflection is related through an integral to the relative velocity or velocity differential between the spring ends. Dampers or shock absorbers dissipate energy and are considered R-elements. Levers and rocker arms operate as transformers. Velocity differentials are represented by 0-junctions, and 1-junctions are used to model summations of forces. These elements are summarized in Table 3.1.

Table 3.1: *Mechanical translation elements*

Element Type	Mechanical Component	A Depiction
R-element	Damper	Figure 2.1 (a)
C-element	Spring	Figure 2.3 (a)
I-element	Mass	Figure 2.5 (a)
Effort Source	External Force	Figure 2.8 (c)
Flow Source	Shaker or Velocity Source	Figure 2.8 (d)
Transformer	Lever or Rocker Arm	Figure 2.10 (a)
0-Junction	Velocity Differential	Figure 2.14 (c)
1-Junction	Sum of Forces	Figure 2.14 (f)

The basic guidelines for synthesizing bond graphs for mechanical translation systems are:

1. *Identify distinct velocities.* For each distinct velocity establish a 1-

junction. Distinct velocities are often associated with masses. Thus, if a mass is specifically associated with a distinct velocity an I-element representing the mass should be placed directly off that 1-junction using a bond. Moreover, if there are any other elements directly related with a specific velocity, they too should appear off the associated 1-junction.

2. *Insert force-generating 1-ports and energy converting 2-ports.* The force-generating 1-ports (i.e., springs and dampers) are placed between appropriate pairs of 1-junctions off of 0-junctions. Recall that the secondary condition for a 0-junction is a summation of flows. In mechanical systems, 0-junctions are used to calculate velocity differentials. If a spring or damper is attached between two moving points, it will experience a velocity differential between its two ends. Additionally, springs and dampers generate equal magnitude forces at each end that are applied in opposite directions. Their ends generally move at different rates. Thus, these elements are attached off of 0-junctions to indicate the common magnitude force and distinct velocities at each end. Transformers and gyrators can also appear between distinct velocities connected by levers or pulleys.

3. *Assign power directions.* Use the general guidelines for assigning power directions detailed in Section 3.2. The remaining bonds with unassigned power direction can be assigned somewhat arbitrarily much like in dynamics where a sign convention is chosen and consistently used. However, some forethought can minimize the need for extra work down the line. For example, in the next step we eliminate zero-velocity junctions. This will inevitably lead to junctions with only two bonds. Thus, it is convenient to follow a power-in-power-out convention so that the junction can be readily collapsed as illustrated in Figure 3.2 without having to concern ourselves with sign changes.

4. *Eliminate zero velocities.* To differentiate between the velocities at each end of a spring or damper, zero velocities were explicitly identified as being distinct and a 1-junction was used as a "place holder." The zero-velocity 1-junctions are redundant. Therefore, eliminate zero-velocity 1-junctions and their attached bonds. This aids in simplifying the bond graph.

5. *Simplify.* Simplify the resulting graph by condensing 2-port 0- and 1-junctions into single bonds as depicted in Figure 3.2 (a) and (b) and by

recognizing that if adjacent junctions are of the same kind, those junctions can be condensed into a single junction as depicted in Figure 3.2 (c) and (d).

6. *Assign causality.* Assign causal strokes on bonds to indicate the input-output relations at each port. This step will be discussed in more detail later in this chapter. For now we will focus on synthesizing and simplifying the bond graph without causality. Assigning causality is the last step just before deriving the differential and algebraic equations from the bond graph that represent the system's dynamics.

The guidelines are best illustrated by example. In the examples that follow, we will synthesize the bond graphs for the mechanical systems depicted in Figure 3.3.

Example 3.1

Synthesize the bond graph for the mass-spring-damper system depicted in Figure 3.3 (a).

Solution. The first step is to *identify the distinct velocities*. As shown in Figure 3.4 (b), there are three distinct velocities – the zero velocity of ground, the velocity of the leftmost mass, and the velocity of the rightmost mass. As indicated in (c), there are elements directly associated with each of these velocities. A flow source is used to identify the zero velocity of ground. I-elements are placed directly off the other 1-junctions to represent the two masses, and an effort source is placed off the third 1-junction to account for the external force, $F(t)$.

The next step is to *insert force-generating 1-ports and energy-converting 2-ports*. Between ground and the first mass, there are a damper (R-element) and spring (C-element). Additionally, another spring (C-element) exists between the two masses. Thus, as Figure 3.4 (d) shows, a C-element is placed off a 0-junction between the leftmost and middle 1-junctions as is an R-element. The other C-element is placed between the middle and rightmost 1-junctions.

At this point, you can *assign power directions*. Note that in Figure 3.4 (d) the bonds are generally directed from right to left so that consistently at each 0-junction the relative velocity is calculated as the right-bond velocity minus the left-bond velocity. This could be done in the opposite direction to derive an equivalent bond graph.

Now that the basic bond graph is established, simplification can begin. First, eliminate zero velocity sources which involves deleting the attached junction and bonds directly connected to that junction. Once done, the

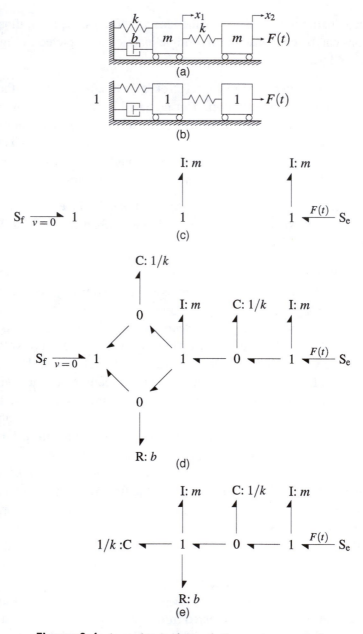

Figure 3.4: *A mechanical translation system example*

resulting simplified bond graph looks like Figure 3.4 (e). This example illustrates all but the final guideline – *assign causality*. We will return to

▌ this later.

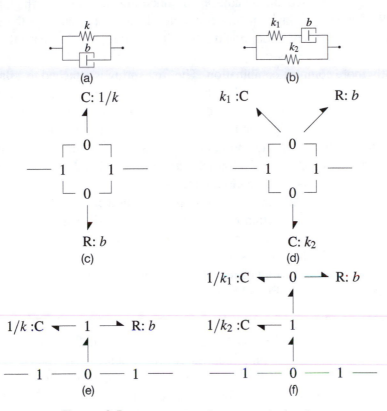

Figure 3.5: *Mass-spring-damper equivalencies*

In mass-spring-damper systems, springs and dampers are often grouped between the same pair of moving points. That is, their ends are attached to the same pair of moving masses or rigid bodies and they share a relative velocity. Figure 3.5 (a) illustrates such a pairing. Because they share a relative velocity, they can appear together off a 1-junction. To illustrate, take a close look at the bond graph segment in Figure 3.5 (c) and its equivalency in (e). The bond graph segment in (c) results from following the guidelines detailed earlier. However, this results in two separate 0-junctions that appear between the same pair of 1-junctions. The C-element appears off of one and the R-element off the other. Each element generally has a distinct force. The secondary condition for the 0-junction is the sum of the velocities which can also be thought of as a velocity differential calculation. Though the two 0-junctions have distinct efforts, they sum the same velocity differential. Hence, instead of using two separate 0-junctions to sum the same differential, a single 0-junction is used and an adjacent 1-junction is attached to which the C- and R-elements are immediately connected to indicate that both the spring and the damper share a common relative differential (i.e., they have the same flow). The second mass-damper system depicted in Figure 3.5 (b) illustrates a slightly more complex combination. The first spring is attached to a damper, and that combination is paired with another spring between the same set of moving points. Figure 3.5 (d) and (f) show the bond graph segments that result from the guidelines and the respective equivalency. Note that in the equivalency, the top spring and damper appear off a 0-junction because the two have a common force. That combination is paired to the other C-element via a 1-junction. The combination of the first spring and damper has the same relative velocity as the second spring alone. Another way to think of it is that the combined deflection of the first spring and damper is equal to the deflection of the second spring.

The example that follows will demonstrate a commonly occurring simplification that results from using the equivalency previously detailed. Example 3.2 will show how this is handled.

Example 3.2

Synthesize the bond graph for the problem depicted in Figure 3.3 (b). This figure depicts a simplified model for a quarter-car suspension system. A quarter-car suspension system is used to approximate the dynamic response of one corner of a four-wheeled vehicle.

Solution. As in the previous example, the first step is to *identify distinct velocities*. As shown in Figure 3.6 (b), the vehicle mass and the contact point with the road have distinct velocities. The road provides a displacement at the contact which has an associated velocity. Thus, a flow source is used at the bottom 1-junction to represent the vertical velocity input from the road. The upper 1-junction is associated with the vertical motion of the vehicle. Thus, the vehicle mass and external gravitational force acting on that mass correspond to that 1-junction.

We then *insert force-generating 1-ports* between the pairs of 1-junctions – in this case the spring (C-element) and damper (R-element). At this stage, we can also *assign power directions*.

The next stage involves simplification. In this problem there is no zero-velocity source that when eliminated would simplify the bond graph. However, there is a common occurrence that can be readily simplified if one understands the physical interpretation. The occurrence is a common relative velocity. Note that both the spring and damper operate between

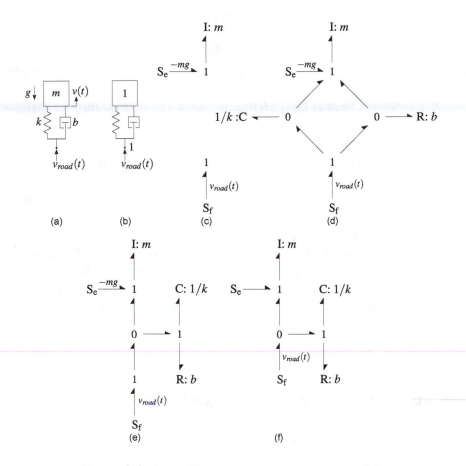

Figure 3.6: *A simplified quarter-car suspension model*

the same set of velocities (i.e., the same set of 1-junctions). Though each appear off of separate 0-junctions, both 0-junctions are connected between the same pair of 1-junctions. In other words, they have the same relative velocity. Recall that apart from having a common force, 0-junctions also sum velocities or, as in this case, represent velocity differentials. Hence, as illustrated in Figure 3.6 (e), a single 0-junction can be used to calculate the velocity differential, and both the C- and R-elements representing the spring and damper can be placed off a common 1-junction to indicate the common velocity differential. As a result, the 1-junction directly attached to the flow source now has only two bonds – one in and one out. The final fully simplified result is shown in Figure 3.6 (f).

3.4 Mechanical Rotation

Mechanical rotation systems are directly analogous to mechanical translation. This results from the fact that the equations of motion for rotation, like those for translation, are derived through the use of Newton's Laws, especially Newton's Second Law. Hence, it should come as no surprise that the bond graph synthesis guidelines for rotation are similar to translation.

Before we get to that though, let us review the rotational system elements. Table 3.2 summarizes those elements. Recall that though bearings are often designed to be very efficient, they dissipate a nominal amount of energy and thus are classified as R-elements. A torsion shaft or spring stores potential energy when its ends are twisted relative to one another. That potential energy is released as the shaft becomes unloaded. Hence, a torsion spring is a C-element. Rotational inertias are composed of mass. Instead of translating, the mass is spun or twisted. As the mass rotates, it stores kinetic energy making the inertia an I-element. Gear pairs and chain-and-sprockets operate as transformers in rotational systems. Angular velocity differentials occur at 0-junctions, and moments (torques) are summed at 1-junctions.

Table 3.2: *Mechanical rotation elements*

Element Type	Mechanical Component	A Depiction
R-element	Bearing	Figure 2.1 (b)
C-element	Torsion Spring	Figure 2.3 (b)
I-element	Rotational Inertia	Figure 2.5 (b)
Effort Source	External Torque	N/A
Flow Source	Angular Velocity Source	N/A
Transformer	Gear Pair and Chain-and-Sprocket	Figure 2.10 (b)
0-Junction	Angular Velocity Differential	N/A
1-Junction	Sum of Moments (or Torques)	N/A

Like mechanical translation systems, we initiate our bond graph synthesis process by identifying places where the system has distinct velocities. These often occur where there is an inertia and/or where elements are rigidly connected. The basic guidelines for synthesizing bond graphs for mechanical rotation systems are:

1. *Identify distinct angular velocities.* For each distinct angular velocity, establish a 1-junction. Distinct angular velocities are often associated

with rotational inertias. If a rotational inertia has a distinct velocity an I-element representing that inertia should be placed directly off the 1-junction. Moreover, if there are any other elements directly related with a specific angular velocity, they too should appear off the junction.

2. *Insert torque generating 1-ports and energy-converting 2-ports.* Torque- or moment-generating 1-ports are placed between appropriate pairs of 1-junctions off of 0-junctions. Similar to mechanical translation systems, torsion springs and sometimes bearings generate equal magnitude moments at each end that are applied in opposite directions. Because they can be attached between two moving points, they may experience an angular velocity differential. Their ends generally move at different rates. The secondary condition on a 0-junction is the sum of angular velocities or an angular velocity differential. Thus, these elements are attached off of 0-junctions to indicate the common magnitude torque and distinct angular velocities at each end. Transformers or gear pairs can also appear between distinct angular velocities.

3. *Assign power directions.* As before, use the general guidelines for assigning power. The remaining bonds can be assigned somewhat arbitrarily. It is convenient to follow a power-in-power-out convention to facilitate collapsing redundant junctions.

4. *Eliminate zero velocities.* As was done with translational systems, to differentiate between the angular velocities at each end of a torsion spring or other twistable elements, zero velocities were identified as being distinct and a 1-junction was used as a "place holder." The zero-angular-velocity 1-junctions are redundant. Therefore, to aid in simplification, eliminate zero-angular-velocity 1-junctions and their attached bonds.

5. *Simplify.* Simplify the resulting graph by condensing 2-port 0- and 1-junctions into single bonds and by recognizing that if adjacent junctions are of the same kind, those junctions can be condensed into a single junction.

6. *Assign causality.* Assign causal strokes on bonds to indicate the input-output relations at each port.

To illustrate the synthesis process, we again exam several examples. We will first examine the rotational system illustrated in Figure 3.7.

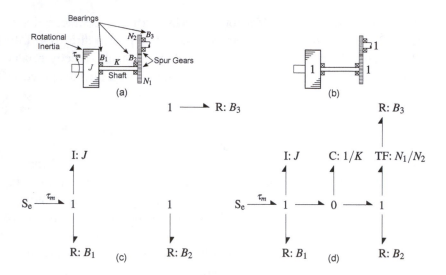

Figure 3.7: *A simple rotational system example*

Example 3.3

Synthesize the bond graph for the rotational system given in Figure 3.7.

Solution. As with mechanical translation systems, we first *identify the distinct velocities*. Figure 3.7 (b) shows that there are three different angular velocities – one at the rotational inertia and one at each spur gear. The rotational inertia is attached at one end of the shaft next to a bearing. In addition, each spur gear is also attached to an adjacent bearing. Furthermore, an external moment, τ_m, is applied directly to the rotational inertia. Therefore, as Figure 3.7 (c) depicts, an I-element, an R-element, and an effort source are attached off the leftmost 1-junction to account for the external torque, rotational inertia, and bearing. R-elements are attached to the remaining 1-junctions to represent the bearings adjacent to each gear. Next we *insert torque-generating 1- and 2-ports* between appropriate pairs of 1-junctions. The shaft, which stores potential energy, is represented by a C-element that goes off a zero junction between the 1-junctions connected to the rotational inertia and first spur gear. The gear pair is represented by a transformer between the 1-junctions accounting for the angular velocities of each gear. We can now *assign power directions*. Note how the unassigned power bonds tend to flow from the source out toward the energy-dissipating bearings at the right end. There are no explicit zero-velocity junctions to consider so we proceed to the subsequent step – *simplification*. This problem has only one redundant junction

– the 1-junction attached to the R-element representing the bearing next to the top spur gear – which can be collapsed so that it ends up connected directly to the end of the transformer as shown in Figure 3.7 (d).

The following example is much like the previous except for inclusion of a motor and an additional gear pair. Depending on how a motor is operated, it can be, in its most simplified terms, treated as an external torque or angular velocity source. As listed in Table 3.2, rotational transformers take on numerous forms including gears and chain-and-sprockets. As you may have seen in other courses, there are numerous types of gears. The system in Figure 3.8 has a spur gear pair and a bevel gear pair. Both can be used to generate a mechanical advantage. The difference is that the latter also changes the axis of rotation by a right angle.

Example 3.4

Synthesize the bond graph for the rotational system in Figure 3.8.

Solution. We first *identify distinct angular velocities* and any associated elements. This system has four distinct angular velocities – one at the motor output shaft (ω_m), one at the left end of the compliant shaft, another at the right end of the compliant shaft, and a final at the rotational inertia (refer to Figure 3.8 (b)). The motor supplies an angular velocity and its short shaft is attached to a gear through a bearing. Hence, as Figure 3.8 (c) depicts, a flow source and R-element are attached to the first junction. There are bearings at each end of the shaft. To account for these, R-elements are connected to the second and third 1-junctions. The rotational inertia is mounted on a short shaft with a bearing, and R- and I-elements are linked to the final junction to represent these. We now *insert torque-generating 1- and 2-ports* as illustrated in Figure 3.8 (d). A spur gear pair connects the motor output shaft to the left end of the compliant torsion shaft. A transformer is placed between the associated 1-junctions. The torsion shaft is represented by a C-element placed off a 0-junction between the 1-junctions representing the velocities at each of the shafts. The bevel gear pair joins the shaft to the rotational inertia (J). This, like the spur gear pair, is represented by a transformer between the two remaining angular velocities. We *assign the power directions* starting from the flow source and going up and around out to the rotational inertia. There are no simplifications to make, and thus we stop here.

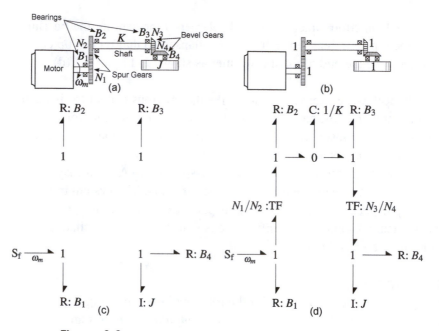

Figure 3.8: *A more complex mechanical rotation example*

3.5   Electrical Circuits

The process for bond graphing electric circuits varies significantly from mechanical systems. In mechanical systems, translational and rotational alike, we first identified distinct velocities and established a 1-junction to represent each. This derives from the fact that a common velocity (a flow) is shared at points where mechanical elements are rigidly attached. In electric circuits where elements are connected, they share a voltage (which is an effort) rather than a current (which is a flow). This implies a fundamental shift in how we synthesize the bond graph. We identify points of common effort rather than points of common flow.

Let us first review the basic elements in electric circuits listed in Table 3.3 for your convenience. Resistors dissipate energy and are therefore R-elements. Potential energy is stored in capacitors making them C-elements, while kinetic energy is stored by inductors which serve as I-elements in electric circuits. Ideal voltage sources and batteries serve as effort sources, while ideal current sources supply flow. Electric transformers operate similar to gear pairs in that the voltages (or efforts) on either side are proportional as are the currents (or flows). If you recall from a Circuits or Physics course, Kirchhoff's laws where used to determine currents and voltages. In circuits, elements connected in series share a common current (a flow). Hence, 1-junctions are used to represent circuit loops where the current is shared and the voltages are summed as done using Kirchhoff's Voltage Law. When in parallel, circuit elements share a voltage and currents are summed at the node. A common-effort or 0-junction is used to represent circuit nodes where Kirchhoff's Current Law is applied.

Table 3.3: *Electric circuit elements*

Element Type	Electrical Component	A Depiction
R-element	Resistor	Figure 2.1 (c)
C-element	Capacitor	Figure 2.3 (c)
I-element	Inductor	Figure 2.5 (c)
Effort Source	Battery or Ideal Voltage Source	Figure 2.8 (a)
Flow Source	Ideal Current Source	Figure 2.8 (b)
Transformer	Transformer	Figure 2.10 (c)
0-Junction	Sum of Currents into a Node	Figure 2.14 (a)
1-Junction	Sum of Voltage Drops around a Loop	Figure 2.14 (d)

Bond graph synthesis of electric circuits differs from mechanical systems significantly. Though many of the steps remain basically the same, the synthesis of electric circuit bond graphs begins by identifying unique efforts as opposed to unique flows.

1. *Identify distinct voltages.* Unlike mechanical systems where we first identify distinct velocities (or flows) and establish a 1-junction, in electric circuits we identify distinct voltage potentials (or efforts) including ground and establish a 0-junction for each. If there are any elements directly related to any of the distinct voltages, we place them, using a bond, directly off the respective 0-junction. Distinct potentials are located between elements.

2. *Insert 1-port circuit elements and energy-converting 2-ports.* Energy-storing and dissipating circuit elements typically generate voltage drops. Therefore, we place them off of 1-junctions between appropriate pairs of 0-junctions representing the voltages at the ends of the element. They appear off of 1-junctions because the secondary condition for a 1-junction dictates a summation of efforts. In electric circuits, this can be thought of as a voltage drop calculation. Additionally, energy converting 2-ports like transformers can cause a change in voltage and can

appear between a pair of 0-junctions representing the distinct potentials at either end.

3. *Assign power directions.* Remember to use the general guidelines for assigning power. The remaining bonds are assigned arbitrarily, but it is convenient to follow a power-in-power-out convention to facilitate collapsing redundant junctions.

4. *Eliminate explicit ground.* To facilitate proper placement of 1-ports, we explicitly included ground as a distinct voltage in the first step. In circuits we often measure voltages relative to ground. At this point we can eliminate the 0-junction associated with ground and all bonds directly attached.

5. *Simplify.* Simplify the resulting graph by condensing 2-port 0- and 1-junctions into single bonds and by collapsing adjacent junctions that are of the same type.

6. *Assign causality.* Assign causal strokes on bonds to indicate the cause-and-effect relations at each port. We will return to this step soon.

Let us review some examples for illustration. The first example problem will be the simple circuit illustrated in Figure 3.9 (a).

Example 3.5

Synthesize the bond graph for the circuit illustrated in Figure 3.9 (a).

Solution. First, we *identify the distinct voltage potentials* and, for each, establish a 0-junction as illustrated in Figure 3.9 (b). At this stage, we also want to identify any elements directly associated with those junctions. The only element would be a ground which is represented in Figure 3.9 (c) as an effort source with 0 voltage. Next we *insert the 1-port circuit elements*. This problem has no 2-ports. The 1-ports appear off of 1-junctions between the appropriate pair of 0-junctions. The *power directions are then assigned*. Recall that, though the power directions can be assigned somewhat arbitrarily, careful selection can further facilitate later simplification. We can then *eliminate the explicit ground* which entails removing the bottom 0-junction and any bonds and elements directly attached. Then *simplify*. The bond graph can be simplified by collapsing trains of bonds and junctions to arrive at the final bond graph illustrated in Figure 3.9 (d). Note that bond graphs for electric circuits can be done often times by inspection. The circuit is made up of three branches connected in parallel (i.e., the three branches share a common voltage). Hence, the

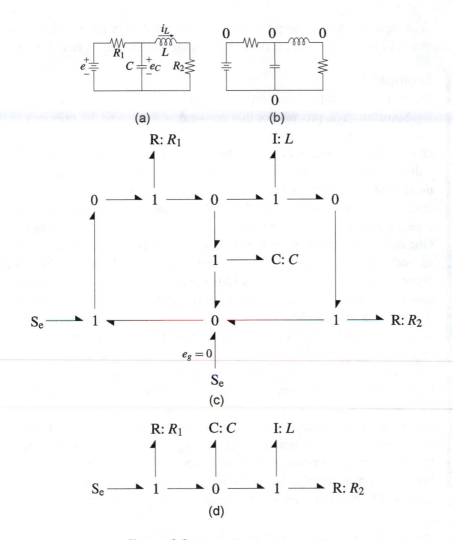

Figure 3.9: *A simple circuit example*

bond graph can be thought of as three branches off the 0-junction. The leftmost branch consists of a battery and resistor connected in series (i.e., they have a common current). Thus the effort source and R-element appear off a 1-junction together. The central branch is just a capacitor (or C-element). The rightmost branch is an inductor and resistor in series (again a common current). The representative I- and R-elements appear off a 1-junction together.

The next example is slightly more complicated than the previous. As shown in Figure 3.10 (a) the second circuit example includes a transformer.

Example 3.6

Derive the differential equations for the circuit in Figure 3.10 (a).

Solution. The process for this problem is much like the previous except that we now have not just 1-ports to insert into our bond graph but also a 2-port – the transformer. The first thing is to identify the distinct voltages (refer to Figure 3.10 (b)). Also, at this point you want to *identify an element directly associated with a distinct voltage* (refer to Figure 3.10 (c)). In this case there is just the explicit ground. We *insert the 1- and 2-port circuit elements.* A nuance results when inserting the transformer. One end of the transformer is between the inductor and ground and the second between the resistor and ground. That is, each end lies between distinct pairs of 0-junctions. As illustrated in the figure, 1-junctions are inserted between each pair of 0-junctions, and the transformer connects the two 1-junctions. Then, the *power directions are assigned.* The simplification process begins by *eliminating explicit ground.* We further *simplify* by collapsing trains of bonds. The resulting simplified bond graph is given in Figure 3.10 (d). As noted before, circuit bond graphs can sometimes be done readily by inspection. The circuit is composed of two loops attached by a transformer. The simplified bond graph is made of two branches. The left branch is the voltage source, inductor, and the left side of the transformer that share a common flow. The right branch is the right side of the transformer, the resistor, and the capacitor in series. The right and left branches appear off of 1-junctions because each branch shares a common current. The two branches interface through the transformer.

Recall the equivalencies that occur in mass-spring-damper systems when we have springs and dampers connected between the same pair of moving points. In electric circuits we have similar equivalencies when circuit elements or branches are connected in parallel. Figure 3.11 illustrates several examples. Example (a) is just an inductor and a resistor in parallel. When elements in a circuit are connected in parallel, they share a common voltage drop and therefore should appear off a 0-junction together. However, if we follow the guidelines for electric circuit bond graphs, we will arrive at a bond graph segment similar to that illustrated in Figure 3.11 (c). Note that the inductor and resistor are represented by an I- and R-element, each of which appears off a distinct 1-junction. Recall that the secondary condition for a 1-junction is the summation of efforts. In electric circuits, this can be thought of

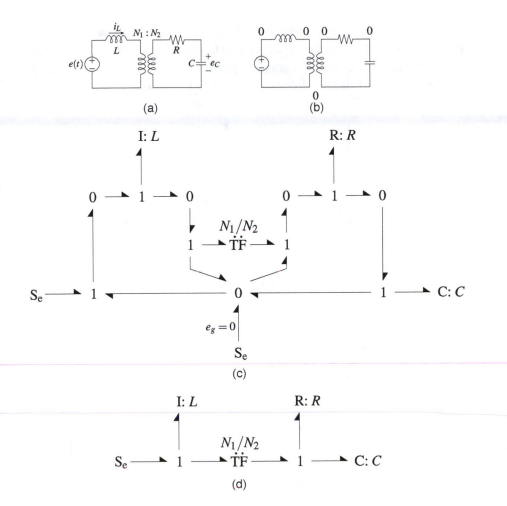

Figure 3.10: *A circuit example with a transformer*

as the calculation of a voltage differential. Hence, though the two 1-junctions represent distinct currents, each calculates the same voltage drop. Therefore, a single 1-junction can be used to which we affix, using a bond, a 0-junction off of which are connected both the I- and the R-elements to represent the fact that the two have a common voltage differential. The second circuit segment in (b) is composed of a resistor and capacitor in series connected in parallel to an inductor. If we follow the guidelines, the bond graph segment given in (d) will result. Nonetheless, the bond graph can be simplified by using an equivalency. The resistor and capacitor share a common current and should appear off a 1-junction; the combination of the two share a voltage drop with the inductor. Thus, the resistor-capacitor combo can be placed off a 0-junction with the inductor. Moreover, a single 1-junction can be used to determine the common voltage differential that exists across the inductor and the resistor-capacitor combo. The resulting simplified equivalency is depicted in (f).

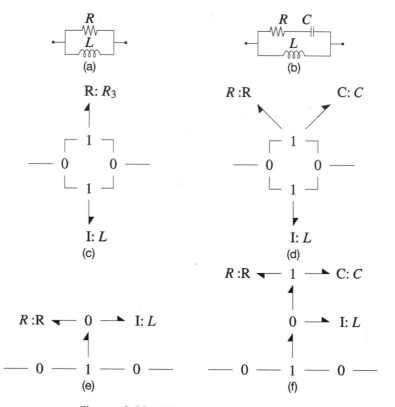

Figure 3.11: *Electric circuit equivalents*

The circuit illustrated in Figure 3.12 has a circuit segment composed of a resistor and inductor connected in parallel. The following example discusses the synthesis of the bond graph for this circuit and how an equivalency can be used to further simplify the graph.

Example 3.7

Synthesize the bond graph for the circuit in Figure 3.12.

Solution. The bond graph is synthesized in much the same manner as the previous two circuit problems. Identify the distinct voltages, insert the 1-port circuit elements between appropriate pairs of 0-junctions off of

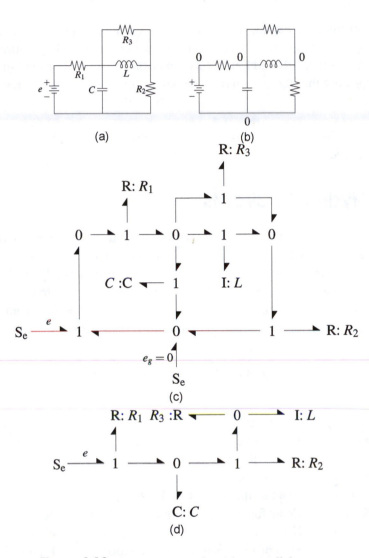

Figure 3.12: *A circuit example with a parallel segment*

one junctions, assign the power directions, eliminate explicit ground, and simplify. It is when we simplify that this problem deviates slightly from the previous two. Note in Figure 3.12 (c) how the R-element representing the resistance R_3 appears off a 1-junction between a pair of 0-junctions, and that the I-element representing the inductor is also off a 1-junction between the same pair of 0-junctions. The two generally do not have the same current and cannot be placed off a single 1-junction. However, a single 1-junction can be used to determine the common voltage differential and then both the R- and I-elements can be placed off a common 0-junction to represent that the two share a voltage drop. If this equivalency is employed the final bond graph shown in (d) results.

In the following section we will discuss the synthesis of bond graphs for hydraulic systems.

3.6 Hydraulic Circuits

Mechanical translation and mechanical rotation systems were presented back-to-back because the two have similar processes for synthesizing bond graphs. This should come as no surprise as both are governed basically by the same rules – Newtonian Mechanics. Oddly enough, hydraulic circuits are very similar to electric circuits. As we will see, the guidelines for synthesizing bond graphs for hydraulic circuits are directly analogous to those for electric circuits.

Table 3.4: *Hydraulic circuit elements*

Element Type	Hydraulic Component	A Depiction
R-element	Valve or Surface Roughness	Figure 2.1 (d)
C-element	Accumulator	Figure 2.3 (d)
I-element	Slug of Fluid	Figure 2.5 (d)
Effort Source	Displacement Pump or Pressure Source	N/A
Flow Source	Centrifugal Pump or Ideal Flow Source	N/A
Transformer	N/A	N/A
0-Junction	Sum of Flow Rates into a Junction	Figure 2.14 (b)
1-Junction	Sum of Pressure Drops around a Loop	Figure 2.14 (e)

First, let us review the basic elements for hydraulic circuits listed in Table 3.4. R-elements in hydraulic circuits are used to represent pressure drops that result from energy losses through valves or because of roughness along the surface of the pipe walls. Accumulators can store potential energy by varying the height (or volume) of fluid. Fluids have density and can store considerable amounts of kinetic energy has they flow through pipes and tubing. A displacement pump is often used to supply a pressure to a hydraulic circuit and can be treated as an ideal pressure source. A centrifugal pump, on the other hand, usually governs the flow through a hydraulic system and thus can be treated as an ideal flow source. A physical junction in a hydraulic circuit where pipes or tubes connect is similar to a node in an electric circuit where branches of the circuit connect. At these physical junctions, flow rates combine (i.e., a sum of flow rates results). Additionally, physical junctions are points of common pressure. Hydraulic circuit elements can be connected "in series" much like electric circuit elements. When they are connected in such a manner the elements share a common volumetric flow rate and pressure drop results across element.

Given the similarities between electric and hydraulic circuits, it should come as no surprise that the guidelines for synthesizing bond graphs of hydraulic circuits is a near carbon copy of those for electric circuits.

1. *Identify distinct pressures.* In hydraulic circuits we identify distinct pressures (or efforts) including atmospheric pressure and establish a 0-junction for each. If there are any elements directly related to any of the distinct pressures, we place them, using a bond, directly off the respective 0-junction. Distinct pressures are located between hydraulic elements.

2. *Insert 1-port hydraulic circuit elements and energy-converting 2-ports.* Energy-storing and dissipating hydraulic circuit elements typically generate pressure drops. Therefore, we place them off of 1-junctions between appropriate pairs of 0-junctions representing the pressures at ends of the element. They appear off of 1-junctions because the secondary condition for a 1-junction dictates a summation of efforts. In hydraulic circuits, this can be thought of as a pressure drop calculation. Additionally, energy converting 2-ports can cause a change in pressure and can appear between a pair of 0-junctions representing the distinct potentials at either end.

3. *Assign power directions.* Remember to use the general guidelines for assigning power. The remaining bonds can be assigned arbitrarily, but it is convenient to follow a power-in-power-out convention to facilitate collapsing redundant junctions.

4. *Eliminate atmospheric pressure.* Much like electric circuits where we typically measure voltage relative to ground, in hydraulic circuits we measure gauge pressure or the pressure differential relative to atmospheric pressure. If we typically use gauge as opposed to absolute

pressure we can eliminate atmospheric pressure. At this point we can eliminate the 0-junction associated with atmospheric pressure and all bonds directly attached.

5. *Simplify.* Simplify the resulting graph by condensing 2-port 0- and 1-junctions into single bonds and by collapsing adjacent junctions that are of the same type.

6. *Assign causality.* Assign causal strokes on bonds to indicate the cause-and-effect relations at each port. We will return to this step in just a few sections.

We will illustrate the process by examining a few key examples, the first of which is illustrated in Figure 3.13.

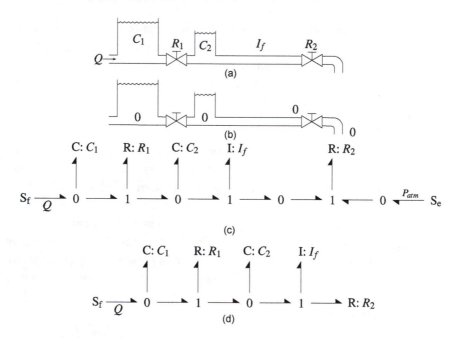

Figure 3.13: *A hydraulic circuit problem*

Example 3.8

Synthesize the bond graph for the hydraulic circuited in Figure 3.13 (a).

Solution. The hydraulic circuit is composed of a flow source, two accumulators, two valves, and a pipe. The first step is to *identify the distinct pressures*. In this problem, distinct pressures exist at the bottoms of each accumulator, at the right end of the long pipe, and at the exit (see Figure 3.13 (b)). For each we establish a 0-junction. As illustrated in Figure 3.13 (c), accumulators are associated with two of the junctions and atmospheric pressure is associated with a third. Thus, C-elements and an effort source are placed off the appropriate 0-junctions. Next, we *insert 1- and 2-port elements*. A valve exists between the two accumulators which causes a pressure drop. It is represented by an R-element off a 1-junction between the two leftmost 0-junctions. The long pipe flows a significant amount of fluid with inertia. Hence, an I-element is placed off a 1-junction between the second and third 0-junctions representing the pressures at the left and right ends of the pipe. Another valve exists between the end of the pipe and the exit which is exposed to atmospheric pressure. This valve is represented by an R-element off a 1-junction between the two rightmost 0-junctions. Now we *assign power directions*. Directions are chosen in a manner that makes sense and facilitates later simplification. Now one can *eliminate atmospheric pressure* which involves removing the associated effort source, the attached 0-junction, and any bond attached to that 0-junction. This allows us to *simplify* by collapsing trains of bonds. Finally, we arrive at the simplified solution in Figure 3.13 (d).

The next example illustrates a complication that can occur in some hydraulic systems – decoupled subsystems. Figure 3.14 (a) illustrates a hydraulic circuit composed of two subsystems. The outlet of the top subsystem is exposed to atmosphere and provides flow into the second. This results in the transfer of flow only and not power between the two subsystems. We illustrate in the following example how this manifests in a bond graph.

Example 3.9
Synthesize the bond graph for the hydraulic circuit in Figure 3.14 (a).

Solution. The hydraulic circuit depicted in Figure 3.14 is really composed of two circuits which are related only through a flow. The flow out of the top circuit is the flow into the bottom. We can think of it in this manner: The flow out the top circuit modulates the flow into the bottom circuit. Because the outlet of the top circuit is exposed to atmosphere, and no pressure intrinsic to the bottom circuit affects the top, only a signal and not power connects the two subsystems. Despite this complication, we still proceed as before. That is, we first *identify the distinct pressures*. The top circuit has distinct pressures at the bottom of the accumulator and at the outlet. The same is true for the bottom circuit. Thus as illustrated in Figure 3.14 (b), four 0-junctions are established to represent these pres-

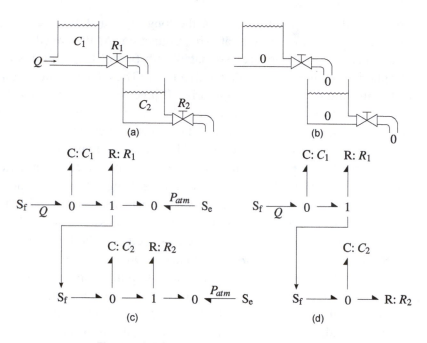

Figure 3.14: *A decoupled hydraulic circuit*

sures. The accumulators are represented by C-elements placed off two of the zero junctions, and atmospheric pressure at the outlets of circuits are represented by effort sources off the other two 0-junctions. Furthermore, each circuit has a flow source feeding into the junctions at the bottoms of the two accumulators. These are represented by flow sources attached to each of those 0-junctions. We *insert the 1-ports* which include two valves, each of which is placed off a 1-junction between pairs of 0-junctions that represent the pressures at the bottoms of the accumulators and at the outlets of the circuits. *Power directions are then assigned.* Assuming that the pressures are measured relative to atmosphere, we can *eliminate the sources representing atmospheric pressure* and *simplify.* At this point you might be tempted to collapse the R-element for the top circuit to the 0-junction. However, since the flow going through the valve is the flow out of the top circuit and into the bottom circuit, we leave the 1-junction and attach a *signal* bond (not a power bond) from the 1-junction to the flow source for the bottom circuit. A signal bond only carries effort or flow in one direction, unlike a power bond which carries effort in one direction and flow back in the other. The signal bond is shown in Figure 3.14 (c).

This type of annotation implies that the signal associated with the top 1-junction (in this case a volumetric flow rate) modulates the bottom flow source. We can now simplify being sure to maintain the top 1-junction and not collapsing the top R-element to the 0-junction. This results in the final bond graph in Figure 3.14 (d).

3.7 Mixed Systems

Many dynamic systems are not purely of a single energy domain but rather mixed systems with subsystems that operate in various domains. Systems can be electromechanical, hydromechanical, rotational and translational, etc. Mixed energy systems can be decomposed into subsystems, each of a single energy domain. In this manner, the subsystems can be handled using the guidelines detailed in the previous sections for mechanical translation, mechanical rotation, electric circuits, and hydraulic circuits. The subsystems are coupled by devices referred to as transducers, which convert energy from one form to another. Some such devices are depicted in Figure 3.15. Typically, transducers are considered to be energy conservative and can be represented by transformers or gyrators depending on how they convert energy. We will study a mixed-system example to illustrate how a transducer can couple energy domains and how a system can be decomposed into subsystems that operate in a single energy domain.

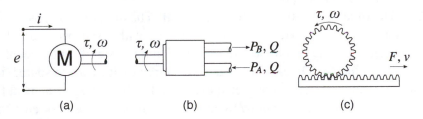

(a)   (b)   (c)

Figure 3.15: *Examples of transducers including (a) an ideal motor, (b) a centrifugal pump, and (c) a rack-and-pinion*

Example 3.10
Recall the PMDC motor schematic from Figure 1.3. The schematic is repeated for convenience in Figure 3.16. Synthesize a bond graph for the system.

   Solution. Many electromechanical devices such as robotic arms, wind turbines, and computer numerical controlled (CNC) machines use motors

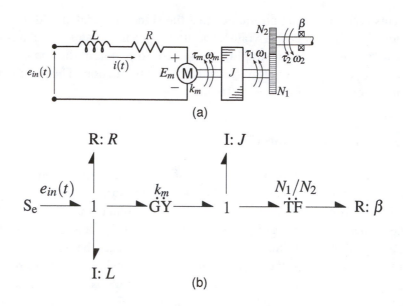

(a)

(b)

Figure 3.16: *A PMDC motor model*

and/or servos. These devices, though electrically driven, are used to position and move mechanical systems. A commonly used device in such systems is a PMDC motor.

The motor in itself is electromechanical. The motor is composed of a rotating coil that is made of a copper winding with resistance and inductance and, depending on the size of the winding, a significant amount of rotational inertia. On one side you have a circuit with voltage source, resistance, and inductance connected in series. The bond graph fragment for the electrical side can be done by inspection. It is composed of an effort source, R-element, I-element, and open bond off a 1-junction.

On the other side you have the rotational inertia of the winding, a gear train, and bearing. The electrical and mechanical sides of the system are coupled by an "ideal motor," which represents the electromechanical coupling between both sides. It relates the torque generated on the mechanical side to the current conducted through the electrical side. Moreover, it relates the electromotive force – the voltage drop across electrical side due to the generation of torque on the mechanical side – to the angular velocity of the rotating coil. A motor constant, k_m, is used to cross-relate the efforts and flows on electrical and mechanical sides,

$$\tau_m = k_m i(t) \text{ and } E_m = k_m \omega_m.$$

Therefore, the "ideal motor" can be represented by a gyrator element with a modulus of k_m.

The mechanical side is characterized by two distinct angular velocities – one at the motor output shaft and the other after the gear pair. The output gear is shaft-mounted through a bearing. We establish two 1-junctions. Off the first we attach an I-element to represent the rotational inertia of the coil. From the second, we attach an R-element to represent the bearing. Between the two 1-junctions we insert a transformer to represent the gear pair. The resulting bond graph can be simplified by collapsing the R-element representing the bearing at the output shaft to the transformer directly.

3.8 State Equation Derivation

Once a bond graph has been synthesized and simplified, one can begin to derive a set of differential equations that describe the dynamic response. Though power directions are assigned somewhat arbitrarily, care must be taken to ensure that the resulting sign convention complies with common practices such as coordinate systems.

Though there is no established process for deriving differential equations from bond graphs, the following is provided as a set of guidelines:

1. *Assign causality.* As has been mentioned before, the first step in deriving the state equations is assigning causality. The bond graph must be annotated with causal strokes to indicate on each bond the end or port where effort is assumed to be an input. Causal strokes are assigned in a systematic manner:

   (a) We *begin with sources* because they have an inherent cause-and-effect that cannot be altered. Once a bond has an assigned causality we check to see if that affects adjacent junctions in such a manner that the cause-and-effect is propagated. As illustrated in Figure 3.17 (a), if an effort source is attached to a 0-junction then it specifies the effort at that junction. Hence, the remaining bonds specify flow into the junction and effort out to the adjacent junctions or elements. Similarly, if a flow source is attached to a

1-junction, the source specifies the flow at the junction, and the remaining bonds must therefore specify efforts into the junction (refer to Figure 3.17 (b)).

(b) Then we *proceed to the energy-storing elements* (i.e., I- and C-elements). Start with any I- or C-element and place it, if possible, in integral causality. If the applied causality affects an adjacent junction, propagate the cause-and-effect accordingly. For instance, a C-element in integral causality attached to a 0-junction will specify the effort at that junction as will an I-element in integral causality specify the flow at a 1-junction (refer to Figure 3.17 (c) and (d)). Note the resulting number of energy-storying elements in integral causality will be equal to the number of independent dynamic states and consequently also the number of first-order differential equations we will derive.

(c) If any bonds remain, *select an R-element and specify its causality.* Usually, after the energy-storing elements, all bonds have an assigned causality. Occasionally, when unassigned bonds remain you must select an R-element with an unassigned causality, specify its causality, and propagate the cause-and-effect. Such instances result in algebraic loops. An example will be provided later to demonstrate.

2. *Label efforts and flows on the energy-storying elements.* Once causality is assigned throughout the bond graph, efforts and flows must be labeled on the energy-storing elements. Recall that for generic linear C-elements the associated state will be a displacement (q), the flow will be the time rate of change of that displacement ($\dot{q}$), and the effort will be the displacement divided by the compliance (q/C). For generic linear I-elements, the associated state will be a momentum (p), the time rate of change of that momentum ($\dot{p}$) will be the effort, and the flow will be the momentum divided by the inertance (p/I). For the power and energy variables of specific domains recall Tables 2.5 and 2.6.

3. *Apply the primary conditions.* To generate the state equations we need to remember the primary and secondary conditions that occur at each junction. These conditions are used to propagate information throughout the bond graph and determine the efforts that sum at 1-junctions or the flows that sum at 0-junctions. As an aid we can distribute this information throughout the bond graph by labeling the efforts and flows that result on each bond when applying the conditions. We apply the

primary conditions and propagate the information. While applying the primary conditions at the junctions some differential equations may result.

4. *Apply the secondary conditions.* We finish by applying the secondary conditions to finalize distributing information and deriving the remaining differential equations.

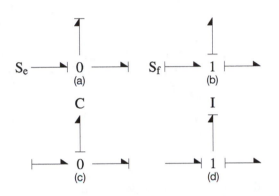

Figure 3.17: *Examples of elements affecting the causality on adjacent junctions*

Sometimes an energy-storing element cannot be placed in integral causality. This implies that the associated state is not independent of the others and that a differential equation is not necessary for that state. Later, we will review some examples where this occurs.

In many problems, once you have assigned causality for the sources and energy-storing elements, the cause-and-effect for complete bond graph has been specified. If any bonds are left unspecified, select a causality and propagate until the remaining bonds are specified. Such an occurrence implies an algebraic loop that will need to be solved in order to derive the equations for your system. This will also be illustrated in an example later.

The following examples illustrate the process detailed above. We begin by examining a mass-spring-damper system.

Example 3.11

Derive the differential equations for the mass-spring-damper system in Figure 3.18.

Solution. The mass-spring-damper system and previously derived bond graph from Example 3.1 are shown in Figure 3.18. To derive the differential equations, we begin by *applying causality*. First the *source is assigned*. The single effort source will specify an effort into the system; thus

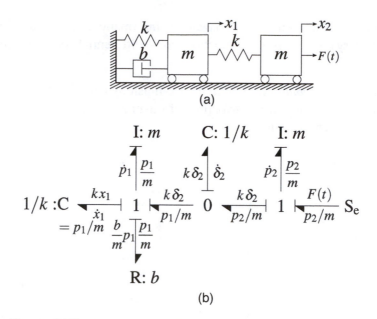

Figure 3.18: *Equation derivation for a mass-spring-damper system*

the causal stroke is placed on the end of the bond pointing into the system. Next we *assign causal strokes on the energy-storing elements*. It does not matter which I- or C-element we start with. However we choose, the result is the same – all energy-storing elements end up in integral causality. The C-elements specify effort into the adjacent junctions, and the I-elements specify flow into their respective junctions. At this stage all bonds have an assigned causality. Then we *label the efforts and flows* on the bonds attached to the energy-storing elements.

Now we *apply the primary condition* at each junction which, at the left 1-junction, leads us to the differential equation

$$\dot{x}_1 = \frac{p_1}{m}. \tag{3.1}$$

Additionally, this forces us to identify the common flow p_1/m at the left 1-junction, the common effort $k\delta_2$ at the 0-junction, and the common flow p_2/m at the right 1-junction. Finally, we *apply the secondary condition* at each junction to derive the remaining three differential equations. At the left 1-junction there is a summation of forces (efforts),

$$-kx_1 - \dot{p}_1 + k\delta_2 - \frac{b}{m}p_1 = 0.$$

However, note that the left, right, and bottom bonds at that junction specify the effort into the junction and that the top bond outputs an effort from the junction. To account for the causality, the above summation should be written as

$$\dot{p}_1 = -kx_1 - \frac{b}{m}p_1 + k\delta_2. \tag{3.2}$$

At the 0-junction the summation of flow results in

$$-\frac{p_1}{m} - \dot{\delta}_2 + \frac{p_2}{m} = 0,$$

but the causality at the junction indicates that $\dot{\delta}_2$ is an output flow from the junction. Thus, to indicate this, the equation should be reorganized as

$$\dot{\delta}_2 = -\frac{p_1}{m} + \frac{p_2}{m}. \tag{3.3}$$

Application of the secondary condition at the right 1-junction produces

$$-k\delta_2 - \dot{p}_2 + F(t) = 0,$$

which becomes

$$\dot{p}_2 = -k\delta_2 + F(t), \tag{3.4}$$

when accounting for the causality at the junction. The efforts $k\delta_2$ and $F(t)$ are inputs used to determine $\dot{p}_2$ as an output. The resultant state equations are in terms of the displacement of the first mass, the momentum of the first mass, the deflection of the second spring, and the momentum of the second mass (i.e., x_1, p_1, δ_2, and p_2).

There is no unique set of states or state equations for any given dynamic system. In a bond graph formulation the states are displacements or deflections and momenta. In a traditional formulation, the commonly used states are displacements and velocities. Regardless of the method used, the resultant system of equations should be equivalent. We will confirm this by deriving the state equations using the free body diagrams illustrated in Figure 3.19. By summing the forces on the left mass we can derive an equivalent representation of Equation 3.2,

$$m\ddot{x}_1 = m\dot{v}_1 = -kx_1 + k(x_2 - x_1) - bv_1,$$

which when solved for $\dot{v}_1$ gives

$$\dot{v}_1 = -\frac{2k_1}{m}x_1 - \frac{b}{m}v_1 + \frac{k}{m}x_2. \tag{3.5}$$

By recognizing the deflection δ_2 is the difference between the displacements x_2 and x_1 and that the momentum $p_1 = mv_1$, we can readily show that $m\dot{v}_1$ is equal to Equation 3.2,

$$\dot{p}_1 = -kx_1 + k\delta_2 - \frac{b}{m}p_1 = -k_1 + k(x_2 - x_1) - bv_1 = m\dot{v}_1.$$

Similarly, we can show that

$$\dot{p}_2 = -k\delta_2 + F(t) = -k(x_2 - x_1) + F(t) = m\dot{v}_2.$$

Solving the above equation for $\dot{v}_2$ gives

$$\dot{v}_2 = \frac{k}{m}x_1 - \frac{k}{m}x_2 + \frac{1}{m}F(t). \tag{3.6}$$

Moreover, we can write Equation 3.1 in terms of the velocity of the first mass

$$\dot{x}_1 = v_1 \tag{3.7}$$

and replace Equation 3.3 with

$$\dot{x}_2 = v_2, \tag{3.8}$$

which substitutes the spring deflection – a differential or relative motion – with the displacement of the second mass. Though not equal, per se, the alternate equations are equivalent and will predict the same dynamic response in terms of a different set of states (x_1, v_1, x_2, and v_2).

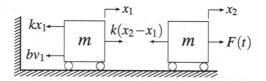

Figure 3.19: *Free body diagrams of the two masses in Figure 3.4*

As the following example illustrates, we follow the same basic process for rotational systems.

Example 3.12

We previously derived the bond graph for Example 3.4. It is repeated

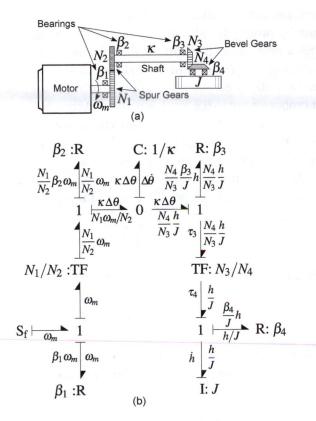

Figure 3.20: *Equation derivation for a rotational system*

in Figure 3.20 for convenience. Derive the differential equations of this problem.

Solution. We begin by *applying the causality starting with the source*. It is attached to a 1-junction. Since it specifies the flow into that junction the remaining attached bonds must specify efforts. Thus the causality is propagated to the attached bonds. Additionally, if the flow is specified into one end of the left transformer, then the flow out of the transformer is, as a result, also specified. Consequently, the flow into the top-left 1-junction is determined by the transformer, and the two remaining bonds at that junction must input efforts. Then, we *assign the causality on the energy-storying elements*. If we begin with the C-element, then the effort at the 0-junction is prescribed. The remaining unprescribed bonds at that junction must supply flow as inputs into the 0-junction. Placing the I-element in integral causality stipulates the flow into the bottom-right 1-junction. The causality propagates all the way out to the top-right 1-junction and prescribes a cause-and-effect for all the remaining bonds. Now we can *label the efforts and flows* on the C- and I-elements.

Note how the prescribed causalities of the flow source and the I-element propagate out to the top 1-junctions. Accordingly, when *applying the primary condition* at the bottom 1-junctions, the flows are determined all the way up the left and right sides of the bond graph up toward the 0-junction. Given the common angular velocity, ω_m, at the bottom 1-junction we know the torque lost in the first bearing, $B\omega_m$. The angular velocity at the left end of the shaft is proportional to ω_m through the first gear ratio

$$\omega_2 = \frac{N_1}{N_2} \omega_m .$$

Given this, we can determine the torque dissipated in the second bearing,

$$\tau_{\beta_2} = \frac{N_1}{N_2} \beta_2 \omega_m .$$

In a similar fashion, we can surmise the angular velocity at the right end of the shaft and the moment lost at the third bearing,

$$\omega_3 = \frac{N_4}{N_3} \omega_4 = \frac{N_4}{N_3} \frac{h}{J} \quad \text{and} \quad \tau_{\beta_3} = \frac{N_4}{N_3} \frac{\beta_3}{J} h .$$

Application of the primary condition at the 0-junction means that the at-tached bonds have the same effort as the C-element, $\kappa \Delta\theta$. Finally, the common angular velocity at the bottom right 1-junction can be used to determine the torque in the fourth bearing,

$$\tau_{\beta_4} = \frac{\beta_4}{J} h .$$

To derive the differential equations, the *secondary condition must be applied*. At the 0-junction, the secondary condition results in an angular velocity differential,

$$\boxed{\Delta\dot{\theta} = \frac{N_1}{N_2} \omega_m - \frac{N_4}{N_3} \frac{h}{J} .}$$

The left and right bonds at the 0-junction are inputs used to calculate $\Delta\dot{\theta}$ as an output. To determine the remaining differential equation, $\dot{h}$, we need to determine the torque transferred through the bevel gear pair. This is specified by the summation of moments at the top-right 1-junction,

$$\tau_3 = \kappa \Delta\theta - \frac{N_4}{N_3}\frac{\beta_3}{J}h.$$

The torques at the left and top bonds are inputs and τ_3 is the output. Knowing τ_3, we can calculate τ_4 using the gear ratio,

$$\tau_4 = \frac{N_4}{N_3}\tau_3 = \frac{N_4}{N_3} = \frac{N_4}{N_3}\left[K\Delta\theta - \frac{N_3}{N_4}\frac{\beta_3}{J}h \right].$$

Now we can sum the moments at the bottom-right 1-junction to determine $\dot{h}$. Summing the input torques at the top and right bonds results in

$$\boxed{\dot{h} = \tau_4 - \frac{\beta_4}{J}h = \frac{N_4}{N_3}\left[\kappa\Delta\theta - \frac{N_4}{N_3}\frac{\beta_3}{J}h \right] - \frac{\beta_4}{J}h.}$$

By recognizing that the angular momentum is $h = J\omega$ one can rewrite the above equation in terms of the more traditionally used state ω,

$$J\dot{\omega} = \frac{N_4}{N_3}\left[\kappa\Delta\theta - \frac{N_4}{N_3}\beta_3\omega \right] - \beta_4\omega.$$

The following illustrates the use of this process on an electric circuit.

Example 3.13

Let us now review an electric circuit example. Recall the circuit and bond graph from Example 3.6. The circuit and annotated bond graph are provided in Figure 3.21 for convenience. Derive the differential equations for the circuit.

Solution. As before, we *assign the causality beginning with the source*. Since the source is an effort and is adjacent to a common-flow junction, the causality does not propagate. We proceed to *apply causality to the energy-storying elements*. Note that if we assign causality to the I-element first, it will propagate from the left 1-junction through the transformer to the right 1-junction, specifying causality on all the remaining bonds in the process. As a result, the C-element, though assigned, still ends up in integral causality. Alternatively, if we begin with the C-element and place it in integral causality first, that does not affect the right 1-junction and propagate. Subsequently, assigning causality to the

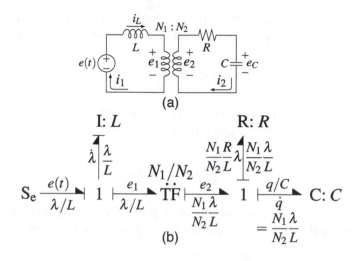

Figure 3.21: *Equation derivation for an electric circuit*

I-element does propagate to the remaining bonds. Regardless of the element we chose to assign first, the resulting causality on each bond is the same, and both energy-storing elements end up in integral causality.

Now we can *label the efforts and flows* for the energy-storing elements. For the inductor, the effort is the time rate of change of the flux linkage ($\dot{\lambda}$) and the flow is the inductor current ($i_L = \lambda/L$). The capacitor voltage (effort) and current (flow) are q/C and $\dot{q}$, respectively.

The differential equations are derived by applying the junction conditions. We *apply the primary condition first.* If we examine the causality at the right 1-junction, the transformer specifies the flow into that junction. This can be traced back to the left 1-junction and ultimately to the I-element. Hence, all the flows in the bond graph can be directly related to the current through the inductor ($i_L = \lambda/L$). As a result we have derived the differential equation for the capacitor charge,

$$\dot{q} = \frac{N_1}{N_2}\frac{\lambda}{L}.$$

The λ equation is derived by applying the secondary condition at the left 1-junction. However, we need to determine the voltage drop across the left side of the transformer (e_1), which is proportional to the voltage e_2 through the windings ratio (N_1/N_2). We can solve for voltage e_2 by summing the voltages at the right 1-junction,

$$e_2 = \frac{N_1}{N_2}\frac{R}{L}\lambda + \frac{q}{C}.$$

Knowing e_2 we can determine e_1:

$$e_1 = \frac{N_1}{N_2}e_2 = \frac{N_1}{N_2}\left[\frac{N_1}{N_2}\frac{R}{L}\lambda + \frac{q}{C}\right].$$

Finally, the voltages can be summed at the left 1-junction to determine the differential equation for λ,

$$\boxed{\dot{\lambda} = e(t) - e_1 = e(t) - \frac{N_1}{N_2}\left[\frac{N_1}{N_2}\frac{R}{L}\lambda + \frac{q}{C}\right].}$$

We can arrive at an equivalent set by applying Kirchhoff's Voltage Law to the two loops. The left loop results in the following summation:

$$e(t) - L\frac{di_L}{dt} - e_1 = 0 \Rightarrow L\frac{di_L}{dt} = e(t) - e_1.$$

The sum of the voltages around the right loop is

$$e_2 - R\dot{q} - \frac{q}{C} = 0 \Rightarrow e_2 = R\dot{q} + \frac{q}{C}.$$

We can use the transformer winding ratio to relate the voltages and currents,

$$\frac{e_1}{e_2} = \frac{N_1}{N_2} = \frac{i_2}{i_1}.$$

The current relation can be used to solve for $\dot{q}$ in terms of the inductor current

$$\boxed{\dot{q} = i_2 = \frac{N_1}{N_2}i_1 = \frac{N_1}{N_2}i_L.}$$

By substituting for $\dot{q}$ in the right loop equation and using the voltage relation, we derive the differential equation of i_L,

$$\boxed{L\frac{di_L}{dt} = e(t) - \frac{N_1}{N_2}e_2 = e(t) - \frac{N_1}{N_2}\left[R\dot{q} + \frac{q}{C}\right] = e(t) - \frac{N_1}{N_2}\left[R\frac{N_1}{N_2}i_L + \frac{q}{C}\right].}$$

As illustrated by the following example, the same basic process can be applied to a hydraulic system.

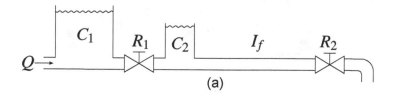

(a)

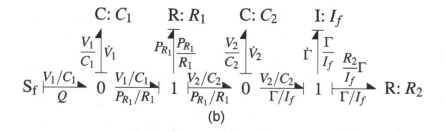

(b)

Figure 3.22: *Equation derivation of a hydraulic circuit problem*

Example 3.14

Derive the differential equations for the circuit from Example 3.8. The schematic and bond graph are repeated in Figure 3.22 for convenience.

Solution. We *assign causality* to derive the differential equations *beginning with the source*. There is a single source – a flow source – attached to an adjacent 0-junction. The causality does not propagate further so we continue by *assigning causality to the energy-storing elements*. Since there are three energy-storing elements in integral causality, we expect three differential equations.

With the causality assigned, we continue by *labeling the efforts and flows* on the three energy-storing elements. In hydraulic systems the states associated with C-elements are volumetric displacements whose time rate of change are flow rates. Hence, the C-elements have flows $\dot{V}_1$ and $\dot{V}_2$. If the constitutive relations are linear, the associated efforts are V_1/C_1 and V_2/C_2, respectively. The I-element represents the fluid inertia in the relatively long pipe. It has a hydraulic momentum whose time rate of change ($\dot{\Gamma}$) is an inertial pressure – an effort. Again, assuming a laminar flow, a linear constitutive relation results, and the flow rate through the pipe is proportional to the momentum (i.e., $Q_{pipe} = \Gamma/I_f$).

The *primary condition can now be applied* to distribute information throughout the bond graph. At the left 0-junction there is a common effort specified by the C-element to be V_1/C_1. The common effort at the other 0-junction is prescribed as V_2/C_2 by the second C-element. The I-element stipulates a common flow of Γ/I_f at the rightmost 1-junction. Given the flow at this 1-junction, we can determine the pressure drop across the second valve,

$$\Delta P_{R_2} = \frac{R_2}{I_f}\Gamma.$$

This provides us sufficient information to derive differential equation for Γ,

$$\dot{\Gamma} = \frac{V_2}{C_2} - \frac{R_2}{I_f}\Gamma \qquad (3.9)$$

To derive the remaining differential equations we need to apply the secondary condition at the leftmost 1-junction. The secondary condition at this junction results in a pressure drop across the valve R_1 (P_{R_1}); it is the difference in pressure between the the two accumulators,

$$P_{R_1} = \frac{V_1}{C_1} - \frac{V_2}{C_2}.$$

Knowing the pressure differential, we can calculate the flow through the valve which is shared by the two other bonds attached to the left 1-junction,

$$Q_{R_1} = \frac{P_{R_1}}{R_1} = \frac{1}{R_1}\left[\frac{V_1}{C_1} - \frac{V_2}{C_2}\right].$$

Now we have all the information necessary to determine the remaining differential equations,

$$\dot{V}_1 = Q - \frac{P_{R_1}}{R_1} = Q - \frac{1}{R_1}\left[\frac{V_1}{C_1} - \frac{V_2}{C_2}\right], \qquad (3.10)$$

and

$$\dot{V}_2 = \frac{P_{R_1}}{R_1} - \frac{\Gamma}{I_f} = \frac{1}{R_1}\left[\frac{V_1}{C_1} - \frac{V_2}{C_2}\right] - \frac{\Gamma}{I_f}. \qquad (3.11)$$

Differential equations for mixed systems can also be derived in the same manner.

Example 3.15

Derive the differential equations for the PMDC motor model in Example 3.10. The system schematic and bond graph are repeated in Figure 3.23.

Solution. *Assign the causality starting with the source* – a voltage input. This has no effect on the attached 1-junction so we continue to *assign causality to the energy-storying elements*. Regardless of the I-element we choose to assign first, both will ultimately be in integral causality. The causality of the inductor propagates out to the right side of the ideal motor (gyrator), and that of the rotational inertia propagates through the gear pair (transformer) out to the bearing (R-element).

Given the causality, we can commence *applying the primary condition*. The inductor, at the left 1-junction, specifies the current at that junction (λ/L). Consequently, the voltage source, resistor, and ideal motor (gyrator) have the same current. At the right 1-junction, the rotational inertia determines the angular velocity which is shared by the ideal motor and the first gear. The motor torque is proportional to the current and the the electromotive force to the motor angular velocity:

$$\tau_m = k_m i(t) = \frac{k_m}{L}\lambda \quad \text{and} \quad E_m = k_m \omega_m = k_m \frac{h}{J}.$$

Recall that the flows on either side of a transformer (gear pair) are directly related. If one is known the other can be calculated using the modulus (gear ratio)

$$\omega_2 = \frac{N_1}{N_2}\omega_1 = \frac{N_1}{N_2}\frac{h}{J}.$$

Once we know the angular velocity of the second gear, we can use the damping constant to determine the moment lost in the bearing and then relate it to the torque at the first gear:

$$\tau_2 = \beta\frac{N_1}{N_2}\frac{h}{J} \quad \text{and} \quad \tau_1 = \frac{N_1}{N_2}\tau_2 = \frac{N_1^2}{N_2^2}\frac{\beta}{J}h.$$

We now have sufficient information to derive the two differential equations.

By *applying the secondary condition* at both 1-junctions, we deduce each differential equation through a summation of efforts:

$$\boxed{\dot{\lambda} = e_{in}(t) - \frac{R}{L}\lambda - \frac{k_m}{J}h} \quad \text{and} \quad \boxed{\dot{h} = \frac{k_m}{L}\lambda - \frac{N_1^2}{N_2^2}\frac{\beta}{J}h}.$$

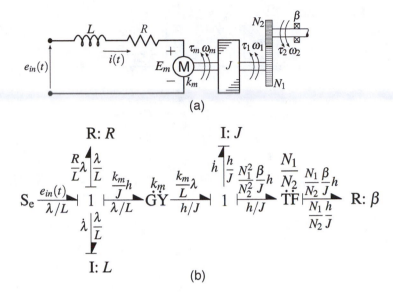

Figure 3.23: *Equation derivation of a mixed system*

3.9 Algebraic Loops and Derivative Causality

All the examples we have examined until now had fully assigned causality after allocating cause-and-effect on the energy-storing elements. That is, no bonds were left unassigned after we propagated the causalities assigned to the energy-storing elements. As we shall see in this section, bond graphs with unassigned bonds have algebraic loops that must be solved to derive the state equations and as a consequence the state-space representation. In other words, some extra effort is necessary to derive the final set of state equations. This phenomenon is best illustrated by examples.

Example 3.16

The mass-spring-damper system depicted in Figure 3.24 is seemingly straightforward. Imagine, however, that the image depicts a simplified model of two railcars being pushed up against a snubber. What if the first railcar were a fully loaded coal car and the second an empty flatbed railcar? The mass of the first railcar would be substantially greater than the second (i.e., $m_1 \gg m_2$). Derive the differential equations assuming that the mass of the second railcar is relatively negligible.

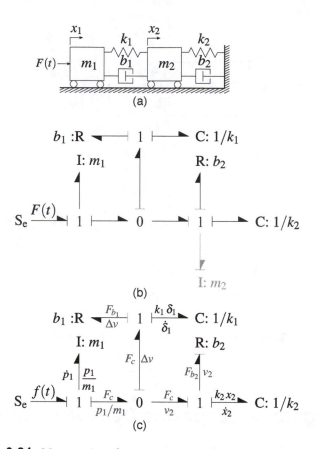

Figure 3.24: *Mass-spring-damper system with potential algebraic loop*

Solution. If the mass of the second car is negligible, one may be inclined to treat it as a massless cart. This presents a dilemma. As Figure 3.24 (b) indicates, the presence of the second mass specifies a causality that propagates through the bond graph to otherwise unassigned bonds. Note the causal strokes which are assigned by propagating the causality of the second I-element. If the second railcar is treated as a massless cart, four bonds are left with unassigned causality. Hence, causality must be specified by selecting one of the remaining unassigned R-elements, assigning its input-output relation, and propagating the resulting cause-and-effect. It does not matter which R-element we choose. If we select the second damper and prescribe that it supplies the flow at the attached one junction, the four unassigned bonds will have the same causality they would were the second mass present in the model (refer to Figure 3.24 (c)).

Now all the bonds have an assigned causality. We can proceed to apply the primary and secondary conditions to derive the differential equations. Or can we? According to the resultant bond graph, to derive the equations we will need to know the force generated by the first damper (F_{b_1}) and the velocity of the second damper (v_2). Upon closer inspection we find that the force generated by the first damper is a function of the velocity of the second and that the velocity of the second is a function of the force of the first. We cannot know one without the other, but neither is prescribed. This is an algebraic loop.

To resolve the loop we must solve for one of the unknowns (i.e., F_{b_1} or v_2) in terms of itself algebraically. To facilitate solving for v_2, we will label the force of the second damper F_{b_2} and the common force at the 0-junction, which is the combined force of the left spring and damper, F_c. We solve for v_2 in steps around the loop:

$$v_2 = \frac{F_{b_2}}{b_2} = \frac{F_c - k_2 x_2}{b_2} = \frac{[F_{b_1} + k_1 \delta_1] - k_2 x_2}{b_2} = \frac{[b_1 \Delta v + k_1 \delta_1] - k_2 x_2}{b_2}$$

$$= \frac{[b_1(p_1/m_1 - v_2) + k_1 \delta_1] - k_2 x_2}{b_2}.$$

The above equation is algebraically manipulated so that the terms with v_2 are gathered on one side of the equation:

$$b_1 v_2 + b_2 v_2 = (b_1 + b_2) v_2 = \frac{b_1}{m_1} p_1 + k_1 \delta_1 - k_2 x_2 .$$

Now we can readily solve for v_2 in terms of the states p_1, δ_1, and x_2:

$$v_2 = \frac{\dfrac{b_1}{m_1} p_1 + k_1 \delta_1 - k_2 x_2}{b_1 + b_2} = \boxed{\frac{b_1}{m_1(b_1 + b_2)} p_1 + \frac{k_1}{b_1 + b_2} \delta_1 - \frac{k_2}{b_1 + b_2} x_2}.$$

Given v_2 we can determine F_{b_1}:

$$F_{b_1} = b_1 \Delta v = b_1 \left[\frac{p_1}{m_1} - v_2 \right]$$

$$= b_1 \left[\frac{p_1}{m_1} - \left(\frac{b_1}{m_1(b_1 + b_2)} p_1 + \frac{k_1}{b_1 + b_2} \delta_1 - \frac{k_2}{b_1 + b_2} x_2 \right) \right]$$

$$= \boxed{\frac{b_1}{m_1} \left(1 - \frac{b_1}{b_1 + b_2} \right) p_1 - \frac{b_1}{b_1 + b_2} k_1 \delta_1 - \frac{b_1}{b_1 + b_2} k_1 x_2}.$$

To complete the differential equations we needed to know the combined force of the first spring and damper (F_c) and their shared relative velocity (Δv). Additionally, we needed to know the velocity of the second damper (v_2). In the process of solving the algebraic loop for v_2 we also determined the relative velocity,

$$\Delta v = \frac{p_1}{m_1} - v_2 = \left(1 - \frac{b_1}{b_1 + b_2}\right)\frac{p_1}{m_1} - \frac{1}{b_1 + b_2}k_1\delta_1 - \frac{1}{b_1 + b_2}k_1 x_2 ,$$

and the combined force,

$$F_c = F_{b_1} + k_1\delta_1 = \frac{b_1}{m_1}\left[1 - \frac{b_1}{b_1 + b_2}\right]p_1 + \left[1 - \frac{b_1}{b_1 + b_2}\right]k_1\delta_1 - \frac{b_1}{b_1 + b_2}k_1 x_2 .$$

The above relations for v_2, Δv, and F_c can be substituted into the following to finish deriving the state equations:

$$\dot{p}_1 = F(t) - F_c, \quad \dot{\delta}_1 = \Delta v, \quad \text{and } \dot{x}_2 = v_2 .$$

The above equations result from applying the secondary condition to the leftmost 1-junction and the primary condition to the remaining 1-junctions using the effort and flow labels we specified to facilitate resolving the loop.

One arbitrary causal assignment was necessary to specify the remaining unassigned bonds. We specified the causality on a single R-element and that propagated through to the remaining bonds. Generally, the number of necessary arbitrary assignments will be equal to the number of algebraic loops that must be solved. If several arbitrary causal assignments are necessary, a problem can become increasingly tedious. Had we chosen from the beginning to account for the second mass regardless of relative size, we would have avoided this complication from the get-go. Many times, algebraic loops can be avoided by simply using a slightly more elaborate model that accounts for elements that we might be prone to ignore.

Aside from algebraic loops, derivative causality can result and require some extra analysis. Those energy-storing elements that are in derivative causality are not dynamically *independent*. They do not contribute an independent state variable to the mathematical model or require a differential equation to describe their time-varying dynamics. The *dependent* state can be determined algebraically from one or more of the state variables associated with the energy-storing elements that are in integral causality. However, the presence of derivative causality implies that the output effort or flow is a function of the derivative of one or more of the independent state variables. Hence, some extra effort is necessary to finalize derivation of the differential equation. Take, for example, the toy car model depicted in Figure 3.25. It appears to have two states (h and p), but, as the following example will illustrate, one of those states is dependent on the other.

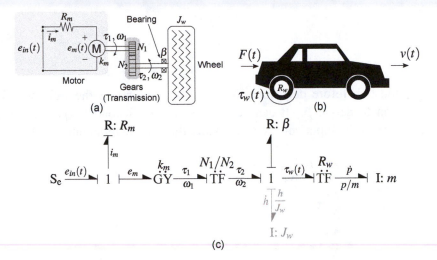

Figure 3.25: *A toy car bond graph model with derivative causality*

Example 3.17

Derive the differential equations for the simplified toy car model in Figure 3.25.

Solution. Closer inspection of the bond graph in Figure 3.25 (c) reveals that only one of the momenta can be an independent state. (Note that i_m, $e_m(t)$, τ_1, ω_1, τ_2, ω_2, $F(t)$, and $v(t)$ are labels provided to facilitate deriving the differential equation and are not external efforts or flows.) The causality has been chosen such that the I-element representing the mass is in integral form to facilitate determining the car velocity as an output ($v(t) = p/m$). The I-element representing the wheel rotational inertia that is depicted in Figure 3.25 (c) is in derivative causality.

By applying the primary condition at the rightmost 1-junction and following the causality out through the transformer, we can show that the angular momentum of the wheel is algebraically related to the momentum of the car,

$$\frac{p}{m} = R_w \frac{h}{J_w}.$$ (3.12)

Hence, a separate differential equation is not necessary to determine the angular momentum (h) since it is algebraically related to the translational momentum (p).

Since there is only one energy-storing element in integral causality, only one differential equation is necessary to model the dynamics,

$$\dot{p} = F(t), \tag{3.13}$$

where

$$F(t) = \frac{\tau_w}{R_w}. \tag{3.14}$$

We must therefore determine the sum of moments at the rightmost 1-junction to derive $\dot{p}$. It can be shown that the second torque (τ_2) is a function of the input voltage ($e_{in}(t)$) and the vehicle momentum (p).

$$
\begin{aligned}
\tau_w &= \tau_2 - \beta \frac{p/m}{R_w} - \dot{h} \\
&= \frac{N_2}{N_1}\tau_1 - \frac{\beta}{R_w m}p - \dot{h} \\
&= \frac{N_2}{N_1}k_m i_m - \frac{\beta}{R_w m}p - \dot{h} \\
&= \frac{N_2}{N_1}k_m \left[\frac{e_{in}(t) - e_m}{R_m}\right] - \frac{\beta}{R_w m}p - \dot{h} \\
&= \frac{N_2}{N_1}k_m \left[\frac{e_{in}(t) - k_m \omega_1}{R_m}\right] - \frac{\beta}{R_w m}p - \dot{h} \\
&= \frac{N_2}{N_1}k_m \left[\frac{e_{in}(t) - k_m(N_2/N_1)\omega_2}{R_m}\right] - \frac{\beta}{R_w m}p - \dot{h} \\
&= \frac{N_1}{N_2}k_m \left[\frac{e_{in}(t) - k_m(N_2/N_1)p/(R_w m)}{R_m}\right] - \frac{\beta}{R_w m}p - \dot{h}
\end{aligned}
$$

Using Equations 3.12 to 3.14 we can substitute for t_w and $\dot{h}$:

$$R_w \dot{p} = \frac{N_1}{N_2}\frac{k_m}{R_m}\left[e_{in}(t) - k_m\frac{N_2}{N_1}\left(\frac{p}{R_w m}\right)\right] - \frac{\beta}{R_w m}p - \frac{J_w}{R_w m}\dot{p}.$$

Note that $\dot{p}$ appears on both sides of the equation, which can be algebraically manipulated to arrive at

$$\boxed{\dot{p} = \left\{\frac{N_1}{N_2}\frac{k_m}{R_m}\left[e_{in}(t) - k_m\frac{N_2}{N_1}\left(\frac{p}{R_w m}\right)\right] - \frac{\beta}{R_w m}p\right\} \div \left(R_w + \frac{J_w}{R_w m}\right).}$$

As with algebraic loops, derivative causality requires extra effort to derive the differential equations. Both necessitate extra algebraic manipulation to arrive at the final differential equations. This section is only a cursory introduction into analysis of systems requiring extended formulation. Nonetheless, it serves to show that such systems are still solvable.

3.10 Summary

▷ As illustrated in Figure 3.1 (a), generally, it is assumed that power flows from the system to energy-storing or dissipating elements.

▷ Usually, it is assumed that power flows from the source to the system. Moreover, effort sources supply effort as an input and flow sources supply flow inputs (refer to Figure 3.1 (b)).

▷ Transformers and gyrators have power through convention. As depicted in Figure 3.1 (c), the power goes in one port and out the other.

▷ Adjacent 0- or 1-junctions can be collapsed into a single junction. Common junction types adjacent to one another are in actuality the same junction and the attached bonds share a common effort or flow (Figure 3.2).

▷ When synthesizing bond graphs for mechanical systems, we first identify distinct velocities and establish 1-junctions. For each 1-junction we identify elements that are directly associated. For example, inertias are commonly associated with distinct velocities. Then we insert effort-generating 1-ports off of 0-junctions or 2-ports between appropriate pairs of 1-junctions. Next, we eliminate zero-velocity sources and simplify.

▷ For circuits (both electric and hydraulic) we first identify distinct potentials (voltages or pressures) and establish 0-junctions. If there are any elements directly associated with these distinct efforts, we place them directly off the associated junction using a bond. We then insert the 1- and 2-ports between pairs of 0-junctions. The 1-ports are placed off of 1-junctions that are inserted between pairs of 0-junctions. Next, we eliminate the ground or reference pressure and simplify.

▷ Mixed systems can be dissected into subsystems, each of which is of a single energy domain. Each subsystem can be analyzed using the asso-

ciated guidelines. The subsystems interface at energy-converting transducers which are modeled as either transformers or gyrators. Some
examples were provided in Figure 3.15.

▷ When deriving differential equations from a bond graph one must first
assign causality beginning with the sources, then the energy-storying
elements, and last, if necessary, the R-elements. At each stage we assign the causality to an element and propagate if the causality affects
adjacent junctions and/or elements. The process proceeds until all the
bonds have an assigned causality. The differential equations result from
applying the primary and secondary conditions at the junctions.

▷ Algebraic loops and derivative causality require extra analysis to derive
the differential equations.

3.11   Review

R3-1   What are the assumed power directions for energy-dissipating and energy storying elements?

R3-2   What are the assumed power directions for energy sources?

R3-3   Explain the power through conventions for transformers and gyrators.

R3-4   Describe in your own words the guidelines for synthesizing bond graphs
of mechanical translation systems.

R3-5   How are the guidelines for mechanical rotation similar to those for
translation?

R3-6   Describe in your own words the guidelines for synthesis of electric
circuit bond graphs.

R3-7   How is the synthesis of hydraulic circuit bond graphs similar to electric
circuits?

R3-8   Explain how the synthesis of bond graphs for a mixed system is handled.

R3-9   Describe in your own words the derivation of differential equations
from a bond graph.

R3-10   Why and how do algebraic loops occur in bond graph models?

R3-11   What does derivative causality imply?

3.12 Problems

For each of the systems specified below, synthesize the bond graph as needed including annotations and causality. Use the bond graph to derive the differential equations.

P3-1 *Mass-Spring-Damper Systems (Figure 1.15).* For (c) you need only derive the differential equations. Recall that the bond graph was synthesized in Example 3.2. The input for (a) is the velocity $v(t)$. The displacement associated with the input velocity is $x(t)$ and is labeled for illustration. The input to (b) is the forcing function $F(t)$. For (c) and (d) the external force is an input. The force due to gravity, if considered, can be treated as an external force and as second input to each system.

P3-2 *A Mass-Spring-Damper System with Massless Body.* Figure 3.26 depicts a mass-spring-damper system with a massless platform that connects the upper spring and damper to the lower spring. Synthesize the bond graph and derive the differential equations.

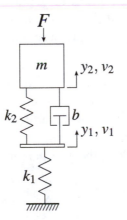

Figure 3.26: *A mass-spring-damper system with massless body*

P3-3 *Rotational Systems (Figure 1.16).* An input torque τ_m is supplied for (a) and (b) while (c) has two input torques.

P3-4 *Electric Circuits (Figure 1.17).* The bond graph for circuit (a) was previously derived in Example 3.5. Derive its differential equations. The input for circuit (a) is the voltage e. Circuit (b) was examined in

Examples 3.6 and 3.13. Derive the differential equations for circuit (c). The input is the external voltage $e(t)$.

P3-5 *A Circuit with Parallel and Series Elements.* A circuit with parallel branches is given in Figure 3.27. Synthesize the bond graph and derive the differential equations. One or more system may involve an algebraic loop.

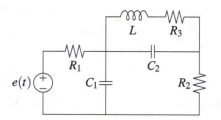

Figure 3.27: *An electric circuit with series elements in a parallel branch*

P3-6 *Hydraulic Circuits (Figure 1.18).* Hydraulic circuit (a) was discussed in Examples 3.8 and 3.14. The bond graph for circuit (b) was discussed in Example 3.9. Circuit (b) has an input volumetric flow source $Q(t)$. For circuit (c) the input is a pressure source, P_{in}.

P3-7 *A Simple Wind Turbine Generator.* A simple wind turbine generator with an input wind speed, $v(t)$, is depicted in Figure 3.28. Synthesize the bond graph and derive the differential equations for the angular momentum, h, of the rotational inertia, J_g, and the flux linkage, λ_g, of the inductor, L_g. Recall that the relation between the torque and angular velocities on each side of the gear pair are

$$\frac{\tau_1}{\tau_2} = \frac{N_1}{N_2} = \frac{\omega_2}{\omega_1}$$

and the motor relations are

$$\tau_m = k_m i_m \text{ and } e_m = k_m \omega_m.$$

Note that the shafts are assumed to be rigid and that τ_m, ω_m, τ_1, ω_1, τ_2, and ω_2 are labels and not sources.

P3-8 *Dependent States.* The electromechanical system in Figure 3.29 has two rotational inertias whose dynamics are not independent of one another. In addition to synthesizing bond graph and deriving the differential equations, use the bond graph to show that one inertia is dependent

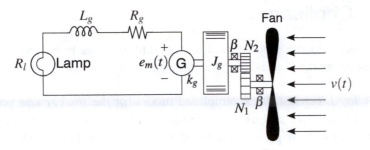

Figure 3.28: *A simple wind turbine generator*

on the other. Furthermore, show how the circumstance changes if the motor inertia is assumed to be negligible.

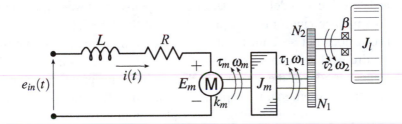

Figure 3.29: *An electromechanical system with an algebraic loop*

P3-9 *Algebraic Loops.* Synthesize the bond graph for the electric circuit in Figure 3.30. You should find that it has an algebraic loop. Resolve the algebraic loop and derive the differential equations.

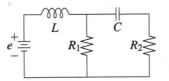

Figure 3.30: *A circuit problem with an algebraic loop*

3.13   Challenges

As you did with the practice problems, synthesize the bond graph, assign causality, annotate it, and derive the differential equations.

C3-1 *A Toy Electric Car.* A simplified model for the toy car was provided in Figure 3.25 and used in Example 3.17. The synthesis of the bond graph that was provided is not detailed in that example. Derive your own bond graph and consider aspects that may have been left out of that model. What other elements might you include that are not incorporated in the schematic depicted in Figure 3.25? How would you change your model to eliminate the derivative causality? Justify your answer.

C3-2 *A Quarter-Car Suspension Model.* Figure 3.6 and Example 3.2 provide and describe a model for a quarter-car suspension. However, this model does not account for the unsprung mass of the wheel, brakes, tire, etc., and it does not account for the effective compliance and damping of the tire wall and compressed air. Figure 1.2 (c) depicts a mass-spring-damper model of the suspension that includes unsprung mass and the effective damping and compliance of the tire. Be sure to incorporate these aspects in your model. The displacement due to the road surface profile has an associated vertical velocity ($v_{road}(t)$) that supplies the input to the system.

C3-3 *An Electric Drill.* The electric drill is powered by a battery which excites a PMDC motor with a selectable gear pair. The drill is used to drive a screw into a wooden frame. Friction between the wood and screw dampens the rotational energy. Assume that the battery supplies a specified input voltage e_{in}.

Chapter 4

The State Space and Numerical Simulation

One of our primary reasons for modeling systems is predicting their dynamic response. It is useful to understand how dynamic states vary with time when excited by inputs. We numerically simulate to predict dynamic responses. Systems can be exposed to a variety of inputs which can generally vary with time. To numerically simulate such responses we must consider a number of things:

▷ What mathematical formulation might we apply to the models we derived in the previous chapter in order to facilitate predicting and analyzing their dynamic responses?

▷ What are typical system inputs and outputs?

▷ How can we use a numerical computing language like MATLAB to simulate dynamic responses?

▷ What are typical responses for common types of inputs?

4.1 Introduction

In Chapter 3 we synthesized bond graphs and used them to systematically derive differential equations for dynamic systems. Because we focused strictly on linear systems, the differential equations took the form of linear combinations of the states and inputs. Moreover, the equations derived from bond graphs are first-order differential equations. These characteristics allow one to represent the differential equations in what is termed the *state-space form* – the first of two primary methods we will explore for representing linear systems. The state-space form is a linear algebraic representation of the differential equations and the system outputs.

Herein, we will examine the formulation and use of state-space representations for some of the systems discussed in the previous chapter. In addition, we shall review common inputs and learn how MATLAB is used to simulate responses of state-space models.

> ▷ *Objectives:*

> 1. To understand how systems of first-order differential equations are converted to state-space representations,

> 2. To study and understand the use of mathematical functions utilized to model commonly occurring system inputs, and

> 3. To simulate dynamic systems represented by state-space equations using MATLAB.

> ▷ *Outcomes:* Upon completion, you should

> 1. be able to reformulate equations derived using bond graphs into state-space representations,

> 2. be able to model a variety of physical inputs using some basic mathematical functions, and

> 3. simulate dynamic responses for simple and moderately complex systems using the state-space formulation.

4.2   The State Space

You may recall from your advanced Mathematics course(s) that a linear set of equations can be represented using matrices and vectors. A linear set of n differential equations can be written in the general form

$$\dot{\mathbf{x}}(t) = \mathbf{A}\mathbf{x}(t) + \mathbf{B}\mathbf{u}(t), \tag{4.1}$$

where $\mathbf{x}(t)$ is the $n \times 1$ *state vector* (i.e., $\mathbf{x}(t) \in \mathbb{R}^n$), $\mathbf{A}$ is an $n \times n$ matrix of coefficients associated with the states (i.e., $\mathbf{A} \in \mathbb{R}^{n \times n}$), $\mathbf{u}$ is the $m \times 1$ *input vector* (i.e., $\mathbf{u}(t) \in \mathbb{R}^m$), and $\mathbf{B}$ is an $n \times m$ matrix of coefficients associated with the inputs (i.e., $\mathbf{A} \in \mathbb{R}^{n \times m}$). The state vector, $\mathbf{x}(t)$, includes each state for which an independent differential equation is necessary, and the input vector, $\mathbf{u}(t)$, includes each of the external sources.

In addition to the system of differential equations, the state-space representation includes the outputs, which are also written as a linear combination of the states and inputs. Given r outputs, the *output vector*, $\mathbf{y}(t) \in \mathbb{R}^r$, is represented as

$$\mathbf{y}(t) = \mathbf{C}\mathbf{x}(t) + \mathbf{D}\mathbf{u}(t), \tag{4.2}$$

where $\mathbf{C}$ and $\mathbf{D}$ are $r \times n$ and $r \times m$ matrices, respectively (i.e., $\mathbf{C} \in \mathbb{R}^{r \times n}$ and $\mathbf{D} \in \mathbb{R}^{r \times m}$). Combined, Equations 4.1 and 4.2 compose the state-space representation. An advantage of the state-space representation is that we can use Linear Algebra to analyze the dynamic system to determine a variety of characteristics. This will be discussed in detail in Chapter 7.

4.3 Composing State-Space Representations

All the system differential equations we have derived in the previous chapter were in the form of first-order, ordinary, differential equations. More specifically each problem was represented by a system of linear differential equations.

You may have noticed that, in the Chapter 3 examples, we often isolated each state from its corresponding coefficient. To convert the individual first-order differential equations into a state-space representation one must identify the coefficients associated with each state and input. When examining the equations, if any state or input is not present, the coefficient for that state or input is zero. Hence, to explicitly convert a linear set of differential equations into the state-space form given by Equations 4.1 and 4.2, one can follow this process:

1. *Identify the state vector, input, and outputs.* The states and their order are not unique, but the minimum number of time-varying states is. Explicitly identify the states and their chosen order by specifying the state vector. Make sure to consistently use the same states and order as you generate the state-space form. Each effort or flow source supplies an input. Though the inputs are unique, the order is not. Thus, specify the input vector and maintain the order throughout the process. The outputs may be less obvious. The question to ask is, "why are we modeling this system?" In particular, "what about the dynamic response are we interested in predicting?" These questions help identify the outputs of interest. Specify the output vector using the dynamic variables we

wish to measure or predict.

2. *Rewrite each state equation as a linear combination of all the states and inputs.* To facilitate identifying the coefficients of the matrices **A** and **B** write each state equation as a summation listing the states and their coefficients and the inputs and their coefficients in the same order as specified by the state and input vectors. Be sure to include all states and inputs. If a differential equation is not a function of a specific state or input, the respective coefficient is 0.

3. *Rewrite the linear set of differential equations as a linear algebraic equation.* Use the coefficients identified in the previous step to write the matrices **A** and **B** and separate out the states as the state vector and the inputs as the input vector.

4. *Write each output as a linear combination of the state and input vectors.* This is conducted in the same manner as detailed in step 2 except that the coefficients are associated with matrices **C** and **D**.

5. *Rewrite the output equation as a linear algebraic equation.* Use the coefficients identified in the previous step to write the matrices **C** and **D** and separate out the states as the state vector and the inputs as the input vector.

The above process is just a set of guidelines to facilitate arriving at the state-space representation.

Remember that the number of energy-storing elements in integral causality will be equal to the number of first-order differential equations or the *system order*. Think of the order of the system as the minimum number of first-order differential equations necessary to represent the system dynamics. Furthermore, it is the order of the system that is unique and not the states that result, which can appear in any given arrangement. However, proper use of bond graphs to derive your differential equations should result in a consistent set. Of the previous examples, Examples 3.11 is one of the few systems represented by more than just two differential equations. Hence, we will use this less trivial example to illustrate how to generate a state-space representation.

Example 4.1

Recollect the PMDC motor from Examples 3.10 and 3.15. Place the differential equations in state-space form.

Solution. The system had two states, λ and h, and one input, e_{in}. The state equations are rewritten below.

$$\dot{\lambda} = -\frac{R}{L}\lambda - \frac{k_m}{J}h + e_{in}(t)$$

$$\dot{h} = \frac{k_m}{L}\lambda - \left[\frac{N_1}{N_2}\right]^2\frac{\beta}{J}h$$

If we choose the states in the same order as they appear in the bond graph, we can specify the state vector as

$$\mathbf{x} = \begin{bmatrix} \lambda \\ h \end{bmatrix},$$

and the input as $u = e_{in}$. We might use the model to predict several things including the output torque or angular velocity. Let us assume that the output angular velocity is of interest. Recall from Figure 3.23 that the output angular velocity is

$$y = \omega_{out} = \frac{N_1}{N_2}\frac{h}{J}.$$

Thus, we have identified the state vector, input, and output. We now rewrite each state equation as a linear combination of each of the states and the input,

$$\begin{array}{rcl} \dot{\lambda} & = & -(R/L)\lambda \qquad\qquad -(k_m/J)h \quad +1\,e_{in}(t) \\ \dot{h} & = & (k_m/L)\lambda \quad -(N_1/N_2)^2(\beta/J)h \quad +0\,e_{in}(t) \end{array}.$$

In this format, the coefficients for matrices $\mathbf{A}$ and $\mathbf{B}$ are self-evident. Thus,

$$\dot{\mathbf{x}} = \begin{bmatrix} \dot{\lambda} \\ \dot{h} \end{bmatrix} = \begin{bmatrix} -R/L & -k_m/J \\ k_m/L & -(N_1/N_2)^2(\beta/J) \end{bmatrix}\begin{bmatrix} \lambda \\ h \end{bmatrix} + \begin{bmatrix} 1 \\ 0 \end{bmatrix}e_{in}(t) = \mathbf{Ax} + \mathbf{Bu}.$$

$$(4.3)$$

The output can be rewritten in terms of both states and the input,

$$y = 0\lambda + \left(\frac{N_1}{N_2}\frac{1}{J}\right)h + 0\,e_{in}(t) \qquad (4.4)$$

$$= \begin{bmatrix} 0 & \dfrac{N_1}{N_2}\dfrac{1}{J} \end{bmatrix}\begin{bmatrix} \lambda \\ h \end{bmatrix} + 0\,e_{in}(t) \qquad (4.5)$$

$$= \mathbf{Cx} + \mathbf{Du}. \qquad (4.6)$$

As expected, the resulting matrices $\mathbf{A}$, $\mathbf{B}$, $\mathbf{C}$, and $\mathbf{D}$ have dimensions 2×2, 2×1, 1×2, and 1×1, respectively, given that there are 2 states, 1 input, and 1 output. Though seemingly trivial, this example helps illustrate the process. The following is substantially more involved.

The following example has more states and should serve as a more thorough illustration.

Example 4.2

In Examples 3.1 and 3.11 we synthesized the bond graph and derived the differential equations for the mass-spring-damper system in Figure 3.3 (a). Place the differential equations in state-space form.

Solution. Recall Equations 3.1, 3.2, 3.3, and 3.4 which represent the dynamics of the mass-spring-damper system. These equations are repeated here for convenience.

$$\dot{x}_1 = \frac{1}{m}p_1$$

$$\dot{p}_1 = -kx_1 - \frac{b}{m}p_1 + k\delta_2$$

$$\dot{\delta}_2 = -\frac{1}{m}p_1 + \frac{1}{m}p_2$$

$$\dot{p}_2 = -k\delta_2 + F(t)$$

Note that the equations are represented in a slightly different fashion that separates the states from the coefficients and lists equations and states in the order of their appearance in the bond graph from left to right. The state vector for our set of equations is

$$\mathbf{x} = \begin{bmatrix} x_1 \\ p_1 \\ \delta_2 \\ p_2 \end{bmatrix},$$

and the single input to the system is the forcing function, $u = F(t)$. The system has 4 states and 1 input. The sequence of the states is not important (e.g., whether x_1 is the first, second, third, or final state does not matter), but once a set and arrangement are chosen, one should not deviate.

To generate the first part of the state-space representation (as in Equation 4.1) we rewrite the above equations accounting for every state and input regardless of its presence or lack thereof. Remember, if a state or input is missing, its coefficient is zero.

$$
\begin{aligned}
\dot{x}_1 &= & 0x_1 & +(1/m)p_1 & +0\delta_2 & & +0p_2 & +0F(t) \\
\dot{p}_1 &= & -kx_1 & -(b/m)p_1 & +k\delta_2 & & +0p_2 & +0F(t) \\
\dot{\delta}_2 &= & 0x_1 & -(1/m)p_1 & +0\delta_2 & +(1/m)p_2 & & +0F(t) \\
\dot{p}_2 &= & 0x_1 & +0p_1 & -k\delta_2 & & +0p_2 & +1F(t)
\end{aligned}
$$

Note how the elements are aligned in columns to emphasize the coefficients of the matrices **A** and **B**. There is a column for each state and the input. From here we can readily convert the system of equations into state-space form,

$$
\dot{\mathbf{x}} = \begin{bmatrix} \dot{x}_1 \\ \dot{p}_1 \\ \dot{\delta}_2 \\ \dot{p}_2 \end{bmatrix} = \begin{bmatrix} 0 & 1/m & 0 & 0 \\ -k & -b/m & k & 0 \\ 0 & -1/m & 0 & 1/m \\ 0 & 0 & -k & 0 \end{bmatrix} \begin{bmatrix} x_1 \\ p_1 \\ \delta_2 \\ p_2 \end{bmatrix} + \begin{bmatrix} 0 \\ 0 \\ 0 \\ 1 \end{bmatrix} F(t) = \mathbf{A}\mathbf{x} + \mathbf{B}u .
$$

(4.7)

Notice that **A** is a 4×4 matrix, that **B** is a 4×1 vector, and that their dimensions correlate with the number of states and the input.

The outputs of the system are those time-varying variables we wish to predict using our model. For this system, we wish to determine the time-varying displacements of each mass (i.e., x_1 and x_2). Because the left side of the first mass is attached to a fixed position by the leftmost spring and damper, the displacement of that mass is equal to the deflection of that spring and damper. Hence the first output is x_1. The rightmost mass is attached to the leftmost mass by a spring. Its overall displacement is equivalent to the displacement of the first mass plus the deflection of the second spring,

$$
x_2 = x_1 + \delta_2 .
$$

Therefore, the model outputs are

$$
\mathbf{y} = \begin{bmatrix} x_1 \\ x_1 + \delta_2 \end{bmatrix},
$$

but recall that the output must be written as a linear combination of the states ($\mathbf{x}$) and the input ($F(t)$). To determine the matrix coefficients we proceed as before with the differential equations. That is, we write out the displacements in terms of all the states and input regardless of their absence from the above equation,

$$
\begin{aligned}
x_1 &= 1x_1 &+0p_1 &+0\delta_2 &+0p_2 &+0F(t) \\
x_2 &= 1x_1 &+0p_1 &+1\delta_2 &+0p_2 &+0F(t)
\end{aligned} .
$$

This helps to identify the matrix coefficients,

$$
\mathbf{y} = \begin{bmatrix} x_1 \\ x_2 \end{bmatrix} = \begin{bmatrix} 1 & 0 & 0 & 0 \\ 1 & 0 & 1 & 0 \end{bmatrix} \begin{bmatrix} x_1 \\ p_1 \\ \delta_2 \\ p_2 \end{bmatrix} + \begin{bmatrix} 0 \\ 0 \end{bmatrix} F(t) = \mathbf{C}\mathbf{x} + \mathbf{D}u .
$$

(4.8)

Before we leave this example, let us revisit the concept of equivalent systems. Recall that this system could alternatively be represented by the equations 3.5, 3.6, 3.7, and 3.8, which are repeated below for convenience.

$$\dot{x}_1 = v_1$$

$$\dot{v}_1 = -\frac{2k}{m}x_1 - \frac{b}{m}v_1 + \frac{k}{m}x_2$$

$$\dot{x}_2 = v_2$$

$$\dot{v}_2 = \frac{k}{m}x_1 - \frac{k}{m}x_2 + \frac{1}{m}F(t)$$

For this set of equations, the alternate state vector is

$$\widetilde{\mathbf{x}} = \begin{bmatrix} x_1 \\ v_1 \\ x_2 \\ v_2 \end{bmatrix}.$$

As with the previous set, we can identify the matrix coefficients by explicitly writing each of the above equations in terms of the new states and input.

$$
\begin{array}{rclllll}
\dot{x}_1 & = & 0x_1 & +1v_1 & +0x_2 & +0v_2 & +0F(t) \\
\dot{v}_1 & = & -(2k/m)x_1 & -(b/m)v_1 & +(k/m)x_2 & +0v_2 & +0F(t) \\
\dot{x}_2 & = & 0x_1 & +0v_1 & +0x_2 & +1v_2 & +0F(t) \\
\dot{v}_2 & = & (k/m)x_1 & +0p_1 & -(k/m)x_2 & +0v_2 & +(1/m)F(t)
\end{array}
$$

These equations can al be represented in state-space form, $\dot{\widetilde{\mathbf{x}}} = \widetilde{\mathbf{A}}\widetilde{\mathbf{x}} + \widetilde{\mathbf{B}}u$,

$$
\begin{bmatrix} \dot{x}_1 \\ \dot{v}_1 \\ \dot{x}_2 \\ \dot{v}_2 \end{bmatrix} = \begin{bmatrix} 0 & 1 & 0 & 0 \\ -2k/m & -b/m & k/m & 0 \\ 0 & 0 & 0 & 1 \\ k/m & 0 & -k/m & 0 \end{bmatrix} \begin{bmatrix} x_1 \\ v_1 \\ x_2 \\ v_2 \end{bmatrix} + \begin{bmatrix} 0 \\ 0 \\ 0 \\ 1/m \end{bmatrix} F(t). \quad (4.9)
$$

Since the outputs are two of the alternative states, x_1 and x_2, the output vector becomes

$$
\widetilde{\mathbf{y}} = \begin{bmatrix} 1 & 0 & 0 & 0 \\ 0 & 0 & 1 & 0 \end{bmatrix} \begin{bmatrix} x_1 \\ v_1 \\ x_2 \\ v_2 \end{bmatrix} + \begin{bmatrix} 0 \\ 0 \end{bmatrix} F(t) = \widetilde{\mathbf{C}}\widetilde{\mathbf{x}} + \widetilde{\mathbf{D}}u . \quad (4.10)
$$

Both equivalent systems generate the same outputs despite having distinct state vectors. As we will see in a later chapter, regardless of the state vector chosen, the equations model the same system and should have the same dynamic characteristics as evidenced by properties such as the eigenvalues or characteristic roots. We will return to this problem in Chapter 7 when we discuss time domain analysis.

Not all the systems we model are single-input systems. The circuit depicted in Figure 4.1 has both a voltage and a current source. The following example illustrates how to deal with multiple inputs and how they manifest in the state-space representation.

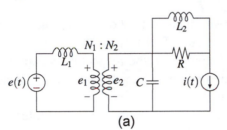

(a)

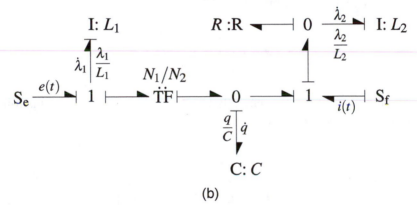

(b)

Figure 4.1: *A circuit with multiple inputs.*

Example 4.3

Using the bond graph given in Figure 4.1 (b) it can be shown that the differential equations for the circuit are

$$\dot{\lambda}_1 = e(t) - \frac{N_1}{N_2}\frac{q}{C}, \quad \dot{q} = \frac{N_1}{N_2}\frac{\lambda}{L_1} - i(t), \quad \text{and} \quad \dot{\lambda}_2 = \left[i(t) - \frac{\lambda_2}{L_2} \right] R,$$

where the states and inputs are

$$\mathbf{x} = \begin{bmatrix} \lambda_1 \\ q \\ \lambda_2 \end{bmatrix} \quad \text{and} \quad \mathbf{u} = \begin{bmatrix} e(t) \\ i(t) \end{bmatrix}.$$

Note that $e(t)$ appears in the first equation and $i(t)$ in the second and third. Hence, the first part of the state-space representation is

$$\dot{\mathbf{x}} = \begin{bmatrix} 0 & -\dfrac{N_1}{N_2}\dfrac{1}{C} & 0 \\ \dfrac{N_1}{N_2}\dfrac{1}{L_1} & 0 & 0 \\ 0 & 0 & -\dfrac{R}{L_2} \end{bmatrix} \begin{bmatrix} \lambda_1 \\ q \\ \lambda_2 \end{bmatrix} + \begin{bmatrix} 1 & 0 \\ 0 & -1 \\ 0 & R \end{bmatrix} \begin{bmatrix} e(t) \\ i(t) \end{bmatrix} = \mathbf{Ax} + \mathbf{Bu}.$$

Because there are two inputs instead of simply one, the matrix $\mathbf{B}$ has two columns and not just one. Elements of the first column are nonzero if the first input ($e(t)$) appears in the respective equation corresponding to each row. The second column elements are nonzero if $i(t)$ appears in the corresponding equation.

For this example let us assume the outputs of interest are the inductor currents and the capacitor voltage,

$$\mathbf{y} = \begin{bmatrix} i_{L_1} \\ v_C \\ i_{L_2} \end{bmatrix} = \begin{bmatrix} \lambda_1/L_1 \\ q/C \\ \lambda_2/L_2 \end{bmatrix}.$$

The outputs can be more generally represented as a linear combination of all the states and the two inputs,

$$\mathbf{y} = \begin{bmatrix} \dfrac{1}{L_1} & 0 & 0 \\ 0 & \dfrac{1}{C} & 0 \\ 0 & 0 & \dfrac{1}{L_2} \end{bmatrix} \begin{bmatrix} \lambda_1 \\ q \\ \lambda_2 \end{bmatrix} + \begin{bmatrix} 0 & 0 \\ 0 & 0 \\ 0 & 0 \end{bmatrix} \begin{bmatrix} e(t) \\ i(t) \end{bmatrix} = \mathbf{Cx} + \mathbf{Du}.$$

Like the matrix $\mathbf{B}$, matrix $\mathbf{C}$ has two columns. Notice $\mathbf{D}$ is a 3×2 matrix of zeros.

4.4 Basic Transient Responses

When assessing or characterizing dynamic responses, we are typically interested in evaluating the transient response of the system. Natural and forced responses are made up of two parts – the *transient* and *steady-state* (Ogata 2004). The transient is the part of the overall response that occurs between the initial and final conditions of the system (Ogata 2004); it is the part of the response that occurs just after the input(s) is (are) applied until the system settles into steady-state. Because we are interested in the transient as well as the steady-state, we typically assume that the inputs are not applied or "turned on" until some initial time $(t = 0)$. As a consequence of this assumption, many of the mathematical functions that we use as inputs will be zero for the time prior to initiation of the systems dynamic response (i.e., $t < 0$).

Certain inputs are commonly used to assess characteristics of dynamic systems. Some such inputs are *impulse, step,* and *ramp functions*. They are also utilized to approximate some commonly occurring physical inputs. Moreover, these three functions in particular have a special relation. As will be demonstrated, the unit step is the integral of the unit impulse, and the unit ramp is the integral of the unit step. Conversely, the reverse relations are derivative instead of integral, that is, the derivative of the unit ramp is the unit step, and derivative of the unit step is the unit impulse.

In Chapter 7 we will discuss in detail the characteristics of time-domain responses including impulse, step, and ramp responses. For now, these functions are introduced to facilitate a latter section where we will discuss simulation of state-space representations.

4.4.1 The Unit Impulse

The unit impulse is an ideal function of infinite height and infinitesimal width. It has an area under the curve of one, hence its name – the "unit" impulse. Image, as depicted in Figure 4.2, a square pulse of unit height and width. Further, imagine that the height is stretched and the width narrowed all while maintaining an area under the curve of unity. The height is stretched to infinity while the width is narrowed to zero. The unit impulse or *Dirac delta function* is represented mathematically as

$$\tilde{\delta}(t) = \left\{ \begin{array}{ll} \infty, & t = 0 \\ 0, & t \neq 0 \end{array} \right. . \tag{4.11}$$

($\tilde{\delta}(t)$ should not be confused with δ, which is used for a relative displacement.) Note that function is infinite at $t = 0$ and zero for all other values of t (i.e., $t \neq 0$).

Though an idealization, it provides useful insight about the properties of dynamic responses and is sometimes used to approximate inputs such as impact forces, which have large magnitudes applied over short periods of time. Moreover, it is commonly used to characterize the transient response.

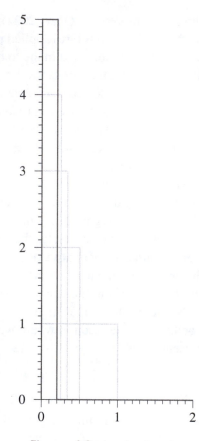

Figure 4.2: *A pulse function.*

4.4.2 The Unit Step

As has been previously mentioned, the unit step or *Heaviside function* illustrated in Figure 4.3 is the integral of the unit impulse function,

$$1(t) = \int \tilde{\delta}(t) \, dt = \left\{ \begin{array}{l} 0, \, t < 0 \\ 1, \, t \geq 0 \end{array} \right. . \tag{4.12}$$

Recall that the area under of the unit impulse function is one. Hence, its integral, the unit step, will have a constant value of one for all $t > 0$. Conversely, the Dirac delta function is the derivative of the Heaviside function

$$\frac{d}{dt} 1(t) = \left\{ \begin{array}{l} \infty, \, t = 0 \\ 0, \, t \neq 0 \end{array} \right. = \tilde{\delta}(t).$$

Everywhere except $t = 0$ the Heaviside function is constant. Thus, at everywhere except $t = 0$ its derivative is zero. At $t = 0$ the Heaviside function has an infinite slope that corresponds with the infinite spike that occurs at zero in the Dirac delta function.

Step functions are utilized to represent constant inputs that are triggered at initial time (e.g., a constant force that is applied just after $t = 0$). They are also utilized to assess *set point problems*. Set point problems are one in which the system is expected to go from some initial unexcited state to a constant stead-state condition usually within a specified time limit. Like the unit impulse, a unit step input is also used to assess a systems transient response.

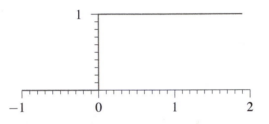

Figure 4.3: *The unit step or Heaviside function.*

4.4.3   The Unit Ramp

If we integrate the unit step, we arrive at the unit ramp,

$$f(t) = \int 1(t)\,dt = \begin{cases} 0, t < 0 \\ t, t \geq 0 \end{cases}, \qquad (4.13)$$

depicted in Figure 4.4. Like the impulse and step functions, the ramp function is zero until initial time, after which it rises at a constant rate. It is referred to as the unit ramp because it has a slope of one. A general ramp function is basically the same but has some slope other than one. The ramp function is a piecewise continuous function. That is, it is composed of pieces that individually have continuous differentials,

$$\frac{d}{dt}f(t) = \begin{cases} d0/dt = 0, t < 0 \\ dt/dt = 1, t \geq 0 \end{cases} = 1(t).$$

As illustrated above, the derivative of the unit ramp function is the unit step function.

Ramp functions are used to represent linearly varying inputs. They are also commonly used to assess how well a system can track an input. *Tracking problems*, unlike set point problems, are one where a system is expected to closely follow a specified input over a duration of time. At steady-state, the system output is expected to track with minimal error the linearly varying input. Unlike the impulse and step functions, the ramp function is commonly used to assess steady-state behavior and not transient characteristics.

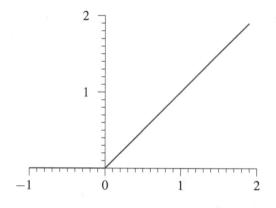

Figure 4.4: *The unit ramp function.*

4.5 State-Space Simulations Using MATLAB

As indicated in Table 4.1, MATLAB includes a series of commands that can be used to numerically simulate basic responses such as the impulse, step, and ramp responses.* Additionally, it provides commands that can be utilized to simulate responses to general inputs and specified initial conditions. Herein, we will explore the use of these commands to simulate responses for various state-space representations.

First, we must define a state-space object that includes matrix information. In the previously discussed example problems, we formulated linear algebraic representations that included four matrices – **A**, **B**, **C**, and **D**. In MATLAB, a state-space object can be defined using the ss command:

```
>> sys = ss(A,B,C,D)
```

---

*. For a comprehensive review of the use and syntax of MATLAB, refer to the "Academia" link at the MathWorks website (www.mathworks.com/academia/).

Table 4.1: *MATLAB commands for LTI response operations.*

Function	Description
ss(A,B,C,D)	defines a state-space object
impulse(sys)	simulates the unit impulse response
step(sys)	simulates the unit step response
lsim(sys)	simulates response to arbitrary inputs
initial(sys,x0)	simulates the initial condition response

This generates an object in MATLAB named sys that contains the state-space system model. Note that >> represents the MATLAB command prompt.

Unit impulse and step responses are obtained through use of the impulse and step commands, respectively:

```
>> impulse(sys)
>> step(sys)
```

There are is no built-in function that will generate the ramp response. However, the lsim command is used to simulate the response of *linear, time-invariant systems* (LTI) to general inputs.[†] The command syntax is

```
>> lsim(sys,u,t)
```

where u is generally a two-dimensional array with columns. Each column represents a system input, and each element in a column the input value at the corresponding time in the same row of the time array t. Thus, the user can define functions for each of the inputs as one-dimensional arrays that are composed of the input values at each instant in time specified in the array t. To simulate the unit ramp response, define a time array and equate the input to the time array so that $u = t$:

```
>> t = [0:0.01:10]';
>> u = t;
>> lsim(sys,u,t)
```

where t is a time array specified as values between 0 and 10 seconds spaced 0.10 seconds apart. The apostrophe character is used to transpose the row array t into a column array.

---

†. A *linear, time-invariant* system is one that satisfies the properties of superposition and homogeneity and whose parameters (not states) are constant and not time-varying.

Some systems have non-zero initial conditions. The commands mentioned thus far assume zero initial conditions. To simulate responses that account for this, an array with the initial conditions can be passed to the lsim command,

```
>> lsim(sys,u,t,x0)
```

where each element in the array x0 corresponds to an initial condition. Moreover, some systems are unforced. For instance, free vibration systems do not have external excitations. Vibrations result because the system is not initially in a state of equilibrium (i.e., the initial conditions are not all zero). Because the system is not driven by inputs and is not at equilibrium it will tend to equilibrium (if stable) as time progresses. To simulate such responses, one can utilize the initial command,

```
>> initial(sys,x0)
```

Note that each of the commands mentioned above have additional options that can be specified. For details regarding these options and the syntax use the help command (e.g., help impulse).

4.6 Applications

To illustrate the use of the aforementioned MATLAB commands, we will simulate the responses for a variety of systems we have previously examined. To begin let us predict the impulse response of the simple quarter-car suspension model from Example 3.2.

Example 4.4

Recall the quarter-car suspension model discussed in Chapter 3. The simplified bond graph is illustrated in Figure 3.6 (f). Typical vehicle mass, damping constant, and spring rate for a family sedan are 1700 kg, 750 N-s/m, and 20 kN/m, respectively. If the vehicle drives over a curb, what is the approximate response?

Solution. To simulate the response, we need to characterize the input and derive the system model. Let us begin by deriving the system model. For the purposes of this model, we will ignore the effect due to gravity. Gravity does not change the characteristics of the dynamic response. It simply changes the point of equilibrium. If we think of $y(t)$ as being the displacement from equilibrium, the force due to gravity only changes the

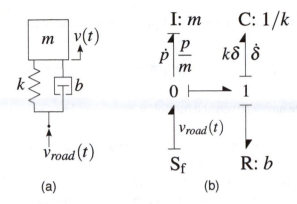

Figure 4.5: *Quarter-car suspension model.*

absolute position about which the response oscillates. Hence, if the effort source representing gravity is removed from the bond graph, the bond attached to the I-element collapses directly to the 0-junction as illustrated in Figure 4.5 (b).

The resulting simplified bond graph has two state equations,

$$\dot{p} = k\delta + b\left[v_{road}(t) - \frac{p}{m}\right] \text{ and}$$

$$\dot{\delta} = v_{road}(t) - \frac{p}{m}.$$

The purpose of this model is to predict the vertical displacement from equilibrium of the sprung mass, $y(t)$. However, we cannot arrive at that displacement through any linear combination of the states, $p(t)$ and $\delta(t)$, and the input, $v_{road}(t)$. To resolve this issue, we can augment the state equations by introducing an additional differential equation recognizing that

$$\dot{y} = \frac{p}{m}.$$

Though not necessary to simulate the system's dynamic response, the additional state equation provides the desired output. This, along with the above equations, forms the first part of our state-space representation,

$$\begin{bmatrix} \dot{p} \\ \dot{\delta} \\ \dot{y} \end{bmatrix} = \begin{bmatrix} -b/m & k & 0 \\ -1/m & 0 & 0 \\ 1/m & 0 & 0 \end{bmatrix} \begin{bmatrix} p \\ \delta \\ y \end{bmatrix} + \begin{bmatrix} b \\ 1 \\ 0 \end{bmatrix} v_{road}(t) = \mathbf{Ax} + \mathbf{Bu}.$$

The output is readily generated from the augmented state vector,

$$y = \begin{bmatrix} 0 & 0 & 1 \end{bmatrix} \begin{bmatrix} p \\ \delta \\ y \end{bmatrix} + 0 v_{road}(t) = \mathbf{C}x + \mathbf{D}u .$$

Now that we have arrived at a model, we can simulate the systems response to an input.

The vertical displacement induced by driving over a curb is much like a step function. The step input is not of unit height, however. Average curb height is in the range of 6 to 8 inches. Thus a height of 20 cm (just under 8 inches) will be utilized. The previous specifications for a typical family sedan will be used with the exception of the vehicle mass. This is a quarter-car suspension model. Thus, one should use the approximate mass suspended by one corner of the vehicle. The weight of a typical front-wheel-drive family sedan is biased toward the front of the vehicle such that approximately 60% of the total weight is suspended by the front suspension. Assuming the vehicle drives over the curb with the front wheels, approximately 30% of the total mass is supported by one corner of the front suspension. Hence, the mass we will use is 30% of 1700 kg or 510 kg. We can input those parameters into MATLAB and define the system model using the above matrices.

```
>> m = 0.30*1700; b = 750; k = 30e3;
>> A = [-b/m,k,0;-1/m,0,0;1/m,0,0];
>> B = [b;1;0];
>> C = [0,0,1];
>> D = 0;
>> sys = ss(A,B,C,D);
```

You might have noticed by now that the input to the above model is the road velocity, $v_{road}(t) = \dot{y}_{road}(t)$, and not the displacement, $y_{road}(t)$. To facilitate the simulation, we can take advantage of a previously introduced relationship between the impulse and step functions. Recall that the impulse is the derivative of the step function. Therefore, the approximate road velocity input that results from driving over the curb is

$$v_{road}(t) = \frac{d}{dt}(0.20 \text{ m}) 1(t) = 0.20 \delta(t) \text{ m/sec} .$$

For linear time invariant systems the effect of scaling an input is the same as scaling the output response. That is, we can generate the unit impulse response using the impulse command and scale it as illustrated in the following:

```
>> [Y,T] = impulse(sys);
>> plot(T,0.20*Y);
>> xlabel('Time (seconds)')
>> ylabel('y(t) (m)')
```

Note that if outputs are specified for the impulse command (as well as the step and lsim command) the plot is suppressed but the results are saved to arrays Y and T, which contain simulated response and time arrays. The approximate response garnered using the above commands is depicted in Figure 4.6.

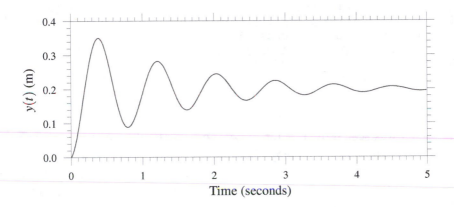

Figure 4.6: *Quarter-car suspension curb response.*

To illustrate the use of step command, we will return to the PMDC motor from Example 4.1.

Example 4.5

As has been mentioned, PMDC motors are ubiquitous in a variety of electromechanical systems. Though the input voltage may vary, there exist many cases where PMDC motors are operated at a constant input voltage. When the voltage is initially applied, the input instantaneously jumps from zero to a constant value just like a step input. A transient response ensues. Plot the response of the system to a 24-volt input.

   Solution. For the purpose of this simulation, let us use a motor for which we can readily attain the pertinent parameters – a Pittman GM8000 series model GM8712S030. The parameters are provided in Table 4.2 (parameters available at www.clickautomation.com/).

Using Equations 4.3 to 4.6 and the MATLAB `step` command, we can simulate the transient response.

```
>> ein_max = 24; L = 8.62e-3; R = 17.2; km = 2.73e-2;
>> J = 9.2e-7; n = 1/187.7; beta = 5.9e-7;
>> A = [-R/L,-km/J;km/L,-n^2*beta/J];
>> B = [1;0];
>> C = [0,n/J];
>> D = 0;
>> sys = ss(A,B,C,D);
>> t = [0:0.001:0.2]';
>> [Y,T] = step(sys,t);
>> wout = 24*Y; % scale the output response
>> u = ein_max*ones(length(t),1); % array of ein_max
>> [AX,H1,H2] = plotyy(T,wout,t,u);
>> xlabel('Time (seconds)');
>> set(get(AX(1),'Ylabel'),'String',...
'\omega_{out}(t) (rad/s)')
>> set(get(AX(2),'Ylabel'),'String','e_{in}(t) (volts)')
```

Note that the `step` command was used to generate the unit step response, which was scaled by a factor of 24. To generate the input voltage array, the `ones` function is employed to create an array of ones with the same dimensions as the time array t, which is then scaled by `ein` the maximum voltage limit. The `plotyy` command is utilized to plot multiple functions with different scales on the same graph. The `set` command can specify the labels for the left and right y-axes. The response $\omega_{out}(t)$ is plotted in Figure 4.7.

Table 4.2: *Pittman GM8712S030 motor parameters.*

Parameter	Variable	Value
Max Input Voltage	$(e_{in})_{max}$	24 V
Inductance	L	8.62 mH
Resistance	R	17.2 Ω
Motor Constant	k_m	2.73×10^{-2} N-m/A or V-s/rad
Rotational Inertia	J	9.2×10^{-7} kg-m^2
Gear Ratio	N_1/N_2	1/187.7
Damping Constant	β	5.9×10^{-7} N-m-s/rad

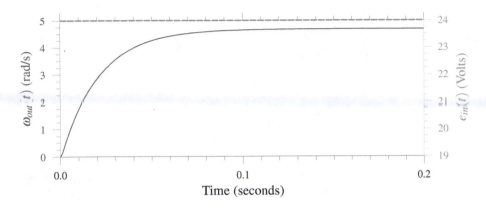

Figure 4.7: *Step response of a Pittman GM8712S030 motor.*

Example 4.6

What would the response of the PMDC motor be if the voltage is ramped from 0 to 24 V in 0.2 seconds and then maintained constant?

Solution. To generate the ramp response for the PMDC motor, an array must be generated that contains the linearly varying values. The Pittman motor from the prior example has a maximum input voltage limit of 24 V. You might imagine a number of ways to generate an array representation of such a function. The code below illustrates one way and assumes that the code for the previous example is still active or that the parameters and system are still defined in the MATLAB workspace.

```
>> t = [0:1/1000:0.3]';
>> u = [0:24/200:24 24*ones(1,100)]';
>> [Y,T] = lsim(sys,u,t);
>> [AX,H1,H2] = plotyy(t,Y,t,u);
>> xlabel('Time (seconds)');
>> set(get(AX(1),'Ylabel'),'String',...
'\omega_{out}(t) (rad/s)')
>> set(get(AX(2),'Ylabel'),'String','e_{in}(t) (Volts)')
```

The time array ranges from 0 to 0.3 in steps of 1/1000. Thus there are 300 divisions between 0 and 0.3. The voltage ramps from 0 to 24 over the first 200 divisions of time and then remains constant for the remaining 100. Figure 4.8 illustrates the predicted response for this input.

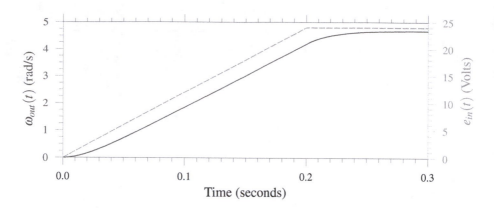

Figure 4.8: *Ramp response of a Pittman GM8712S030 motor.*

Example 4.7

Recall the mass-spring-damper system illustrated in Figure 3.18 whose state-space representation we derived in Example 4.2. The system has an external force that induces motion. Motion, however, can be induced by simply displacing the system from equilibrium. The first mass can be displaced a given amount and then released resulting in free vibration. Simulate the free vibration response for an initial displacement of the leftmost mass of a few centimeters (i.e., $x_1(0) = 0.2$ m = 20 cm). Assume that the masses, damping constants, and spring rates are 10 kg, 20 N-s/m, and 60 N/m, respectively.

   Solution. The free vibration response can be simulated using the following series of commands:

```
>> m = 10; b = 20; k = 60;
>> A = [0,1/m,0,0;-k,-b/m,k,0;0,-1/m,0,1/m;0,0,-k,0];
>> B = [0;0;0;1];
>> C = [1,0,0,0;1,0,1,0];
>> D = [0;0];
>> sys = ss(A,B,C,D);
>> [Y,T,X] = initial(sys,[0.20 0 0 0]);
>> plot(T,Y)
>> xlabel('Time (seconds)')
>> ylabel('Displacement (m)')
>> legend('x_1(t)','x_2(t)')
```

Notice that the MATLAB initial function was utilized to simulate the response. Note that if the first mass is displaced from equilibrium by 20 cm, the second mass will also move the same amount unless the spring between the masses deflects. This is illustrated by the response rendered in Figure 4.9. This system has four states. The responses depicted in Figure 4.9 are referred to as higher-order responses (i.e., greater than first- or second-order). In Chapter 7 we will return to this problem when we discuss the time-domain characteristics of higher-order systems.

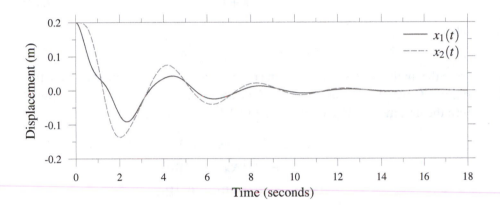

Figure 4.9: *Free vibration of a mass-spring-damper system.*

4.7 State Transformations

As has been previously mentioned, there is no unique set of states or differential equations that can be used to describe a single system. What is unique is the minimum number of states needed to model the dynamics. As was illustrated in Example 4.2, there are at least two equivalent ways to model the mass-spring-damper system from Examples 3.1 and 3.11. Regardless of the states chosen, the characteristics of the system cannot change. We will illustrate this using a state transformation and a few identities from Linear Algebra.

As has been mentioned, there is no unique set of differential equations for a given dynamic system. Equivalent models, however, describe the same dynamic system and must therefore have the same core characteristics including eigenvalues. Take two different models of the same system given in state-space form as

$$\dot{\mathbf{x}} = \mathbf{Ax} + \mathbf{Bu} \tag{4.14}$$

$$\mathbf{y} = \mathbf{Cx} + \mathbf{Du} \tag{4.15}$$

and

$$\dot{\tilde{\mathbf{x}}} = \tilde{\mathbf{A}}\tilde{\mathbf{x}} + \tilde{\mathbf{B}}\mathbf{u} \tag{4.16}$$

$$\mathbf{y} = \tilde{\mathbf{C}}\tilde{\mathbf{x}} + \mathbf{Du} \tag{4.17}$$

where

$$\mathbf{x} = \mathbf{T}\tilde{\mathbf{x}} \quad \text{or} \quad \tilde{\mathbf{x}} = \mathbf{T}^{-1}\mathbf{x}.$$

Note that both models have the same inputs $\mathbf{u}$ and output $\mathbf{y}$. The state transformation $\mathbf{T}$, which also relates the derivatives (i.e., $\dot{\mathbf{x}} = \mathbf{T}\dot{\tilde{\mathbf{x}}}$), can be substituted into the first model (Equations 4.14 and 4.15) to arrive at the second,

$$\begin{aligned}
\dot{\tilde{\mathbf{x}}} = \mathbf{T}^{-1}\dot{\mathbf{x}} &= \mathbf{T}^{-1}(\mathbf{Ax} + \mathbf{Bu}) \\
&= \mathbf{T}^{-1}\mathbf{Ax} + \mathbf{T}^{-1}\mathbf{Bu} \\
&= \mathbf{T}^{-1}\mathbf{AT}\tilde{\mathbf{x}} + \mathbf{T}^{-1}\mathbf{Bu} \\
&= \tilde{\mathbf{A}}\tilde{\mathbf{x}} + \tilde{\mathbf{B}}\mathbf{u}
\end{aligned}$$

and

$$\mathbf{y} = \mathbf{Cx} + \mathbf{Du} = \mathbf{CT}\tilde{\mathbf{x}} + \mathbf{Du} = \tilde{\mathbf{C}}\tilde{\mathbf{x}} + \mathbf{Du}.$$

Thus the transformation matrix also relates the state-space representations,

$$\tilde{\mathbf{A}} = \mathbf{T}^{-1}\mathbf{AT}, \ \tilde{\mathbf{B}} = \mathbf{T}^{-1}\mathbf{B}, \text{ and } \tilde{\mathbf{C}} = \mathbf{CT}.$$

Matrix $\mathbf{D}$ does not change from one representation to the other.

You may recall from Linear Algebra the eigenvalue problem,

$$\mathbf{Ax} = \lambda\mathbf{x} = \lambda\mathbf{Ix} \quad \Rightarrow \quad (\lambda\mathbf{I} - \mathbf{A})\mathbf{x} = \mathbf{0},$$

where $\mathbf{I}$ is the identity matrix. The eigenvalues are defined as those values λ that satisfy the above equation. More precisely, because the above must hold true regardless of the value of $\mathbf{x}$, the eigenvalues are those values of λ that satisfy

$$\det(\lambda\mathbf{I} - \mathbf{A}) = 0. \tag{4.18}$$

They are also known as the characteristic roots of the system. As we shall see in Chapter 7, they can be used to determine properties of the expected system response to various inputs.

Similar matrices have the same eigenvalues. A matrix $\widetilde{\mathbf{A}}$ is said to be similar to matrix $\mathbf{A}$ if there exists a *similarity transformation* such that

$$\widetilde{\mathbf{A}} = \mathbf{T}^{-1}\mathbf{A}\mathbf{T}. \tag{4.19}$$

This can be confirmed from the eigenvalue problem. Take two matrices $\widetilde{\mathbf{A}}$ and $\mathbf{A}$ with bases $\widetilde{\mathbf{x}}$ and $\mathbf{x}$ related through a transformation matrix, $\mathbf{x} = \mathbf{T}\widetilde{\mathbf{x}}$ (or $\widetilde{\mathbf{x}} = \mathbf{T}^{-1}\mathbf{x}$). If the matrices are similar, it can be shown that

$$\widetilde{\mathbf{A}}\widetilde{\mathbf{x}} = \lambda\widetilde{\mathbf{x}} \tag{4.20}$$

given λ are the solutions to the the following eigenvalue problem:

$$\mathbf{A}\mathbf{x} = \lambda\mathbf{x}. \tag{4.21}$$

This is accomplished by substituting Equations 4.19 and 4.21 into Equation 4.20 and using the basis transformation,

$$\widetilde{\mathbf{A}}\widetilde{\mathbf{x}} = \mathbf{T}^{-1}\mathbf{A}\mathbf{T}\widetilde{\mathbf{x}} = \mathbf{T}^{-1}\mathbf{A}\mathbf{x} = \mathbf{T}^{-1}\lambda\mathbf{x} = \lambda\mathbf{T}^{-1}\mathbf{x} = \lambda\widetilde{\mathbf{x}}.$$

Hence λ are the solutions for both eigenvalue problems (i.e., Equations 4.20 and 4.21) and thus matrices $\widetilde{\mathbf{A}}$ and $\mathbf{A}$ are similar. Such similarities arise between distinct models of the same system as illustrated by the following example.

Example 4.8

Models can be transformed from one set of states to another through state transformations. Recall the state vector $\mathbf{x} = \begin{bmatrix} x_1 & p_1 & \delta_2 & p_2 \end{bmatrix}^T$ used in Equations 4.7 and 4.8 and the alternate state vector $\widetilde{\mathbf{x}} = \begin{bmatrix} x_1 & v_1 & x_2 & v_2 \end{bmatrix}^T$ used in Equations 4.9 and 4.10. The two sets of states are related through a transformation matrix $\mathbf{T}$,

$$\mathbf{x} = \begin{bmatrix} 1 & 0 & 0 & 0 \\ 0 & m & 0 & 0 \\ -1 & 0 & 1 & 0 \\ 0 & 0 & 0 & m \end{bmatrix} \widetilde{\mathbf{x}} = \mathbf{T}\widetilde{\mathbf{x}} \quad \text{or} \quad \widetilde{\mathbf{x}} = \begin{bmatrix} 1 & 0 & 0 & 0 \\ 0 & 1/m & 0 & 0 \\ 1 & 0 & 1 & 0 \\ 0 & 0 & 0 & 1/m \end{bmatrix} \mathbf{x} = \mathbf{T}^{-1}\mathbf{x}.$$

The transformation matrix can be used to convert from one set of state equations to another:

$$
\begin{bmatrix} \dot{x}_1 \\ \dot{v}_1 \\ \dot{x}_2 \\ \dot{v}_2 \end{bmatrix} = \begin{bmatrix} 1 & 0 & 0 & 0 \\ 0 & 1/m & 0 & 0 \\ 1 & 0 & 1 & 0 \\ 0 & 0 & 0 & 1/m \end{bmatrix} \begin{bmatrix} \dot{x}_1 \\ \dot{p}_1 \\ \dot{\delta}_2 \\ \dot{p}_2 \end{bmatrix} = \mathbf{T}^{-1}\left(\mathbf{Ax} + \mathbf{Bu}\right)
$$

$$
= \begin{bmatrix} 1 & 0 & 0 & 0 \\ 0 & 1/m & 0 & 0 \\ 1 & 0 & 1 & 0 \\ 0 & 0 & 0 & 1/m \end{bmatrix} \left\{ \begin{bmatrix} 0 & 1/m & 0 & 0 \\ -k & -b/m & k & 0 \\ 0 & -1/m & 0 & 1/m \\ 0 & 0 & -k & 0 \end{bmatrix} \begin{bmatrix} x_1 \\ p_1 \\ \delta_2 \\ p_2 \end{bmatrix} + \begin{bmatrix} 0 \\ 0 \\ 0 \\ 1 \end{bmatrix} F(t) \right\}
$$

$$
= \begin{bmatrix} 0 & 1/m & 0 & 0 \\ k/m & -b/m^2 & k/m & 0 \\ 0 & 0 & 0 & 1/m \\ 0 & 0 & -k/m & 0 \end{bmatrix} \begin{bmatrix} x_1 \\ p_1 \\ \delta_2 \\ v_2 \end{bmatrix} + \begin{bmatrix} 0 \\ 0 \\ 0 \\ 1/m \end{bmatrix} F(t)
$$

$$
= \begin{bmatrix} 0 & 1/m & 0 & 0 \\ k/m & -b/m^2 & k/m & 0 \\ 0 & 0 & 0 & 1/m \\ 0 & 0 & -k/m & 0 \end{bmatrix} \begin{bmatrix} 1 & 0 & 0 & 0 \\ 0 & m & 0 & 0 \\ -1 & 0 & 1 & 0 \\ 0 & 0 & 0 & m \end{bmatrix} \begin{bmatrix} x_1 \\ v_1 \\ x_2 \\ v_2 \end{bmatrix} + \begin{bmatrix} 0 \\ 0 \\ 0 \\ 1/m \end{bmatrix} F(t)
$$

$$
= \begin{bmatrix} 0 & 1 & 0 & 0 \\ -2k/m & -b/m & k/m & 0 \\ 0 & 0 & 0 & 1 \\ k/m & 0 & -k/m & 0 \end{bmatrix} \begin{bmatrix} x_1 \\ v_1 \\ x_2 \\ v_2 \end{bmatrix} + \begin{bmatrix} 0 \\ 0 \\ 0 \\ 1/m \end{bmatrix} F(t) = \widetilde{\mathbf{A}}\widetilde{\mathbf{x}} + \widetilde{\mathbf{B}}\mathbf{u}.
$$

Additionally, we can confirm that since both $\mathbf{A}$ and $\widetilde{\mathbf{A}}$ are similar matrices, they should have the same set of eigenvalues. Assume that the system parameters are $m = 100$ kg, $b = 25$ N-s/m, and $k = 50$ N/m. The eigenvalues can be readily determined using MATLAB.

```
>> m = 100; b = 25; k = 50;
>> A = [0 1/m 0 0; -k -b/m k 0; 0 -1/m 0 1/m; 0 0 -k 0];
>> T = [1 0 0 0; 0 m 0 0; -1 0 1 0; 0 0 0 m];
>> Aalt = inv(T)*A*T;
>> eig(A)

ans =

  -0.0902 + 1.1341i
  -0.0902 - 1.1341i
  -0.0348 + 0.4381i
  -0.0348 - 0.4381i

>> eig(Aalt)
```

```
ans =

  -0.0902 + 1.1341i
  -0.0902 - 1.1341i
  -0.0348 + 0.4381i
  -0.0348 - 0.4381i
```

4.8 Summary

▷ For linear systems, the differential equations and outputs can be written as a linear combination of the states and inputs using Linear Algebra. This type of formulation is called the state-space representation (Equations 4.1 and 4.2).

▷ State-space models are composed by identifying the state, input, and output vectors. The individual first-order differential equations and output equations are written as linear combinations of the states and inputs. This facilitates identifying and separating the coefficients, states, and inputs in each equation.

▷ At $t = 0$, the unit impulse function (or Dirac delta function), $\tilde{\delta}(t)$, has infinite height and infinitesimal width. The function is referred to as unit impulse because the integral under the curve is one.

▷ The unit step function (or Heaviside function), $1(t)$, is the integral of the unit impulse. The function is unity for all values of time greater than zero ($t > 0$).

▷ The unit ramp function is the integral of the unit impulse. For values of $t > 0$ the function increases at a constant rate of unity.

▷ Because the impulse, step, and ramp functions are related through integration and differentiation, so are the output responses to these inputs.

▷ MATLAB provides a variety of commands for defining and simulating the responses of state-space models, including commands to define a state-space object and to simulate responses to an impulse, a step, an arbitrary function, or an initial condition.

▷ State-space representations are not unique. Several models can be derived to represent the same system in terms of distinct sets of states.

A state transformation can be used to transfer from one set of states to another. Regardless of the state vector chosen, the eigenvalues of the system are unique.

4.9 Review

R4-1 Explain what the state space is. Describe the composition of the differential and output equations in state-space form.

R4-2 Describe the process outlined for deriving the state-space representation.

R4-3 Describe in words with the aid of mathematics the unit impulse, step, and ramp functions.

R4-4 How are the three functions related?

R4-5 What are some of the MATLAB commands for defining state-space objects and simulating basic responses.

R4-6 Describe how to transform a state-space representation from one set of states to another.

R4-7 What do similar matrices have in common?

4.10 Problems

Derive the state-space representations for the following problems.

P4-1 *Mass-Spring-Damper Systems (Figure 1.15).* Recall that the bond graph was synthesized in Example 3.2 for (a). The input for (a) is the velocity $v(t)$, and the outputs are $x_1(t)$ and $x_2(t)$. The displacement associated with the input velocity is $x(t)$ and is labeled for illustration. The input to (b) is the forcing function $F(t)$, and, like (a), the outputs are $x_1(t)$ and $x_2(t)$. For (c) and (d) the external force is an input and the displacements of the respective masses are the outputs. The force due to gravity, if considered, can be treated as an external force and as second input to each system.

P4-2 *Rotational Systems (Figure 1.16).* An input torque τ_m is supplied for (a) and (b) while (c) has two input torques. For each rotational system, the inputs are used to generate an angular velocity ω_{out} at the output shaft.

P4-3 *Electric Circuits (Figure 1.17).* The input for circuit (a) is the voltage e. The outputs of interest are the inductor current i_L and the capacitor voltage v_c. Circuit (b) was examined in Examples 3.6 and 3.13. The input is the external voltage $e(t)$ and the outputs are the currents through each inductor and the capacitor voltage. Note that circuit (c) was already done in Example 4.3.

P4-4 *Hydraulic Circuits (Figure 1.18).* Hydraulic circuit (a) was discussed in Examples 3.8 and 3.14. The bond graph for circuit (b) was discussed in Example 3.9. Circuits (a) and (b) have an input volumetric flow source $Q(t)$, and the models will be used to estimate the time-varying output flow from the second valve. For circuit (c) the input is a pressure source, P_{in}, and the output of interest is the flow out of the system like the previous hydraulic circuit.

P4-5 *Permanent Magnet Direct Current (PMDC) Motor with Gear Reduction (Figure 1.3).* The voltage $e_{in}(t)$ is an input used to generate the desired angular velocity ω_2 as an output. The labels $i(t)$, E_m, τ_m, ω_m, τ_1, ω_1, τ_2, and ω_2 identify some the distinct efforts and flows and are provided to facilitate analysis. Only $e_{in}(t)$ is an external source or system input.

Simulate or solve the following problems using MATLAB.

P4-6. *Impulse and Step Responses of a Rotational System.* Recall the rotational system illustrated in Figure 1.16 (b). Assume that the rotational inertias are $J_1 = 0.05$ kg-m^2 and $J_2 = 0.1$ kg-m^2. The bearing constant and shaft rigidity are $\beta = 0.02$ N-m-s/rad and $\kappa = 10$ N-m/rad. The viscous friction coefficient is $\beta_{fric} = 0.1$ N-m-s/rad. Plot the step and impulse responses. The gear ratios are $N_1/N_2 = 5$ and $N_3/N_4 = 2$.

P4-7. *Sinusoidal Response of a Mass-Spring-Damper System.* For the system depicted in Figure 1.15 (a), assume that the masses, damping constants, and spring rates are 100 kg, 25 N-s/m, and 50 N/m, respectively. Plot the unit step, impulse, and ramp responses of the two masses. Also plot the responses for a sinusoidal input velocity $v(t) = 2\sin 3t$ m/s.

P4-8. *Time-Varying Response of a Wind Turbine.* Recollect the Pittman motor modeled in Example 4.5. Imagine that the motor is operated in reverse as a generator and used for the wind turbine generator depicted in Figure 3.28. Derive the state-space model, and use it to simulate the response to random wind gusts between 0 and 5 m/s. (*Hint:* Use the rand command to generate a random number between 0 and 1 and scale it by the maximum velocity, 5 m/s, and consider how frequently wind gusts potentially change. There is no one right answer.) Assume the fan modulus is $k_{fan} = 2$ N-s and remember that the Pittman motor parameters are provided in Table 4.2. The lamp resistance is $R_{lamp} = 20\ \Omega$. The output is the voltage drop across the lamp.

P4-9. *Step Response of an Electric Circuit.* Synthesize the bond graph for the circuit depicted in Figure 4.10. Derive the differential equations and place them in state-space form. The input is the voltage $e(t)$ and the outputs are the voltage drop across the capacitor and the current through the second inductor. Plot the responses to an input of 3 V. Assume that the system parameters are $L_1 = 10$ mH, $N_1/N_2 = 2$, $C = 60\ \mu$F, $L_2 = 20$ mH, and $R = 30\ \Omega$.

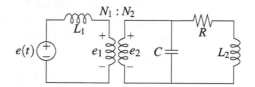

Figure 4.10: *An electric circuit with a transformer.*

P4-10. *Sinusoidal Response of an Electric Circuit.* Recall the electric circuit in Figure 1.17 (c). Assume the system parameters are $L_1 = 10$ mH, $C = 60\ \mu$F, $L_2 = 20$ mH, and $R = 30\ \Omega$. The system outputs are the inductor currents and the capacitor voltage. Plot the responses of each if the current source supplies a constant 200 mA current (i.e., $i(t) = 200$mA) and the voltage source supplies a sinusoidal voltage $= e(t) = 5\sin 10t$ V.

P4-11. *State Transformation of a Mass-Spring-Damper System.* Recall the state-space representations derived in Problem P4-1. Use a state transformation to rewrite the differential equations for (d) in terms of the vertical displacements and velocities of the two masses.

4.11 Challenges

Derive the state-space models for the problems below and simulate the responses using the parameters provided.

C4-1 *A Toy Electric Car.* The purpose of the toy car model is to predict the time-varying speed that results when the voltage from the batteries is applied across the motor terminals. Note that the motor is attached to a gear train at the output. You can reference Figure 3.25, Example 3.17, and Practice Problem P4-5 for suggestions on how you might model this system. Table 4.3 provides parameters you may need to model the system.

Table 4.3: *Toy car parameters.*

Parameter	Variable	Value
Battery Voltage	e_{in}	3 V
Motor Resistance	R_m	1.6 Ω
Motor Inductance	L_m	345 μH
Motor Constant	k_m	1.57×10^{-3} V-s/rad or N-m/A
Motor Inertia	J_m	4.12×10^{-7} kg-m^2
Motor Damping	β	2.80×10^{-7} N-m-s/rad
Gear Ratio	N_1/N_2	1/30
Wheel Inertia	J_w	5.38×10^{-6} kg-m^2
Wheel Radius	R_w	2 cm
Car Mass	m	155 g

C4-2 *A Quarter-Car Suspension Model.* Figure 3.6 and Example 3.2 provide and describe a model for a quarter-car suspension. However, this model does not account for the unsprung mass of the wheel, brakes, tire, etc., and it does not account for the effective compliance and damping of the tire wall and compressed air. Figure 1.2 (c) depicts a mass-spring-damper model of the suspension that includes unsprung mass and the effective damping and compliance of the tire. Be sure to incorporate these aspects in your model. Moreover, the model will be used to estimate the vertical motion of the vehicle as a response to input road displacements that have an associated vertical velocity ($v_{road}(t)$). Using the state-space model, plot the impulse responses of the velocities of the sprung and unsprung masses assuming that the sprung mass, unsprung mass, suspension spring constant, shock absorber damping constant, tire stiffness, and tire damping are $m_s = 400$ kg, $m_{us} = 100$ kg, $k_s = 30$ kN/m, $b_s = 750$ N-s/m, $k_t = 200$ kN/m, and $b_t = 125$ N-s/m, respectively.

C4-3 *An Electric Drill.* The electric drill is powered by a battery which excites a PMDC motor with a selectable gear pair. Assume for this problem that the drill is used to machine a hole into a steel plate. Cutting fluid is used to lubricate the metal-to-metal contact and reduce the friction. As a result, some viscous damping occurs at the outer circumference of the drill bit as it cuts through the steel plate. Assume that the battery supplies a specified input voltage e_{in} that generates an output torque τ_{out} applied to the screw. Use the motor parameters provided in Table 4.2. Select a drill bit of your choosing. Model it as a slender cylinder made of an appropriate steel. Use these details to determine the rigidity or effective torsion stiffness of the drill bit. Research friction models for typical surface-to-surface contacts with lubricating fluids to find a suitable damping constant.

Chapter 5

Laplace Transforms

Knowing now how to derive models in the form of systems of linear differential equations, we transition our attention to the analysis and solution of such models. Though the equations are differential, the Laplace transform provides a means of converting equations in the time domain to algebraic equations in the s-domain where the models become algebraic. Consider the following questions:

$\triangleright$ What are some commonly recurring functions in dynamic systems and their Laplace transforms?

$\triangleright$ How can Laplace transforms be used to solve for a dynamic response in time domain?

$\triangleright$ What purpose does the Laplace transform play in analyzing the time domain response?

$\triangleright$ What are the relationships between the time domain and s-domain?

5.1   Introduction

In the previous chapter we synthesized bond graphs to systematically derive the differential equations. The resulting state-space models can be used to simulate dynamic responses with the aid of MATLAB. With the aid of Laplace transforms and, as we will see later, Linear Algebra, the resulting differential equations can be used to analyze and solve for system responses. Laplace transforms are one of the most important tools for analyzing and calculating responses of linear systems. The fundamental advantage of using Laplace transforms is that they refashion a differential equation problem into

151

an algebraic problem, enabling us to take advantage of the many tools from Algebra that we are already familiar with.

In this chapter we will review the Laplace transform. In particular, we will examine the transforms of commonly recurring mathematical functions used in the modeling of physical dynamic systems. As will be discussed, several Laplace transform theorems exist which enable one to solve differential equations and assess initial and final conditions in the time domain. Laplace transforms are often the tool of choice for solving, in closed form, the time-domain differential equations. To derive such solutions, one must often implement partial fraction expansions to decompose the system in the *s*-domain into components that are readily inverse Laplace transformed.

The objectives and outcomes herein are:

▷ *Objectives:*

1. To review the transforms of commonly occurring functions used for modeling dynamic systems,

2. To review some of the more often used theorems that aid in the analysis and solution of dynamic systems, and

3. To understand algebraic analysis of Laplace transforms.

▷ *Outcomes:* Upon completion, you should

1. be able to transform and inverse transform common functions,

2. be able to conduct partial fraction expansions to find inverse Laplace transforms,

3. be able to use MATLAB to analyze Laplace transforms, and

4. be able to solve linear differential equations using Laplace transforms.

5.2   Complex Numbers, Variables, and Functions

In preparation for the analyses that will be done later in this chapter and others using Laplace transforms, we will review herein some basic properties of complex numbers. This material is covered in more detail in most Engineering Analysis texts. The following is a summary of some of the more pertinent arithmetic and theorems involving complex numbers.

5.2.1 Complex Numbers

When evaluating the roots of polynomial you inevitably have to deal with complex numbers. A complex number is generally of the form

$$z = x + jy$$

where $j = \sqrt{-1}$. As portrayed in Figure 5.1, complex numbers can be thought of as a vector where x is the real part and y is the imaginary part. The complex plane is composed of the real ($\Re$) and imaginary ($\Im$) axes.*

Figure 5.1 illustrates some geometric interpretations of the components of a complex number. In particular, the magnitude of the complex number is the square root of the sum of the squares of the real and imaginary parts,

$$|z| = \sqrt{x^2 + y^2},$$

and the angle of the pseudo vector z is the inverse tangent of the imaginary part over the real part,

$$\theta = \tan^{-1} \frac{y}{x}.$$

The above identities can be used to convert complex numbers represented in rectangular form to polar form,

$$\begin{aligned}
z &= x + jy \\
&= |z| \cos\theta + j|z| \sin\theta \\
&= |z|(\cos\theta + j\sin\theta) \\
&= |z|\underline{/\theta} \\
&= |z|e^{j\theta}.
\end{aligned}$$

The complex conjugate of $z = x + jy$ is

$$\bar{z} = x - jy.$$

Note, as illustrated by Figure 5.2, the complex conjugate has the same real part as z but the imaginary part is equal in magnitude but opposite in sign:

$$\begin{aligned}
z &= z + jy = |z|(\cos\theta + j\sin\theta) = |z|\underline{/\theta} \\
\bar{z} &= x - jy = |z|(\cos\theta - j\sin\theta) = |z|\underline{/-\theta}.
\end{aligned}$$

---

*. $\Re$ is shorthand for the real part and $\Im$ for the imaginary part.

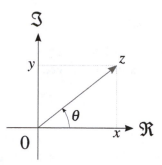

Figure 5.1: *Complex plane depiction of a complex number z.*

Complex conjugate pairs can be thought of as mirror images about the real axis. Furthermore, note that

$$z\bar{z} = (x+jy)(x-jy) = x^2 - jxy + jxy + y^2 = x^2 + y^2,$$

and that

$$\frac{1}{z} = \frac{1}{z}\frac{\bar{z}}{\bar{z}} = \frac{\bar{z}}{x^2 + y^2}.$$

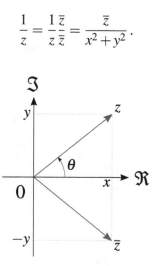

Figure 5.2: *Complex plane depiction of the complex conjugate.*

5.2.2 Euler's Theorem

Recall from Chapter 2 the general Taylor Series expansion of a single variable function (Equation 2.18). The Taylor Series expansions of $\cos\theta$ and $\sin\theta$

about zero are

$$\cos\theta = \cos 0 - \frac{\sin 0}{1!}\theta - \frac{\cos 0}{2!}\theta + \frac{\sin 0}{3!}\theta + \frac{\cos 0}{4!}\theta - \frac{\sin 0}{5!}\theta - \frac{\cos 0}{6!}\theta + \cdots$$

$$= 1 - \frac{\theta^2}{2!} + \frac{\theta^4}{4!} - \frac{\theta^6}{6!} + \cdots$$

and

$$\sin\theta = \sin 0 + \frac{\cos 0}{1!}\theta - \frac{\sin 0}{2!}\theta - \frac{\cos 0}{3!}\theta + \frac{\sin 0}{4!}\theta + \frac{\cos 0}{5!}\theta - \frac{\sin 0}{6!}\theta + \cdots$$

$$= \theta - \frac{\theta^3}{3!} + \frac{\theta^5}{5!} - \frac{\theta^7}{7!} + \cdots.$$

Therefore, it can be shown that

$$\cos\theta + j\sin\theta = \left(1 - \frac{\theta^2}{2!} + \frac{\theta^4}{4!} + \cdots\right) + j\left(\theta - \frac{\theta^3}{3!} + \cdots\right)$$

$$= 1 + j\theta - \frac{\theta^2}{2!} - j\frac{\theta^3}{3!} + \frac{\theta^4}{4!}$$

$$= 1 + j\theta + j^2\frac{\theta^2}{2!} + j^3\frac{\theta^3}{3!} + j^4\frac{\theta^4}{4!}$$

$$= 1 + j\theta + \frac{(j\theta)^2}{2!} + \frac{(j\theta)^3}{3!} + \frac{(j\theta)^4}{4!}.$$

You may also recall the Taylor Series expansion of an exponential function about zero,

$$e^x = 1 + \frac{e^0}{1!}x + \frac{e^0}{2!}x^2 + \frac{e^0}{3!}x^3 + \cdots$$

$$= 1 + x + \frac{x^2}{2!} + \frac{x^3}{3!} + \cdots.$$

Comparing the two prior expansions, it becomes evident that

$$\cos\theta + j\sin\theta = e^{j\theta}.$$

This identity is known as *Euler's theorem*. The complex conjugate of $e^{j\theta}$ is

$$e^{-j\theta} = \cos\theta - j\sin\theta.$$

The complex exponentials $e^{j\theta}$ and $e^{-\theta}$ are employed to find $\cos\theta$ and $\sin\theta$:

$$e^{j\theta} + e^{-j\theta} = (\cos\theta + j\sin\theta) + (\cos\theta - j\sin\theta)$$

$$= 2\cos\theta$$

$$\frac{e^{j\theta} + e^{-j\theta}}{2} = \cos\theta$$

and

$$e^{j\theta} - e^{-j\theta} = (\cos\theta + j\sin\theta) - (\cos\theta - j\sin\theta)$$
$$= 2j\sin\theta$$

$$\frac{e^{j\theta} - e^{-j\theta}}{2j} = \sin\theta\,.$$

5.2.3 Complex Algebra

Complex algebra does not differ much from algebra involving only real numbers. If two complex numbers,

$$z = x + jy \quad \text{and} \quad w = u + jv,$$

are said to be equal then it stands to reason that $x = u$ and $y = v$. To add complex numbers, sum the real and imaginary components much like vectors:

$$z + w = (x + jy) + (u + jv) = (x + u) + j(y + v)\,.$$

Subtraction is conducted in the same manner:

$$z - w = (x + jy) - (u + jv) = (x - u) + j(y - v)\,.$$

Multiplication of complex numbers is a little more involved. When a real number is multiplied by a complex number, the real and imaginary components are scaled by a factor equal to the real number:

$$az = a(x + jy) = ax + jay\,,$$

where a is a real number. When complex numbers are multiplied, we must recall that $j^2 = -1$ and multiply terms much like we do polynomials:

$$zw = (x + jy)(u + jv)$$
$$= xu + jxv + jyu + j^2yv$$
$$= (xu - yv) + j(xv + yu)\,.$$

Complex numbers can also be multiplied in complex form. Given that $z = |z|\underline{/\theta}$ and $w = |w|\underline{/\phi}$, it can be shown that

$$zw = |z||w|\underline{/\theta + \phi}\,. \tag{5.1}$$

Multiplication and division by j rotates the vector representation of the complex number by $90°$ in the complex plane (refer to Figure 5.3). Multiplication of a vector z by j yields

$$jz = j(x+jy) = jx + j^2y = -y + jx.$$

Alternatively, noting that j is in and of itself a vector in the complex plane of unit length pointing upward along the imaginary axis,

$$j = 0 + j1,$$

multiplication by j in polar form results in

$$jz = 1\underline{/90°}\ |z|\underline{/\theta} = |z|\underline{/\theta + 90°}.$$

Thus, the vector representation in the complex plane is rotated counter-clockwise by $90°$ as illustrated in Figure 5.3.

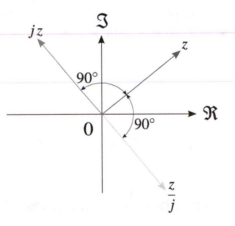

Figure 5.3: *Multiplication and division of a complex number by j.*

Division of complex numbers in polar form can be readily done by dividing the first vector length by the second and determining the difference rather than the sum of the angles:

$$\frac{z}{w} = \frac{|z|\underline{/\theta}}{|w|\underline{/\phi}} = \frac{|z|}{|w|}\underline{/\theta - \phi}. \tag{5.2}$$

In rectangular form, division is less convenient and relies on use of the complex conjugate:

$$\begin{aligned}
\frac{z}{w} &= \frac{x+jy}{u+jv} = \left(\frac{x+jy}{u+jv}\right)\left(\frac{u-jv}{u-jv}\right) \\
&= \frac{xu - jxv + jyu + yv}{u^2 + v^2} \\
&= \frac{xu + yv}{u^2 + y^2} + j\frac{yu - xv}{u^2 + y^2}.
\end{aligned} \tag{5.3}$$

Division by j has a similar effect to multiplication by j, but the rotation as depicted in Figure 5.3 is clockwise instead:

$$\frac{z}{j} = \frac{x+jy}{j} = \frac{x+jy}{j}\frac{j}{j} = \frac{jx - y}{j^2} = y - jx.$$

In polar form

$$\frac{z}{j} = \frac{|z|\angle\theta}{1\angle 90°} = |z|\angle\theta - 90°.$$

The nth power or root of a complex number is best determined in polar form. The nth power of a complex number z is

$$z^n = (|z|\angle\theta)^n = |z|^n\angle n\theta,$$

whereas the nth root is

$$z^{1/n} = (|z|\angle\theta)^{1/n} = |z|^{1/n}\angle\theta/n.$$

5.2.4 Complex Variables and Functions

Thus far we have discussed complex numbers that have constant real and imaginary parts. A complex variable, however, is not constant and is composed of real and imaginary parts that may vary. Such a variable is denoted herein as

$$s = \sigma + j\omega$$

where σ is the real part and ω the imaginary.

A complex function generally varies with s, e.g., $G(s)$. As will be seen in subsequent chapters, complex functions, like Laplace transforms, are commonly used in analysis of linear systems. Laplace transforms are used to solve linear ordinary differential equations. When using Laplace transforms to analyze and solve linear systems, the equations that result are generally of the form

$$G(s) = \frac{K(s+z_1)(s+z_2)\cdots(s+z_m)}{(s+p_1)(s+p_2)\cdots(s+p_n)}.$$

Such functions, as will be discussed in the next chapter, are referred to as transfer functions. The complex function $G(s)$ equals zero at $s = -z_1$, $s = -z_2, \ldots, s = -z_m$. The values of s for which the function is zero are referred to as the *zeros* of the complex function. The values of s for which the denominator is zero (or the complex function is infinite) are referred to as the *poles* (i.e., $s = -p_1$, $s = -p_2, \ldots, s = -p_n$). Note that the denominator can be composed of repeated poles (e.g., $(s+p)^k$).

5.3 The Laplace Transform

The Laplace transform is commonly used to analyze and solve linear systems. The primary advantage of its use is that differential equations in the time domain become algebraic in the s-domain. Once in the s-domain, the many algebraic tools you have learned over the years can be brought to bear. The complex function (or transfer function) can be decomposed through partial fraction expansions into terms that are readily converted back to the time domain. In this section, we will explore mathematical functions that are commonly used in the modeling of physical systems. Unlike other methods for solving ordinary differential equations, the Laplace transform approach accounts for initial conditions and can concurrently solve for both the particular and complimentary solutions.

As you may recall from a prior course in Engineering Analysis or Differential Equations, the Laplace transform of a function $g(t)$ is

$$\mathscr{L}[g(t)] = \mathsf{g}(s) = \int_0^\infty e^{-st} g(t)\, dt. \tag{5.4}$$

Though in many texts the generic mathematical function is more commonly referred to as $f(t)$, $g(t)$ is used to prevent confusion with f, which, until now, has been used to denote a generalized flow. Note that the Laplace transform is defined over the range $0 < t < \infty$. This makes physical sense because it is assumed that the dynamic response initiates at time $t = 0$ and that prior to that time, the system is at equilibrium. Thus, the function $g(t)$ is assumed to be zero for $t < 0$; otherwise, the dynamic system would not be at equilibrium at initial time. The resultant function, $\mathsf{g}(s)$, is a complex function of the Laplace operator s. Notice that a distinct font is used to distinguish the Laplace transform from its respective time domain function. The inverse Laplace transform,

$$\mathscr{L}^{-1}[g(s)] = g(t),\tag{5.5}$$

is the reverse procedure for transforming the s-domain algebraic function back to its time domain equivalent.

The Laplace transform is a linear operation. Thus, it follows that

$$\mathscr{L}[ag_1(t) + bg_2(t) + cg_3(t)] = a\mathscr{L}[g_1(t)] + b\mathscr{L}[g_2(t)] + c\mathscr{L}[g_3(t)]$$
$$= ag_1(s) + bg_2(s)cg_3(s).$$

Though the common practice is to utilize a Laplace transform table like that in Table 5.1 to convert from the time domain function to the s-domain, it is a useful exercise to examine or review the transforms of some common functions to garner better intuition for later analysis.

5.4 Common Functions and Their Transforms

Herein we will review the transforms of some of the most commonly recurring functions in System Dynamics.

Existence of the Laplace transform. Generally, for a Laplace transform of a function to exist, we require that integral be bounded as time approaches infinity. Stated otherwise, the integral must converge. The following are sufficient but not necessary conditions to ensure the existence of a Laplace transform:

1. $g(t)$ is a piecewise continuous function on the interval $0 < t < \infty$.

2. $g(t)$ is of *exponential order*. That is to say there exist real-valued positive constants A and t such that

$$|g(t)| \le Ae^{at} \text{ for all } t \ge T.$$

The first condition indicates that the function can be discontinuous and comprised of pieces that are continuous as shown in Figure 5.4. Examples include discontinuous repeating functions like rectangular and sawtooth waves. Though the second condition basically requires that the function be bounded, due to the first condition, the function can be composed, piecewise, by components that are not bounded over all time. This will be explored in more detail in Section 5.5.

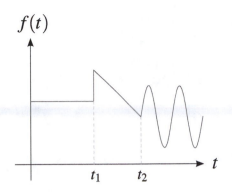

Figure 5.4: *Piecewise continuous function.*

Exponential functions. The decaying exponential recurs often in a variety of dynamic responses. Figure 5.5 portrays a decaying exponential. Note that at initial time, $t = 0$, the function is equal to A. The function decays and asymptotically approaches zero as time approaches infinity (i.e., $Ae^{-at} \to 0$ and $t \to \infty$). The Laplace transform of

$$g(t) = \begin{cases} 0, t < 0 \\ Ae^{-at}, t \geq 0 \end{cases}$$

is

$$\mathscr{L}\left[Ae^{-at}\right] = \int_0^\infty e^{-st}\left(Ae^{-at}\right)dt = A \int_0^\infty e^{-st}e^{-at}dt$$

$$= A \int_0^\infty e^{-(s+a)t}dt = -A \left[\frac{e^{-(s+a)t}}{s+a}\right]_0^\infty$$

$$= -A \left[\frac{0-1}{s+a}\right] = A\frac{1}{s+a}.$$

Notice that the integrated resultant includes a decaying exponential, $e^{-(s+a)t}$, assuming $s+a > 0$. This must be assumed to ensure the existence of the Laplace transform.

Step functions. Recall the unit step function (Equation 4.12) introduced in Chapter 4. A general step function is of the form

$$A1(t) = \begin{cases} 0, t < 0 \\ A, t \geq 0 \end{cases}.$$

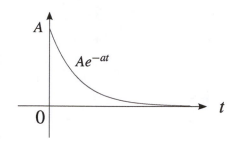

Figure 5.5: *A decaying exponential function.*

This function is constant over the range 0 to ∞, which are the limits for the Laplace transform integral. Thus, the Laplace transform of a general step function is

$$\mathscr{L}\left[A\,1(t)\right] = A \int_0^\infty e^{-st} dt = -A \left.\frac{e^{-st}}{s}\right|_0^\infty = -A\frac{(0-1)}{s} = \frac{A}{s}.$$

Ramp functions.   The unit ramp function was also previously introduced in Chapter 4. Generally, a ramp function may have a slope other than one,

$$g(t) = \begin{cases} 0, \, t < 0 \\ At, \, t \geq 0 \end{cases}.$$

To find its Laplace transform,

$$\mathscr{L}[At] = \int_0^\infty e^{-st} At\, dt = A \int_0^\infty e^{-st} t\, dt,$$

we must do integrations by parts,

$$\int_a^b u\, dv = uv \Big|_a^b - \int_a^b v\, du,$$

where in this case $u = t$ and $dv = e^{-st}$. The resultant integral is

$$\mathscr{L}[At] = A \int_0^\infty e^{-st} t\, dt = A \left[\left.\frac{te^{-st}}{-s}\right|_0^\infty - \int_0^\infty \frac{e^{-st}}{-s} dt \right]$$

$$= A \left[\frac{0-0}{-s} - \left.\frac{e^{-st}}{s^2}\right|_0^\infty \right] = A \left[-\frac{0-1}{s^2} \right]$$

$$= \frac{A}{s^2}.$$

Note that when integrating by parts, the uv term includes the function te^{-st}. It may not be obvious that the limits of this function are both zero. This function is the multiplication of two functions (t and e^{-st}), each of which is different at the limits zero and infinity. For the purposes of this discussion assume that s is positive and real valued. As Figure 5.6 indicates, the function t starts at zero and steadily progresses to infinity as time increases. In contrast, e^{-st} begins at one and decays rapidly to zero. The exponential function decays at a faster rate than the ramp function approaches infinity. Hence the actual function initially increases but rapidly decays back down to zero.

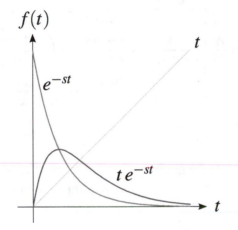

Figure 5.6: *Comparison of functions t and e^{-st}.*

Sinusoidal functions.   Sinusoidal functions commonly occur in a variety of dynamic systems including AC circuits and vibrating structures. To derive the Laplace transform of a sinusoid, we must employ Euler's theorem. Take for instance a sine function of the form

$$g(t) = \begin{cases} 0, \, t < 0 \\ A \sin \omega t, \, t \ge 0 \end{cases} .$$

Recall from Euler's theorem that

$$\sin \omega t = \frac{1}{2j} \left(e^{j\omega t} - e^{-j\omega t} \right) .$$

Thus, the Laplace transform is

$$
\begin{aligned}
\mathscr{L}\left[A\sin\omega t\right] &= \frac{A}{2j}\int_0^\infty e^{-st}\left(e^{j\omega t}-e^{-j\omega t}\right)dt \\
&= \frac{A}{2j}\int_0^\infty \left(e^{-(s-j\omega)t}-e^{-(s+j\omega)t}\right)dt \\
&= \frac{A}{2j}\left[-\frac{e^{-(s-j\omega)t}}{s-j\omega}+\frac{e^{-(s+j\omega)t}}{s+j\omega}\right]_0^\infty \\
&= \frac{A}{2j}\left[-\frac{0-1}{s-j\omega}+\frac{0-1}{s+j\omega}\right]=\frac{A}{2j}\left[\frac{1}{s-j\omega}-\frac{1}{s+j\omega}\right] \\
&= \frac{A}{2j}\left[\left(\frac{1}{s-j\omega}\right)\left(\frac{s+j\omega}{s+j\omega}\right)-\left(\frac{1}{s+j\omega}\right)\left(\frac{s-j\omega}{s-j\omega}\right)\right] \\
&= \frac{A}{2j}\left[\frac{s+j\omega-s+j\omega}{s^2+\omega^2}\right]=\frac{A}{2j}\left[\frac{2j\omega}{s^2+\omega^2}\right] \\
&= A\frac{\omega}{s^2+\omega^2}.
\end{aligned}
$$

Notice how the complex conjugates were used to find a common denominator and simplify the solution. You can show through a similar process that the Laplace transform of $\cos\omega t$ is

$$
\mathscr{L}\left[A\cos\omega t\right]=A\frac{s}{s^2+\omega^2}.
$$

5.5 Advanced Transforms and Theorems

Some functions are more complex than those discussed thus far. Some are piecewise continuous or have a time delay. Moreover, to solve linear differential equations using Laplace transforms, we need to review the Laplace theorems for integration and differentiation. Additionally, Laplace theorems can be used to facilitate determining initial or final values of a function in the time domain.

Multiplication by e^{-at}. We will see in later sections and chapters that decaying sinusoids like $e^{-at}\sin\omega t$ commonly occur in underdamped systems. This raises the question: What is the Laplace transform of such a function? More generally: What is the Laplace transform of a function of the form

$e^{-at}g(t)$? The Laplace transform is

$$\mathcal{L}\left[e^{-at}g(t)\right] = \int_0^\infty e^{-st}e^{-at}g(t)\,dt$$

$$= \int_0^\infty e^{-(s+a)t}g(t)\,dt$$

$$= g(s+a).$$

The result is similar to the definition of the Laplace transform (Equation 5.4) except s has to be substituted by $s+a$.

Therefore, if we wish to find the Laplace transform of decaying sinusoid like $e^{-at}\sin \omega t$ we recall that

$$\mathcal{L}[\sin \omega t] = \frac{\omega}{s^2 + \omega^2}$$

and we replace s with $s+a$ to arrive at

$$\mathcal{L}[e^{-at}\sin \omega t] = \frac{\omega}{(s+a)^2 + \omega^2}.$$

Similarly, we can show that

$$\mathcal{L}[e^{-at}\cos \omega t] = \frac{s+a}{(s+a)^2 + \omega^2}.$$

Shifting versus translation of a function.   To obtain the transform of piecewise continuous functions, we must first understand the concept of translation of a function. In the discipline of Signals and Systems, many of the signals analyzed are, for all intents and purposes, perpetually repeating functions. With such functions, it is not as important what the initial time is. Shifting or delaying a signal an amount a in time is like sliding the function along the time axis and is equivalent to evaluating the mathematical function at $t-a$ instead of t; $g(t)$ shifted to the right by a is equal to $g(t-a)$. Since the repeating signal is effectively perpetual, the signal is generally non-zero for $t < 0$. This is different from System Dynamics where the interest lies in analyzing the transient and steady-state responses, and we assume the system to be at some initial condition, for example, equilibrium, prior to $t = 0$.

Figure 5.7 (a) shows a shifted sinusoid. In System Dynamics and when utilizing Laplace transforms, we assume the function to be zero prior to initial time, like the unit step function plotted in Figure 5.7 (b). The unit step is like a constant value of one that is "truncated" or "zeroed" prior to $t = 0$. In

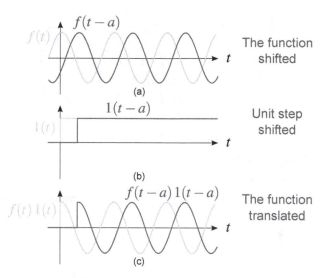

Figure 5.7: *Shifting and translating a function.*

our prior discussion and derivations of Laplace transforms we have assumed every function to be of the form

$$g(t) = \begin{cases} 0, t < 0 \\ \tilde{g}(t), t \geq 0 \end{cases}$$

where $\tilde{g}(t)$ is only defined for values of time greater than zero. Mathematically, what we effectively did is take every function and multiply it by a unit step, for example,

$$g(t) = \begin{cases} 0, t < 0 \\ Ae^{-at}, t \geq 0 \end{cases} = \left(Ae^{-at}\right) 1(t).$$

To translate a function, however, we must truncate and shift the function as shown in Figure 5.7 (c). This involves shifting the original function and multiplying by a shifted unit step, $g(t-a)1(t-a)$. The Laplace transform of a translated function

$$\mathscr{L}[g(t-a)1(t-a)] = \int_0^\infty e^{-st} g(t-a)1(t-a)\, dt$$

can be evaluated by first recognizing that the lower limit of the integral can be changed from 0 to a without loss of generality. This is because the translated function will be zero until $t = a$. Additionally, a substitution can be used to facilitate integration, $\tilde{t} = t - a$ ($d\tilde{t} = dt$). In the shifted time scale, $\tilde{t}$, the bounds of integration are 0 to ∞ (i.e., at $t = a$, $\tilde{t} = 0$). The resultant integral is

$$\mathcal{L}[g(t-a)1(t-a)] = \int_a^\infty e^{-st}g(t-a)1(t-a)\,dt$$
$$= \int_0^\infty e^{-s(\tilde{t}+a)}g(\tilde{t})1(\tilde{t})\,d\tilde{t}$$
$$= e^{-as}\int_0^\infty e^{-s\tilde{t}}g(\tilde{t})\,d\tilde{t}$$
$$= e^{-as}g(s).$$

Notice that the last form of the integral is compose of the term e^{-as} times what appears to be the Laplace transform written in terms of $\tilde{t}$ instead of t. Also, the unit step function $1(\tilde{t})$ is constant and equal to one over the range of the integral. Hence, translating a function by an amount a in the time domain is equivalent to multiplying by e^{-as} in the s-domain.

A rectangular pulse.   Given the above discussion of translated functions, let us look at a commonly occurring function composed of two translated step functions of equal amplitude but opposite sign,

$$g(t) = A\,1(t-t_1) - A\,1(t-t_2) = A\,[1(t-t_1) - 1(t-t_2)]\,.$$

The two step functions sum to create a rectangular pulse like the one illustrated in Figure 5.8 (a). Each step function is translated or delayed. Using the previously derived transform for translated function, the Laplace transform of $g(t)$ is

$$\mathcal{L}[g(t)] = A\left[\frac{e^{-t_1 s}}{s} - \frac{e^{-t_2 s}}{s}\right] = \frac{A}{s}\left[e^{-t_1 s} - e^{-t_2 s}\right]\,.$$

Example 5.1

Find the Laplace transform for the piecewise continuous function in Figure 5.9.

Solution.   As we saw with the rectangular pulse we can compose piecewise continuous functions using translation. The function starts as a ramp and then plateaus. Mathematically, this entails canceling the effect of the ramp by opposing it with a ramp of opposite slope. The opposing ramp, however, is translated to $t = t_1$. The overall function is therefore

$$g(t) = At\,1(t) - A\,(t-t_1)1(t-t_1)\,.$$

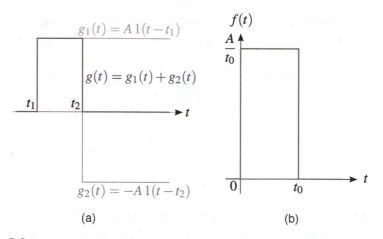

Figure 5.8: *A rectangular pulse composed from the summation of two step functions.*

Applying the transforms for ramp and translated functions, the resultant Laplace transform is

$$\mathcal{L}[g(t)] = \frac{A}{s^2} - \frac{Ae^{-t_1 s}}{s^2} = \frac{A}{s^2}\left[1 - e^{-t_1 s}\right].$$

Impulse functions.   Recall the unit impulse (or Dirac delta function) illustrated in Figure 4.2 and given in Equation 4.11. The general impulse function is created by taking a rectangular pulse function like one previously analyzed. This pulse, however, has a height of A/t_0 and a width of t_0 and begins at $t = 0$:

$$f(t) = \begin{cases} \dfrac{A}{t_0}, & 0 < t < t_0 \\ 0, & t < 0 \text{ and } t_0 < t \end{cases}.$$

The Laplace transform of such a pulse function is derived by defining the overall function piecewise using two step functions,

$$g(t) = \frac{A}{t_0}1(t) - \frac{A}{t_0}1(t - t_0) = \frac{A}{t_0}\left[1(t) - 1(t - t_0)\right]].$$

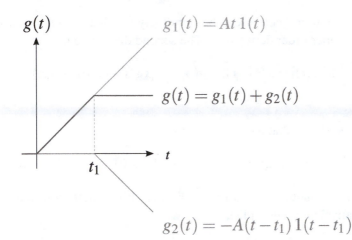

Figure 5.9: *Composition of a piecewise continuous function.*

The identities for step and translated functions can be utilized to derive the transform for the rectangular pulse,

$$\mathscr{L}[g(t)] = \frac{A}{t_0} \left[\frac{1}{s} - \frac{e^{-t_0 s}}{s} \right] = \frac{A}{t_0 s} \left[1 - e^{-t_0 s} \right].$$

The Laplace transform of an impulse is determined by taking the limit of the above transform as t_0 approaches zero. Remember that the impulse is attained by taking a rectangular pulse and stretching the height while narrowing the width and maintaining the area under curve constant. Hence, the Laplace transform is

$$\mathscr{L}[A\,\tilde{\delta}(t)] = \lim_{t_0 \to 0} \frac{A}{t_0 s} \left[1 - e^{-t_0 s} \right] = \lim_{t_0 \to 0} \frac{\dfrac{d}{dt_0} A \left[1 - e^{-t_0 s} \right]}{\dfrac{d}{dt_0} t_0 s}$$

$$= \lim_{t_0 \to 0} \frac{A s e^{-t_0 s}}{s} = \frac{A s}{s} = A.$$

The area under the impulse function is equal to the Laplace transform.

Differentiation theorem.   The Laplace transform of the first derivative of a function $g(t)$ is

$$\mathscr{L}[\dot{g}(t)] = s\mathscr{L}[g(t)] - g(0) = s g(s) - g(0).$$

The differentiation theorem can be iteratively applied to derive the second, third, and higher-order derivatives. The second derivative is

$$\mathscr{L}[\ddot{g}(t)] = s\mathscr{L}[\dot{g}(t)] - \dot{g}(0) = s[s\mathbf{g}(s) - g(0)] - g(0)$$
$$= s^2\mathbf{g}(s) - sg(0) - \dot{g}(0),$$

and it can be shown that the nth derivative is

$$\mathscr{L}[g^{(n)}(t)] = s^n\mathbf{g}(s) - s^{n-1}g(0) - s^{n-2}\dot{g}(0) - \cdots - g^{(n-1)}(0).$$

Apart from accounting for initial conditions, differentiation in the time domain is equivalent to multiplication by s.

Example 5.2

Using the differentiation theorem, derive the Laplace transforms of the untranslated unit step and impulse functions from the unit ramp function.

Solution. In the previous chapter, we discovered that the impulse, step, and ramp functions are related through integration and differentiation. We can apply the differentiation theorems for the first and second derivatives of the unit ramp function to find the Laplace transforms of the step and impulse functions. The transform of the unit step function should be

$$\mathscr{L}[1(t)] = \mathscr{L}\left[\frac{d}{dt}t\right] = s\mathscr{L}[t] - t\Big|_0 = s\frac{1}{s^2} - 0 = \frac{1}{s}.$$

For the unit impulse, the Laplace transform should be

$$\mathscr{L}[\tilde{\delta}(t)] = \mathscr{L}\left[\frac{d^2}{dt^2}t\right]$$
$$= s^2\mathscr{L}[t] - st\Big|_0 - \left[\frac{d}{dt}t\right]_0$$
$$= s^2\frac{1}{s^2} - 0 - 0$$
$$= 1.$$

Integration theorem.   The transform of the integral of a function is

$$\mathscr{L}\left[\int g(t)\,dt\right] = \frac{\mathbf{g}(s)}{s} + \frac{g^{(-1)}(0)}{s}$$

where

$$g^{(-1)}(0) = \int g(t)\,dt \text{ at } t = 0.$$

For a definite integral, the transform is simply

$$\mathscr{L}\left[\int_0^t g(t)\,dt\right] = \frac{g(s)}{s}.$$

Hence, integration in the time domain is basically equivalent to division by s or multiplication by $1/s$ in the s-domain.

Example 5.3

Beginning with the unit impulse function, find the Laplace transforms of the unit step and ramp functions.

Solution. To derive the Laplace transform of the unit step, we apply the integration theorem to the unit impulse function,

$$\mathscr{L}[1(t)] = \mathscr{L}\left[\int \tilde{\delta}(t)\,dt\right] = \frac{\mathscr{L}[\delta(t)]}{s} = \frac{1}{s}.$$

The integration theorem can be applied a second time to find the transform for the unit ramp,

$$\mathscr{L}[t] = \frac{\mathscr{L}[1(t)]}{s} = \frac{1/s}{s} = \frac{1}{s^2}.$$

Final and initial value theorems.   The final value theorem enables one to determine the steady-state condition of a time domain function using its Laplace transform. The theorem states

$$\lim_{t \to \infty} g(t) = \lim_{s \to 0} s\,g(s) \tag{5.6}$$

assuming that the transform $g(t)$ and its derivative exist and that $\lim_{t \to \infty}$ also exists. Similarly, the initial value theorem can be used to determine the initial condition using the Laplace transform,

$$g(0+) = \lim_{s \to \infty} s\,g(s) \tag{5.7}$$

where $g(0+)$ is the value of $g(t)$ at $t = 0$ approaching from $+\infty$.

Table 5.1 summarizes the transforms that have been discussed herein.

Table 5.1: *Commonly used Laplace transforms.*

	$g(t)$	$g(s)$
1	$\tilde{\delta}(t)$	1
2	$1(t)$	$\dfrac{1}{s}$
3	t	$\dfrac{1}{s^2}$
4	e^{-at}	$\dfrac{1}{s+a}$
5	$t e^{-at}$	$\dfrac{1}{(s+a)^2}$
6	$\sin \omega t$	$\dfrac{\omega}{s^2+\omega^2}$
7	$\cos \omega t$	$\dfrac{s}{s^2+\omega^2}$
8	$\sinh \omega t$	$\dfrac{\omega}{s^2-\omega^2}$
9	$\cosh \omega t$	$\dfrac{s}{s^2-\omega^2}$
10	$e^{-at}\sin \omega t$	$\dfrac{\omega}{(s+a)^2+\omega^2}$
11	$e^{-at}\cos \omega t$	$\dfrac{s+a}{(s+a)^2+\omega^2}$
12	$\dot{g}(t)$	$s\mathbf{g}(s)-g(0)$
13	$\ddot{g}(t)$	$s^2\mathbf{g}(s)-sg(0)-\dot{g}(0)$
14	$\displaystyle\int g(t)\,dt$	$\dfrac{\mathbf{g}(s)}{s}+\dfrac{g^{(-1)}(0)}{s}$
15	$e^{-at}g(t)$	$\mathbf{g}(s+a)$
16	$g(t-a)\,1(t-1)$	$e^{-as}\mathbf{g}(s)$

5.6   Inverse Laplace Transforms

Laplace transforms are one of the primary means of solving linear, time-invariant differential equations. The differentiation theorems are utilized to convert a differential equation (or set of differential equations) into an algebraic equation. The resulting algebraic equation is manipulated to solve for the output. Then the resultant transformation must be decomposed through partial fraction expansions to attain components that can be readily inverse Laplace transformed. In the previous chapter, we derived the system models as sets of n linear, time-invariant first-order differential equations. In the subsequent chapter we will discover that we can alternatively derive the system model as a transfer function. As we will see, a transfer function is an algebraic output/input representation that results from applying Laplace transforms to the system model or directly to the individual constitutive relations of the bond graph elements. For now we examine some relatively simple cases by converting n first-order differential equations into an nth-order differential equation and then applying Laplace transforms.

A transform can generally be of the form of one polynomial of s over another:

$$g(s) = \frac{B(s)}{A(s)} = \frac{b_0 s^m + b_1 s^{m-1} + \cdots + b_{m-1} s + b_m}{a_0 s^n + a_1 s^{n-1} + \cdots + a_{n-1} s + a_n}.$$

To find the inverse Laplace transform and determine the response in the time domain, the polynomials must be factors and transform decomposed through partial fraction expansions. The resultant components depend on the type of roots the denominator has. As previously mentioned, the roots of the numerator are the zeros of the system and those of the denominator are poles. The poles dictate the type of terms that will result in the partial fraction expansion and ultimately in the inverse Laplace transform. The roots of a polynomial can be generally complex. Thus, the poles can be purely real, purely imaginary, complex, and can even repeat.

Partial fraction expansions with distinct real poles.   If the poles are real and distinct, the denominator can be factored and the partial fraction expansion will be of the form

$$G(s) = \frac{B(s)}{A(s)} = \frac{K(s+z_1)(s+z_2)\cdots(s+z_m)}{(s+p_1)(s+p_2)\cdots(s+p_n)} = \frac{r_1}{s+p_1} + \frac{r_2}{s+p_2} + \cdots + \frac{r_n}{s+p_n}$$

where the coefficients r_k (for $k = 1,\ldots,n$) are referred to as the *residues* and $s = -p_k$ are the poles. Each residue is a constant and does not vary with s.

Thus, to find each residue, we can multiply both sides of the above equation by $(s + p_k)$ and strategically set $s = -p_k$:

$$\left[(s + p_k) \frac{B(s)}{A(s)} \right]_{s=-p_k} = \left[\frac{r_1}{s + p_1}(s + p_k) + \frac{r_2}{s + p_2}(s + p_k) + \cdots \right.$$

$$\left. + \frac{r_k}{s + p_k}(s + p_k) + \cdots + \frac{r_n}{s + p_n}(s + p_k) \right]_{s=-p_k}$$

$$= r_k .$$

Note that all the terms on the right side of the equation go to zero except for the kth term where $(s + p_k)$ cancels with the denominator leaving only r_k. In summary, the kth residue is

$$r_k = \left[(s + p_k) \frac{B(s)}{A(s)} \right]_{s=-p_k} .$$

Example 5.4

Recall the quarter-car suspension simulated in Example 4.4. Assume for this particular example that the mass, damping constant, and spring rate are 500 kg, 8,000 N-s/m, and 30,000 N/m, respectively. Derive the response of the system to a unit impulse displacement.

Solution. From Example 4.4, the system differential equations are

$$\dot{p} = k\delta + b \left[v_{road}(t) - \frac{p}{m} \right]$$

$$\dot{\delta} = v_{road}(t) - \frac{p}{m} \text{ and}$$

$$\dot{y} = \frac{p}{m} .$$

To solve for the impulse response using Laplace transforms, we need to solve for $\mathscr{L}[y(t)] = y(s)$. For the unit impulse response, we assume that the initial conditions are all zero. Ergo, the transforms of the above differential equations are

$$sp(s) = k\Delta(s) + b \left[v_{road}(s) - \frac{p(s)}{m} \right],$$

$$s\Delta(s) = v_{road}(s) - \frac{p(s)}{m}, \text{ and}$$

$$sy(s) = \frac{p(s)}{m} .$$

To solve for y(s), we must determine p(s):

$$y(s) = \frac{p(s)}{ms}.$$

To do so, we utilize the first equation and substitute for $\Delta(s)$ using the second:

$$sp(s) = k\Delta(s) + b\left[v_{road}(s) - \frac{p(s)}{m}\right]$$

$$= \frac{k}{s}\left[v_{road}(s) - \frac{p(s)}{m}\right] + b\left[v_{road}(s) - \frac{p(s)}{m}\right]$$

$$= \left[b + \frac{k}{s}\right]v_{road}(s) - \left[b + \frac{k}{s}\right]\frac{p(s)}{m}$$

$$ms^2 p(s) = m[bs + k]v_{road}(s) - [bs + k]p(s)$$

$$[ms^2 + bs + k]p(s) = m(bs + k)\,v_{road}(s)$$

$$p(s) = \frac{m(bs + k)}{ms^2 + bs + k}\,v_{road}(s)$$

$$= \frac{m(bs + k)}{ms^2 + bs + k}\,sy_{road}(s).$$

Substituting for p(s) into the y(s) equation yields

$$y(s) = \frac{1}{ms}\frac{m(bs + k)}{ms^2 + bs + k}\,sy_{road}(s) = \frac{bs + k}{ms^2 + bs + k}\,y_{road}(s).$$

Given the system parameters and input unit impulse (i.e., $\mathscr{L}[y_{road}(t)] = 1$), y($s$) is

$$y(s) = \frac{8,000s + 30,000}{500s^2 + 8,000s + 30,000} = \frac{16s + 60}{s^2 + 16s + 60} = \frac{16s + 60}{(s + 6)(s + 10)}.$$

To determine y(t) we must find the partial fraction expansion

$$y(s) = \frac{16s + 60}{s^2 + 16s + 60} = \frac{r_1}{s + 6} + \frac{r_2}{s + 10}.$$

We can solve for the residues as previously described:

$$r_1 = \left[(s + 6)\frac{16s + 60}{(s + 6)(s + 10)}\right]_{s=-6} = \frac{16s + 60}{s + 10}\bigg|_{s=-6} = -9$$

and

$$r_2 = \left[(s + 10)\frac{16s + 60}{(s + 6)(s + 10)}\right]_{s=-10} = \frac{16s + 60}{s + 6}\bigg|_{s=-10} = 25.$$

Thus, $y(s)$ is

$$y(s) = -9\frac{1}{s+6} + 25\frac{1}{s+10}$$

and its inverse Laplace transform using identity 4 in Table 5.1 is

$$y(t) = -9e^{-6t} + 25e^{-10t}.$$

Though presented hereafter as a means of solving for the residues in the case of repeated roots, the method that follows is a general approach that can be used for real distinct roots and, as shown hereafter, for complex roots.

Partial fraction expansion with repeated poles.   The method presented for distinct poles does not work for repeated poles without some modification. There exists, however, a more generalized method that works regardless of the type of poles. Consider a system like the following with two repeated poles; the partial fraction expansion is

$$G(s) = \frac{B(s)}{A(s)} = \frac{K(s+z_1)(s+z_2)\cdots(s+z_m)}{(s+p_1)(s+p_2)^2\cdots(s+p_n)}$$

$$= \frac{r_1}{s+p_1} + \frac{r_2}{s+p_2} + \frac{r_3}{(s+p_2)^2} + \cdots + \frac{r_n}{s+p_n}.$$

To solve for the n residues, we will need n equations. To derive the n equations we first multiply both sides of the equation by the denominator $A(s)$:

$$A(s)\frac{B(s)}{A(s)} = (s+p_1)(s+p_2)^2\cdots(s+p_n)\frac{K(s+z_1)(s+z_2)\cdots(s+z_m)}{(s+p_1)(s+p_2)^2\cdots(s+p_n)}.$$

What results is a polynomial equation,

$$K(s+z_1)(s+z_2)\cdots(s+z_m) = (s+p_2)^2\cdots(s+p_n)r_1$$
$$+ (s+p_1)(s+p_2)\cdots(s+p_n)r_2 + \cdots$$
$$+ (s+p_1)(s+p_2)\cdots(s+p_{n-1})r_n.$$

The coefficients can be solved by multiplying the terms to generate polynomials and gathering on each side of the equation terms of the same order of s. One can derive n equations to solve for $r_1, \ldots, r_n$ by equating the coefficient sums on either side of the equation with common orders of s. This is best illustrated by example.

Example 5.5

Find the partial fraction expansion of

$$G(s) = \frac{s+4}{(s+1)(s+2)^2(s+3)}.$$

Solution. The partial fraction expansion will be of the form

$$G(s) = \frac{r_1}{s+1} + \frac{r_2}{s+2} + \frac{r_3}{(s+2)^2} + \frac{r_4}{s+3}.$$

Multiplying through by the denominator of $G(s)$ results in

$$
\begin{aligned}
s+4 = {} & (s+2)^2(s+3)r_1 \\
& + (s+1)(s+2)(s+3)r_2 \\
& + (s+1)(s+3)r_3 \\
& + (s+1)(s+2)^2r_4 \\
= {} & (s^3 + 7s^2 + 16s + 12)r_1 \\
& + (s^3 + 6s^2 + 11s + 6)r_2 \\
& + (s^2 + 4s + 3)r_3 \\
& + (s^3 + 5s^2 + 8s + 4)r_4 \\
= {} & (r_1 + r_2 + r_4)s^3 \\
& + (7r_1 + 6r_2 + r_3 + 5r_4)s^2 \\
& + (16r_1 + 11r_2 + 4r_3 + 8r_4)s \\
& + (12r_1 + 6r_2 + 3r_3 + 4r_4).
\end{aligned}
$$

Equating like powers of s on each side of the equations yields four equations to solve for the four unknowns:

$$
\begin{aligned}
s^3: & \quad 0 = r_1 + r_2 + r_4 \\
s^2: & \quad 0 = 7r_1 + 6r_2 + r_3 + 5r_4 \\
s^1: & \quad 1 = 16r_1 + 11r_2 + 4r_3 + 8r_4 \\
s^0: & \quad 4 = 12r_1 + 6r_2 + 3r_3 + 4r_4
\end{aligned}
$$

The four unknowns can be determined a number of ways including using Linear Algebra:

$$
\begin{bmatrix}
1 & 1 & 0 & 1 \\
7 & 6 & 1 & 5 \\
16 & 11 & 4 & 8 \\
12 & 6 & 3 & 4
\end{bmatrix}
\begin{bmatrix}
r_1 \\ r_2 \\ r_3 \\ r_4
\end{bmatrix}
=
\begin{bmatrix}
0 \\ 0 \\ 1 \\ 4
\end{bmatrix},
$$

which when solved gives

$$\begin{bmatrix} r_1 \\ r_2 \\ r_3 \\ r_4 \end{bmatrix} = \begin{bmatrix} 1.5 \\ -1 \\ -2 \\ -0.5 \end{bmatrix}.$$

The partial fraction expansion is therefore

$$G(s) = \frac{1.5}{s+1} - \frac{1}{s+2} - \frac{2}{(s+2)^2} - \frac{0.5}{s+3}.$$

Example 5.6

Find the unit impulse response of the quarter car suspension from Example 5.4 if the spring rate is 32,000 N/m.

Solution. To solve for unit impulse response, we can reuse the transform previously derived in Example 5.4,

$$y(s) = \frac{bs+k}{ms^2+bs+k} y_{road}(s) = \frac{8,000s+32,000}{500s^2+8,000s+32,000}$$
$$= \frac{16s+64}{s^2+16s+64} = \frac{16s+64}{(s+8)^2}.$$

The partial fraction expansion is

$$\frac{16s+64}{(s+8)^2} = \frac{r_1}{s+8} + \frac{r_2}{(s+8)^2}.$$

Multiplying through by the denominator provides

$$16s+64 = (s+8)r_1 + r_2 = r_1 s + 8r_1 + r_2,$$

which we use to generate two equations to solve for the unknowns r_1 and r_2:

$$s^1: \quad 16 = r_1$$
$$s^0: \quad 64 = 8r_1 + r_2$$

Substituting $r_1 = 16$ into the second equation renders $r_2 = -64$. As a result, $y(s)$ is

$$y(s) = \frac{16}{s+8} - \frac{64}{(s+8)^2}.$$

The inverse Laplace transform can be derived using items 4 and 5 in Table 5.1,

$$y(t) = 16e^{-8t} - 64te^{-8t} = 16e^{-8t}(1 - 4t).$$

Partial fraction expansion with complex poles.   Complex poles always occur in complex conjugate pairs. When deriving the Laplace transform of an attenuating sinusoid (e.g., a decaying exponential multiplied by a sinusoid) we utilized the identity derived for multiplication by e^{-at}. The resulting denominator of the Laplace transforms (items 10 and 11 in Table 5.1), $(s+a)^2 + \omega^2$, is second-order, and its roots are generally complex conjugates. With second-order denominators, if the roots are complex, we must *complete the square*. This involves manipulating the second-order polynomial so that it is in the form of $(s+a)^2 + \omega^2$. If in the process of finding the partial fraction expansion there are complex conjugate poles, one must complete the square and recognize that the expansion will have terms that are the transforms of exponentials multiplied by sinusoids. Consider a system with one pair of complex conjugate poles, two real poles, and two zeros. Its partial fraction expansion will be

$$G(s) = \frac{B(s)}{A(s)} = \frac{K(s+z_1)(s+z_2)}{(s+p_1)(s+p_2)[(s+a)^2 + \omega^2]}$$

$$= \frac{r_1}{s+p_1} + \frac{r_2}{s+p_2} + \frac{r_3 s + r_4}{(s+a)^2 + \omega^2}.$$

Notice that the numerator of the third term is an order of one less than the denominator. In other words, the numerator is a first-order polynomial and the denominator is second order. The four unknowns can be solved for in the same general manner as that used for repeated roots by multiplying through by the denominator,

$$\begin{aligned} K(s+z_1)(s+z_2) = {} & (s+p_2)\left[(s+a)^2 + \omega^2\right]r_1 \\ & + (s+p_1)\left[(s+a)^2 + \omega^2\right]r_2 \\ & + (s+p_1)(s+p_2)(r_3 s + r_4). \end{aligned}$$

The terms are multiplied; those with like powers of s are combined; and terms with equivalent powers of s on each side of the equation are equated. This is demonstrated in the following examples.

Example 5.7

Find the partial fraction expansion of

$$G(s) = \frac{s+4}{(s+1)(s^2+2s+5)}.$$

Solution. To find the partial fraction expansion, we must first recognize that the roots of

$$s^2+2s+5$$

are complex conjugates

$$s^2+2s+5 = (s+1+j\omega)(s+1-j\omega).$$

To find the partial fraction expansion, we need to complete the square,

$$s^2+2s+5 = (s^2+2s+1)+4 = (s+1)^2+2^2.$$

That makes the partial fraction expansion

$$G(s) = \frac{s+4}{(s+1)(s^2+2s+5)} = \frac{r_1}{s+1} + \frac{r_2 s + r_3}{(s+1)^2+2^2}.$$

Example 5.8

Find the unit impulse response of the quarter car suspension from Example 5.4 if the spring rate is 34,000 N/m.

Solution. Using the transform from Example 5.4, $y(s)$ is

$$y(s) = \frac{bs+k}{ms^2+bs+k} = \frac{8,000s+34,000}{500s^2+8,000s+34,000} = \frac{16s+68}{s^2+16s+68}.$$

The denominator has complex roots, so we need to complete the square,

$$s^2+16s+68 = (s^2+16s+64)+4 = (s+8)^2+2^2.$$

The partial fraction expansion becomes

$$y(s) = \frac{16s+68}{s^2+16s+68} = \frac{16s+68}{(s+8)^2+2^2}$$

$$= \frac{16s+128-128+68}{(s+8)^2+2^2} = \frac{16(s+8)-60}{(s+8)^2+2^2}$$

$$= 16\frac{s+8}{(s+8)^2+2^2} - 30\frac{2}{(s+8)^2+2^2}.$$

To derive the impulse response, use identities 10 and 11 from Table 5.1,

$$y(t) = 16e^{-8t}\cos 2t - 30e^{-8t}\sin 2t = 2e^{-8t}(8\cos 2t - 15\sin 2t).$$

5.7 Poles, Zeros, Partial Fraction Expansions, and MATLAB

MATLAB can be used to facilitate analyzing transforms including finding poles, zeros, and partial fraction expansions. Table 5.2 provides some of the more commonly used commands for defining and analyzing transfer functions. To utilize the related commands, we must first define our transform. As we have seen in some examples in this chapter, transforms often take the form of a ratio of polynomials in s. As will be discussed in the the next chapter, this ratio of polynomials is called the transfer function. A transfer function can be specified in MATLAB using the tf command by specifying the numerator (num) and denominator (den).

```
>> G = tf(num,den)
```

Polynomials in MATLAB are defined as one-dimensional arrays composed of the polynomial coefficients in descending order of s, where every order must be accounted for even if the coefficient is zero. For example, the polynomial $G(s) = s^3 + 2s + 10$ would be specified in MATLAB as follows:

```
>> G = [1 0 2 10]
```

Note that 0 was used as a placeholder for the s^2 term.

Table 5.2: *MATLAB commands for transfer function operations.*

Function	Description
tf(num,den)	defines transfer function using polynomials
pole(sys)	computes poles of a LTI system object
zero(sys)	computes zeros of a LTI system object
zpk(Z,P,K)	defines transfer function with zeros, poles, and gain
conv(A,B)	calculates polynomial multiplication
residue(num,den)	computes partial fraction expansion

Once a transform (or transfer function) is specified in MATLAB, a few commands can be used to find the poles, zeros, and partial fraction expansion. The poles are found using the `pole` command.

```
>> pole(G)
```

Similarly, the `zero` command is utilized to determine the zeros.

```
>> zero(G)
```

The transfer function can be defined in terms of the zeros, poles, and overall gain instead of the numerator and denominator utilizing the function.

```
>> sys = zpk(Z,P,K)
```

The function `conv` can used for multiplication of polynomials (e.g., A and B), and the partial fraction expansion can be obtained using the `residue` command.

```
>> conv(A,B)
>> [R,P,K] = residue(num,den)
```

Assuming distinct poles, the `residue` command finds the partial fraction expansion such that the elements of the arrays R, P, and K are the residues $(R_1, \ldots, R_n)$, poles $(P_1, \ldots, P_n)$, and remainder $K(s)$ in the transform

$$G(s) = \frac{R_1}{s - P_1} + \frac{R_2}{s - P_2} + \cdots + \frac{R_n}{s - P_n} + K(s).$$

A remainder will only exist if the numerator is of an order greater than the denominator. Additionally, the poles identified using this command can generally be complex. In the case of a repeated pole P_k with multiplicity m, the expansion includes terms of the form

$$\frac{R_k}{s - P_k} + \frac{R_{k+1}}{(s - P_k)^2} + \cdots + \frac{R_{k+m}}{(s - P_k)^m}.$$

Example 5.9

Using MATLAB, find the poles, zeros, and partial fraction expansion of

$$G(s) = \frac{s^2 + 9s + 20}{s^4 + 4s^3 + 10s^2 + 12s + 5}.$$

Solution. To find the poles we employ the `pole` command.

```
>> num = [1 9 20];
>> den = [1 4 10 12 5];
>> sys = tf(num,den);
>> p = pole(sys)

p =

   -1.0000 + 2.0000i
   -1.0000 - 2.0000i
   -1.0000 + 0.0000i
   -1.0000 - 0.0000i
```

The poles are $p_1 = -1 + j2$, $p_2 = -1 - j2$, $p_3 = -1$, and $p_4 = -1$ — complex conjugate poles and a repeated pole. The zero command provides the roots of the numerator.

```
>> z = zero(sys)

z =

   -5.0000
   -4.0000
```

We can redefine the system using the poles and zeros and the overall gain, which for this system is unity.

```
>> sys = zpk(z,p,1)

sys =

       (s+5) (s+4)
  ----------------------
  (s+1)^2 (s^2 + 2s + 5)

Continuous-time zero/pole/gain model.
```

We can confirm that the resultant matches our original transfer function using the tf and conv commands.

```
>> tf(conv([1 5],[1 4]),conv(conv([1 1],[1 1]),[1 2 5]))

ans =

       s^2 + 9 s + 20
```

```
- - - - - - - - - - - - - - - - - - - - - - - - - - - - -
s^4 + 4 s^3 + 10 s^2 + 12 s + 5

Continuous-time transfer function.
```

The complete expansion is obtained using the residue command.

```
>> [R,P,K] = residue(num,den)

R =

   -0.8750 + 0.5000i
   -0.8750 - 0.5000i
    1.7500
    3.0000


P =

   -1.0000 + 2.0000i
   -1.0000 - 2.0000i
   -1.0000
   -1.0000


K =

        []
```

Because two of the poles are complex, the corresponding residues are also complex. The remainder is zero because the numerator is second order and the denominator is fourth order. The partial fraction expansion is

$$
\begin{aligned}
G(s) &= \frac{-0.875 + j0.5}{s - (-1 + j2)} + \frac{-0.875 - j0.5}{s - (-1 - j2)} - \frac{1.75}{s - (-1)} + \frac{3}{s - (-1)} \\
&= \frac{-0.875 + j0.5}{s + 1 - j2} + \frac{-0.875 - j0.5}{s + 1 + j2} - \frac{1.75}{s + 1} + \frac{3}{s + 1} \\
&= \frac{-0.875 + j0.5}{s + 1 - j2} \left[\frac{s + 1 + j2}{s + 1 + j2} \right] + \frac{-0.875 - j0.5}{s + 1 + j2} \left[\frac{s + 1 - j2}{s + 1 - j2} \right] \\
&\quad - \frac{1.75}{s + 2} + \frac{3}{s + 1} \\
&= -\frac{1.75s + 3.75}{s^2 + 2s + 5} - \frac{1.75}{s + 2} + \frac{3}{s + 1}.
\end{aligned}
$$

5.8 Summary

▷ Laplace transforms are used to convert differential equations in the time domain to algebraic equations in the s-domain.

▷ Linear differential equations become polynomials in the s-domain.

▷ Sinusoids can be represented using complex exponential functions. As such, they can be manipulated using basic algebraic principles.

▷ A Laplace transform can be decomposed through partial fraction expansions into terms that can be readily inverse Laplace transformed using Laplace transform primitives.

▷ Laplace transforms lead to transfer function models. A transfer function is an algebraic construct that represents the output/input relation in the s-domain.

▷ The zeros are defined as the roots of the polynomial in the numerator of a transfer function.

▷ The poles are defined as the roots of the polynomial in the denominator of a transfer function.

▷ MATLAB includes a series of functions to represent transfer functions, compute poles and zeros, conduct polynomial multiplication, and compute partial fraction expansions.

5.9 Review

R5-1 Explain the relationship between the rectangular and polar representations of complex numbers.

R5-2 How are sinusoids represented using complex exponentials? What theorem is employed to do so?

R5-3 Describe basic complex algebra operations such as addition, subtraction, multiplication, and division.

R5-4 Why do we make use of the Laplace transform to analyze differential equations?

R5-5 Describe how to translate a function and explain the difference between shifting and translating.

R5-6 How do we make use of partial fraction expansions to derive inverse Laplace transforms?

R5-7 Explain what poles and zeros are.

R5-8 List and describe the MATLAB functions available for representing, analyzing, and manipulating transfer function representations.

5.10 Problems

P5-1 *Real and Imaginary Parts of Complex Numbers.* Calculate the real and imaginary parts of

(a) $\dfrac{3-j5}{2+j4}$,

(b) $(3-j5)(2+j4)$, and

(c) $\dfrac{5+j6}{2-j4}+\dfrac{7+j4}{4+j7}$.

P5-2 *Laplace Transforms of Functions.* Find the Laplace transforms of the following functions:

(a) te^{-4t}

(b) $t^2\cos 2t$

(c) $te^{-2t}\sin 3t$

P5-3 *Laplace Transforms of Piecewise Continuous Plots.* Determine, analytically, the Laplace transforms for the functions plotted in Figure 5.10.

P5-4 *Inverse Laplace Transforms of Functions.* Analytically, find the inverse Laplace transforms of the following transfer functions:

(a) $\dfrac{s+4}{s^2+11s+30}$

(b) $\dfrac{s^2+5s+6}{(s+1)^2}$

(c) $\dfrac{1}{s^2(s^2+9)}$

(d) $\dfrac{s+4}{s^2+16s+80}$

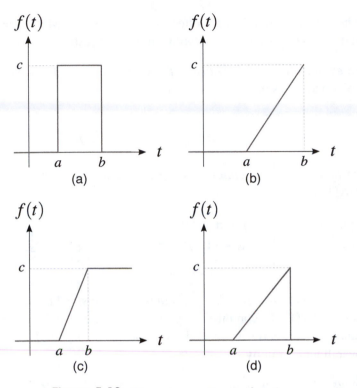

Figure 5.10: *Piecewise continuous functions.*

P5-5 *Laplace Transform Solutions to Initial Value Problems.* Analytically, determine the solutions to the following differential equations:

(a) $\dot{x} + x = 3\cos t$, given $x(0) = 1$

(b) $\ddot{x} + 9x = 0$, given $x(0) = 4$ and $\dot{x}(0) = 0$

(c) $\ddot{x} + 4\dot{x} + 13 = \cos 4t$, given $x(0) = \dot{x}(0) = 0$

P5-6 *Differential Equation with a Rectangular Pulse.* Derive the solution of the following differential equation:

$$\ddot{x} + 4\dot{x} + 13 = f(t),$$

where $f(t)$ is the function plotted in Figure 5.10 (a) where $a = 1, b = 3$, and $c = 2$ and where the initial conditions are zero.

P5-7 *Differential Equation with a Ramp Input.* Solve for the response $x(t)$ to the differential equation:

$$\ddot{x} + 16\dot{x} + 80 = f(t),$$

where $f(t)$ if the function plotted in Figure 5.10 (b) where $a = 1$, $b = 2$, and $c = 1$ and where the system is initially at rest.

P5-8 *Poles and Zeros of a Transfer Function.* Using MATLAB, find the poles and zeros of the transfer function

$$G(s) = \frac{2s^2 + 4s + 5}{6s^5 + 5s^4 + 4s^3 + 3s^2 + 2s + 1}.$$

P5-9 *Polynomial Multiplication.* Complete the following operations using MATLAB:

(a) $(s^2 + 10s + 12)(s^4 + 3s^2 + 11s + 7)$
(b) $(s^3 + 23s^2 + 48s + 54)(s^5 + 49s^4 + 31s^3 + 47s^2 + 7s + 67)$
(c) $(s^4 + 16s^2 + 64)(s^2 + 2s + 5)$

P5-10 *Partial Fraction Expansion of a Pole-Zero Transfer Function.* With the aid of MATLAB, compute the partial fraction expansion for a transfer function with zeros at -4 and -6, poles at $-2 \pm j$ and $-5 \pm j4$, and overall gain of unity.

P5-11 *Sinusoidal Response of a Simple Mass-Spring-Damper System.* Given the simple mass-spring-damper in Figure 5.11, solve, analytically, for the response $x(t)$ if the mass, damping constant, spring rate, and input force are 1 kg, 4 N-s/m, 13 N/m, and $2 \cos 10t$ N, respectively.

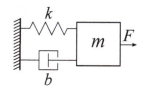

Figure 5.11: *A simple mass-spring-damper.*

5.11 Challenges

C5-1 *A Toy Electric Car.* Given the differential equations you previously derived for the toy electric car in Chapter 4, apply Laplace transforms to find the transfer function representation assuming that the input is the battery voltage and the output is the car velocity.

C5-2 *A Quarter-Car Suspension Model.* Use the Laplace theorems to convert the differential equations previously derived into a transfer function relating the displacement of the vehicle mass to the input road velocity.

C5-3 *An Electric Drill.* Use Laplace transforms to convert the previously acquired differential equations into a transfer function relating the angular velocity at the bit tip to the input voltage.

Chapter 6

Impedance Bond Graphs

Now that we can synthesize a bond graph and derive a state-space representation, let us consider an alternative formulation. Algebra and Linear Algebra provide an extensive set of tools for analyzing linear systems of equations. Could you derive an equivalent algebraic representation that can be used to examine and simulate the system response for each model directly from the bond graph? Consider the following:

▷ What connections are there between Linear Algebra and Differential Equations?

▷ How can you transform a system of differential equations into an algebraic formulation?

▷ What advantages, if any, would an algebraic model have over differential equations?

▷ Are there any concepts from electric circuits that you could generalize to facilitate synthesizing such models?

6.1 Introduction

Laplace transforms have long been used to aid in analysis and design of dynamic systems and controls. Laplace transforms can be used to derive transfer functions. A *transfer function* is the ratio of the Laplace transform of the output to the input assuming that initial conditions are zero. As will be shown in later chapters, transfer functions are a useful means of analyzing the dynamic response of linear systems and designing compensators or controls to alter their response. They can be used to determine stability and predict transient response characteristics. By using Laplace transforms to convert differential equations in the time domain to algebraic equations in s-domain, one can use the many analytical and numeric tools readily available for examining algebraic equations. Because the equations become purely algebraic, integral and derivative causality no longer play a role. Hence, causality becomes unnecessary.

Therefore, it is often useful to model dynamic systems using transfer functions as opposed to differential equations. Thus, in this chapter, it will be shown how transfer functions can be systematically derived using bond graphs. As will be shown, one approach to derive transfer functions is to generate them from the differential equations resulting from the bond graph. Alternatively, an *impedance bond graph* representation, which will be introduced in this chapter, can be used to directly derive the transfer functions.

The chapter objectives and outcomes are:

▷ *Objectives:*

1. To understand relations between state-space and transfer-function representations of linear, time-invariant systems,

2. To understand the synthesis and use impedance bond graphs, and

3. To be able to use alternative methods to derive transfer functions for systems that may require advanced formulation due to sign changes or complex bond graph structures.

▷ *Outcomes:* Upon completion, you will

1. be able to transition between state-space and transfer-function representations of dynamic systems,

2. be able to synthesize impedance bond graphs of mechanical, electrical, and hydraulic systems, and

3. be able to derive transfer functions for dynamic systems using bond graphs as an aid.

6.2 Laplace Transform of the State-Space Equation

As you might recall from a Differential Equations or and Advanced Engineering Mathematics course, the Laplace transform can be applied to a state-space representation to generate transfer functions. As has been shown in previous chapters, bond graphs can be used to derive a system of first-order differential equations that describe the dynamics of the physical system of interest. It has been shown that, assuming the system of interest has linear constitutive relations, these equations can be conveniently represented in the state-space form

$$\dot{\mathbf{x}} = \mathbf{A}\mathbf{x}(t) + \mathbf{B}\mathbf{u}(t)$$
$$\mathbf{y} = \mathbf{C}\mathbf{x}(t) + \mathbf{D}\mathbf{u}(t).$$

Laplace transforms can be applied to the state-space representation to generate the transfer functions for the system. Assuming initial conditions are all zero (i.e., $\mathbf{x}(0) = \mathbf{0}$), the Laplace transforms of the above equations are

$$s\mathbf{X}(s) = \mathbf{A}\mathbf{X}(s) + \mathbf{B}\mathbf{U}(s) \text{ and} \tag{6.1}$$
$$\mathbf{Y}(s) = \mathbf{C}\mathbf{X}(s) + \mathbf{D}\mathbf{U}(s). \tag{6.2}$$

Ultimately, we wish to get the transfer functions, $\mathbf{G}(s)$, relating the outputs, $\mathbf{Y}(s)$, to the inputs, $\mathbf{U}(s)$, such that

$$\mathbf{Y}(s) = \mathbf{G}(s)\mathbf{U}(s).$$

Thus we solve Equation 6.1 for $\mathbf{X}(s)$,

$$\mathbf{X}(s) - \mathbf{A}\mathbf{X}(s) = (s\mathbf{I} - \mathbf{A})\mathbf{X}(s) = \mathbf{B}\mathbf{U}(s) \Rightarrow \mathbf{X}(s) = (s\mathbf{I} - \mathbf{A})^{-1}\mathbf{B}\mathbf{U}(s), \tag{6.3}$$

and substitute into Equation 6.2,

$$\mathbf{Y}(s) = \mathbf{C}(s\mathbf{I} - \mathbf{A})^{-1}\mathbf{B}\mathbf{U}(s) + \mathbf{D}\mathbf{U}(s) = \left[\mathbf{C}(s\mathbf{I} - \mathbf{A})^{-1}\mathbf{B} + \mathbf{D}\right]\mathbf{U}(s),$$

to derive $\mathbf{G}(s)$,

$$\mathbf{G}(s) = \mathbf{C}(s\mathbf{I} - \mathbf{A})^{-1}\mathbf{B} + \mathbf{D}. \tag{6.4}$$

Given a system with m outputs and r inputs, the resulting matrix of transfer functions, $\mathbf{G}(s)$, will be $m \times r$.

Example 6.1

Take for example the simple mass-spring-damper system depicted in Figure 6.1. Derive the transfer function for the system.

Solution. Assuming that the output of interest is the displacement of the mass, $x(t)$, the state equations resulting from the bond graph in state-space form are

$$\begin{bmatrix} \dot{x} \\ \dot{p} \end{bmatrix} = \begin{bmatrix} 0 & 1/m \\ -k & -b/m \end{bmatrix} \begin{bmatrix} x \\ p \end{bmatrix} + \begin{bmatrix} 0 \\ 1 \end{bmatrix} F(t) \tag{6.5}$$

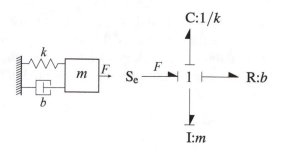

Figure 6.1: *A simple mass spring damper system.*

and

$$y = x = \begin{bmatrix} 1 & 0 \end{bmatrix} \begin{bmatrix} x \\ p \end{bmatrix} + 0 \cdot F(t). \tag{6.6}$$

Equation 6.4 can be applied to generate the transfer function relating the output, $x(s)$, to the input, $F(s)$.

$$G(s) = \begin{bmatrix} 1 & 0 \end{bmatrix} \begin{bmatrix} s & -1/m \\ k & s+b/m \end{bmatrix}^{-1} \begin{bmatrix} 0 \\ 1 \end{bmatrix} + 0$$

$$= \begin{bmatrix} 1 & 0 \end{bmatrix} \frac{1}{s^2+(b/m)s+k/m} \begin{bmatrix} s+b/m & 1/m \\ -k & s \end{bmatrix} \begin{bmatrix} 0 \\ 1 \end{bmatrix} + 0$$

$$= \frac{1/m}{s^2+(b/m)s+k/m}$$

$$= \frac{1}{ms^2+bs+k}$$

These results can be confirmed in MATLAB using the Symbolic Math Toolbox as shown below.

```
>> syms s m b k
>> A = [0 1/m; -k -b/m];
>> B = [0; 1];
>> C = [1 0];
>> D = 0;
>> G = C*inv(s*eye(2)-A)*B+D

G =

1/(s^2*m+s*b+k)
```

The function syms is used in MATLAB to define symbolic variables.

Note that the transfer function derived above is the same that would result by applying the differentiation theorem to the second-order form of the differential equation and assuming the initial conditions are zero. The system can be represented by the second-order differential equation

$$m\ddot{x} + b\dot{x} + kx = F(t).$$

Applying Laplace transforms and assuming zero initial conditions yields

$$ms^2 x(s) + bsx(s) + kx(s) = (ms^2 + bs + k)x(s) = F(s)$$

$$\frac{x(s)}{F(x)} = \frac{1}{ms^2 + bs + k}.$$

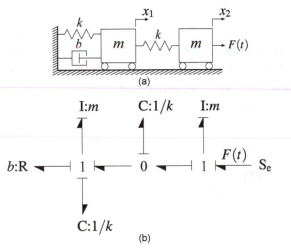

Figure 6.2: *Mass-spring-damper problem from Example 3.1.*

Example 6.2

Remember the mass-spring-damper system from Example 3.11. The system and resultant bond graph are repeated in Figure 6.2 for convenience. Figure 6.2 (b) shows four energy storing elements in integral causality. The outputs of interest are the positions of the two masses, x_1 and x_2. Convert the state-space model to transfer functions relating each of the displacements to the input force.

Solution. The system's dynamics are represented by Equations 3.1,

3.2, 3.3, and 3.4, which can be written in state-space form as

$$
\begin{bmatrix} \dot{x}_1 \\ \dot{p}_1 \\ \dot{\delta} \\ \dot{p}_2 \end{bmatrix} = \begin{bmatrix} 0 & 1/m & 0 & 0 \\ -k & -b/m & k & 0 \\ 0 & -1/m & 0 & 1/m \\ 0 & 0 & -k & 0 \end{bmatrix} \begin{bmatrix} x_1 \\ p_1 \\ \delta \\ p_2 \end{bmatrix} + \begin{bmatrix} 0 \\ 0 \\ 0 \\ 1 \end{bmatrix} F(t)
$$

and

$$
\begin{bmatrix} x_1 \\ x_2 \end{bmatrix} = \begin{bmatrix} 1 & 0 & 0 & 0 \\ 1 & 0 & 1 & 0 \end{bmatrix} \begin{bmatrix} x_1 \\ p_1 \\ \delta \\ p_2 \end{bmatrix} + \begin{bmatrix} 0 \\ 0 \end{bmatrix} F(t).
$$

Because there are two outputs of interest, two transfer functions will result,

$$
\mathbf{G}(s) = \begin{bmatrix} x_1(s)/F(s) \\ x_2(s)/F(s) \end{bmatrix},
$$

which can be determined analytically or with the help of a symbolic math software. A MATLAB script follows that shows the solutions for the transfer functions of the system using Equation 6.4.

```
>> syms s m b k
>> A = [0 1/m 0 0; -k -b/m k 0; 0 -1/m 0 1/m; 0 0 -k 0];
>> B = [0; 0; 0; 1];
>> C = [1 0 0 0; 1 0 1 0];
>> D = [0; 0];
>> G = C*inv(s*eye(4)-A)*B+D;
>> simplify(G)

ans =

              k/(s^4*m^2+3*k*s^2*m+b*s^3*m+s*b*k+k^2)
   (s^2*m+s*b+2*k)/(s^4*m^2+3*k*s^2*m+b*s^3*m+s*b*k+k^2)
```

The transfer functions are

$$
\frac{x_1(s)}{F(s)} = \frac{k}{m^2 s^4 + bms^3 + 3kms^2 + bks + k^2}
$$

and

$$
\frac{x_2(s)}{F(s)} = \frac{ms^2 + bs + 2k}{m^2 s^4 + bms^3 + 3kms^2 + bks + k^2}.
$$

Though the above examples are rather straightforward, in general, systems will have more than two to four states and generating the transfer function(s) from the state-space representation without the aid of a symbolic math software can become rather tedious. However, for many cases, *impedance bond graphs* can be used to facilitate direct derivation of the transfer functions without requiring the intermediate step of deriving the differential equations. To represent dynamic systems using impedance, we must assume that the constitutive relations of the components are linear or linearizable. As will be shown in the following section, some basic Laplace transform properties can be applied to derive the impedances of energy storing and dissipating elements with linear constitutive relations.

6.3   Basic 1-Port Impedances

To synthesize an impedance bond graph, the energy storing and dissipating elements must be replaced by their equivalent impedance. Thus, the impedances for commonly occurring electrical and mechanical elements are presented herein.

To derive the impedances, two Laplace transform theorems will be used, namely, the differentiation theorem,

$$\mathscr{L}\left[\frac{d}{dt}f(t)\right] = s\mathrm{f}(s) - f(0), \tag{6.7}$$

and the integration theorem,

$$\mathscr{L}\left[\int f(t)\,dt\right] = \frac{1}{s}\mathrm{f}(s) + \frac{1}{s}\int_{t=0} f(t)\,dt. \tag{6.8}$$

As has been defined earlier, a transfer function assumes initial conditions are zero. If the initial conditions are assumed to be all zero Equations 6.7 and 6.8 simplify to

$$\mathscr{L}\left[\frac{d}{dt}f(t)\right] = s\mathrm{f}(s) \tag{6.9}$$

and

$$\mathscr{L}\left[\int f(t)\,dt\right] = \frac{1}{s}\mathrm{f}(s), \tag{6.10}$$

respectively. Since, ultimately, the goal is to derive transfer functions, Equations 6.9 and 6.10 are used to transform the linear constitutive relations previously introduced for generalized R-, C-, and I-elements in Chapter 2.

As you may recall from an Electrical Circuits course, the *impedance* of and element is defined as the ratio of the voltages to the current in the s-domain (i.e., $Z = e(s)/i(s)$). With respect to bond graphs, the impedance of an element is generally defined as the ratio of the effort to the flow in the s-domain, $e(s)/f(s)$. Recall that for R-elements, the effort and flow are directly related:

$$e(t) = Rf(t).$$

By taking the Laplace transform of both sides of the above equation, the following relationship results in the s-domain:

$$e(s) = Rf(s),$$

and if the impedance is defined as the ratio of the effort, $e(s)$, to the flow, $f(s)$, then, as you might expect, the impedance of a linear R-element is simply the resistance, R:

$$Z_R = \frac{e(s)}{f(s)} = R.$$

For linear C-elements, the effort is related to the flow through integration:

$$e(t) = \frac{1}{C} \int f(t)dt.$$

Applying the Laplace transform assuming zero initial conditions results in

$$e(s) = \frac{1}{Cs}f(s),$$

which can be solved for the impedance,

$$Z_C = \frac{e(s)}{f(s)} = \frac{1}{Cs}.$$

For a linear I-element, effort is related to the derivative of the flow,

$$e(t) = I\frac{df(t)}{dt}.$$

After applying Equation 6.7, the above equation becomes

$$e(s) = Is\,f(s),$$

which results in an impedance of

$$Z_I = \frac{e(s)}{f(s)} = Is.$$

The impedances for R-, C-, and I-elements in electrical circuits, hydraulic circuits, mass-spring-damper systems, and mechanical rotation systems can be derived in much the same fashion. Table 6.1 summarizes the impedances for R-, C-, and I-elements.

Table 6.1: *Element impedances.*

Element	Effort-Flow Relation	Impedance
R	$e(t) = R f(t) \leftrightarrow e(s) = R f(s)$	$Z_R = R$
C	$e(t) = \dfrac{1}{C} \int f(t)dt \leftrightarrow e(s) = \dfrac{1}{Cs} f(s)$	$Z_C = \dfrac{1}{Cs}$
I	$e(t) = I \dfrac{df(t)}{dt} \leftrightarrow e(s) = Is f(s)$	$Z_I = Is$
damper	$F(t) = b v(t) \leftrightarrow F(s) = b v(s)$	$Z_R = b$
spring	$F(t) = k \int v(t)dt \leftrightarrow F(s) = \dfrac{k}{s} v(s)$	$Z_C = \dfrac{k}{s}$
mass	$F(t) = m \dfrac{dv(t)}{dt} \leftrightarrow F(s) = ms v(s)$	$Z_I = ms$
bearing	$\tau(t) = \beta \omega(t) \leftrightarrow T(s) = \beta \Omega(s)$	$Z_R = \beta$
spring	$\tau(t) = \kappa \int \omega(t)dt \leftrightarrow T(s) = \dfrac{\kappa}{s} \Omega(s)$	$Z_C = \dfrac{\kappa}{s}$
inertia	$\tau(t) = J \dfrac{d\omega(t)}{dt} \leftrightarrow T(s) = Js \Omega(s)$	$Z_I = Js$
resistor	$e(t) = R i(t) \leftrightarrow e(s) = R i(s)$	$Z_R = R$
capacitor	$e(t) = \dfrac{1}{C} \int i(t)dt \leftrightarrow e(s) = \dfrac{1}{Cs} i(s)$	$Z_C = \dfrac{1}{Cs}$
inductor	$e(t) = L \dfrac{di(t)}{dt} \leftrightarrow e(s) = Ls\, i(s)$	$Z_I = Ls$
valve	$P(t) = R_f Q(t) \leftrightarrow P(s) = R_f Q(s)$	$Z_R = R_f$
accumulator	$P(t) = \dfrac{1}{C_f} \int Q(t)dt \leftrightarrow P(s) = \dfrac{1}{C_f s} Q(s)$	$Z_C = \dfrac{1}{C_f s}$
fluid slug	$P(t) = I_f \dfrac{dQ(t)}{dt} \leftrightarrow P(s) = I_f s\, Q(s)$	$Z_I = I_f s$

6.4 Impedance Bond Graph Synthesis

Impedance bond graphs are synthesized similar to regular bond graphs with a few exceptions. First, the R-, C-, and I-elements are replaced with their equivalent impedance. As will be shown, to facilitate "bookkeeping" and simplification, an element's actual impedance can be substituted by a symbolic impedance parameter (i.e., "Z_I" in place of "Is"). Once the overall transfer functions are determined, the symbolic parameter can then be replaced with the actual impedance and the resulting transfer function can be further simplified. Second, causality is not a concern when deriving transfer functions. Thus, it is unnecessary to annotate an impedance bond graph with causality. Third and final, sources are not of impedances and are therefore not included in an impedance bond graph. Instead, the bond where the source element would be attached is annotated with the Laplace transform of the effort or flow associated with the source. These exceptions are best illustrated with a few examples.

Example 6.3

Recall the mass-spring-damper problem from Example 6.2 whose schematic and bond graph are repeated in Figure 6.3 for your convenience. Generate the impedance bond graph for this system.

Solution. The bond graph is readily derived using the methods described in Chapter 3 for mechanical translation. The resulting bond graph can be used to derive the impedance bond graph. The mass-spring-damper system is composed of 4 energy-storing elements and 1 energy dissipating element – 5 impedances in all.

The existing bond graph in Figure 6.3 (b) can be easily modified to produce the impedance bond graph. To do so, the R-, C-, and I-elements are replaced with their equivalent impedances. The causality is unnecessary so the causal strokes on the bonds are removed. Finally, the effort source is eliminated, but the bond to which it was attached is labeled with the Laplace transform of the effort associated with that source, $\mathscr{L}[F(t)] = \mathsf{F}(s)$. The actual impedances could be symbolically represented to facilitate derivation of transfer functions and simplification. For example, Z_1 could replace ms and the transfer functions could be derived in terms of their symbolic representations and simplified. At the end, the actual impedances can be substituted into the transfer function. This will be fully illustrated in a later example.

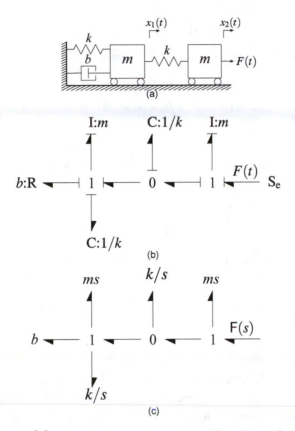

Figure 6.3: *Mass-spring-damper system from Example 6.2.*

Example 6.4

As another example, let us recall the circuit for Example 3.5 depicted in Figure 6.4. Synthesize the impedance bond graph.

   Solution. Note, though in Example 6.3 the impedance bond graph was derived from the regular bond graph, it is not necessary to first derive the regular bond graph. The impedance bond graph can be derived directly using the methods outlined in the previous chapters modified using the exceptions mentioned above:

▷ Use impedances in place of R-, C-, and I elements;

▷ Do not include causal strokes; and

▷ Do not use source elements; just label the respective bonds with the Laplace transforms of the source variables.

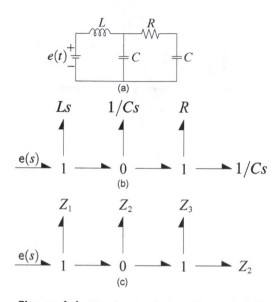

Figure 6.4: *Simple circuit from Example 3.5.*

By inspection, the circuit is made of three branches connected in parallel (common voltage). The three branches are connected off a 0-junction. The left branch is made up of a voltage source and an inductor in series (common current) so they appear together off a 1-junction. The center branch is simply a capacitor, and the right branch is a resistor and capacitor in series (common current) off a 1-junction. Note that the impedance bond graph in part (b) shows the actual impedances, while the impedance bond graph in part (c) uses symbolic substitutions labeled as Z_1 to Z_3. Because both capacitors have the same capacitance, the impedance Z_2 is repeated. As will be shown in later examples, using symbolic substitutions can facilitate algebraic derivation of transfer functions.

As will be shown, transfer functions can be derived directly from the impedance bond graph, but to do so, one must be familiar with the methods to simplify and condense the bond graph through equivalencies. Therefore, the following sections will detail commonly occurring equivalencies that can be used to simplify the impedance bond graph.

6.5 Junctions, Transformers, and Gyrators

As you might recall from an Electronic Circuits course, impedances mathematically sum in much the same manner as resistors. That is, impedances connected in series (or that share a common current) sum like resistors in series, and impedances in parallel (or that share a common voltage) sum like resistors in parallel. These basic premises hold true for impedance bond graphs, except in bond graphs one does not consider things "in series" or "in parallel" but rather as having common effort or common flow. We therefore begin by examining 0- and 1-junctions.

The junction primary and secondary conditions hold true in the s-domain. Recall that elements connected off a 1-junction are characterized by common flow and a summation of efforts. In the s-domain, these two properties are

$$f_1 = f_2 = \ldots = f \ \text{(common flow)}$$

and

$$\sum_{j=1}^{n} e_j = 0 \ \text{(summation of efforts)}.$$

These can be applied to the 1-junction in Figure 6.5 (a):

$$f_1 = f_2 = f$$

and

$$e - e_1 - e_2 = 0 \ \text{or} \ e = e_1 + e_2.$$

The efforts e_1 and e_2 can be calculated from the impedance relations in terms of the common flow, $f(s)$:

$$Z_1 = \frac{e_1(s)}{f_1(s)} = \frac{e_1(s)}{f(s)} \rightarrow e_1(s) = Z_1 f(s)$$

and

$$Z_2 = \frac{e_2(s)}{f_2(s)} = \frac{e_2(s)}{f(s)} \rightarrow e_2(s) = Z_2 f(s),$$

which can be substituted in the effort equation above to derive the equivalent impedance, Z_{eq}:

$$e = Z_1 f + Z_2 f = (Z_1 + Z_2) f \rightarrow Z_{eq} = \frac{e(s)}{f(s)} = Z_1 + Z_2.$$

The 1-junction in Figure 6.5 (a) can be simplified to a single equivalent impedance as illustrated in the figure. In general, for N impedances with common flow, the equivalent impedance is the sum of the impedances attached to the 1-junction,

$$Z_{eq} = \sum_{j=1}^{N} Z_j \text{ (1-junction).} \tag{6.11}$$

Impedances with common flow (connected off a 1-junction) sum like resistors in series.

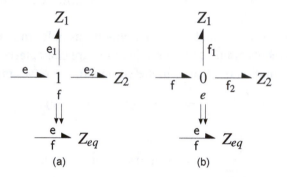

Figure 6.5: *Junction equivalent impedances.*

The equivalent impedance for a 0-junction can be derived in much the same manner. A 0-junction has two properties,

$$e_1 = e_2 = \ldots = e \text{ (common effort)}$$

and

$$\sum_{j=1}^{N} f_j = 0 \text{ (summation of flows).}$$

If the above relations are applied to the 0-junction in Figure 6.5 (b), the results are

$$e_1 = e_2 = e$$

and

$$f - f_1 - f_2 = 0 \text{ or } f = f_1 + f_2.$$

The flows f_1 and f_2 can be calculated in terms of the common effort, $e(s)$, from the impedance relations,

$$Z_1 = \frac{e_1(s)}{f_1(s)} = \frac{e(s)}{f_1(s)} \rightarrow f_1(s) = \frac{e(s)}{Z_1}$$

and

$$Z_2 = \frac{e_2(s)}{f_2(s)} = \frac{e(s)}{f_2(s)} \rightarrow f_2(s) = \frac{e(s)}{Z_2},$$

which can be substituted in the flow equation above to derive the equivalent impedance, Z_{eq}:

$$f = \frac{e(s)}{Z_1} + \frac{e(s)}{Z_2} = \left(\frac{1}{Z_1} + \frac{1}{Z_2}\right)e(s) = \frac{Z_1 + Z_2}{Z_1 Z_2}e(s) \rightarrow Z_{eq} = \frac{e(s)}{f(s)} = \frac{Z_1 Z_2}{Z_1 + Z_2}.$$

For N impedances off a 0-junction, the equivalent impedance is

$$Z_{eq} = \frac{1}{\displaystyle\sum_{j=1}^{N}\frac{1}{Z_j}} \quad \text{(0-junction)}. \tag{6.12}$$

Impedances off a junction with common effort sum like resistors in parallel.

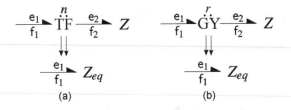

Figure 6.6: *Transformer and gyrator equivalent impedances.*

In many bond graphs there exist transformers and gyrators that convert the attached impedance. The equivalent impedances for these are also readily derived through algebraic relations. Given the transformer depicted in Figure 6.6 (a), which is a transformer in tandem with an impedance, the efforts and flows are related through the modulus

$$e_1(s) = n\,e_2(s) \quad \text{and} \quad f_2(s) = nf_1(s).$$

Hence, by applying the impedance relation between $e_2(s)$ and $f_2(s)$ (i.e., $Z = e_2(s)/f_2(s)$),

$$e_1(s) = n\,e_2(s) = nZf_2(s) = nZ\,(nf_1(s)) = n^2Zf_1(s),$$

the equivalent impedance for a transformer in tandem with an impedance Z is then

$$Z_{eq} = \frac{e_1(s)}{f_1(s)} = n^2Z \quad \text{(transformer)}. \tag{6.13}$$

Table 6.2: *Junctions, transformers, and gyrators.*

Item	Bond Graph	Impedance
1-Junction	Figure 6.5 (a)	$Z_{eq} = Z_1 + Z_2$
0-Junction	Figure 6.5 (b)	$Z_{eq} = \dfrac{Z_1 Z_2}{Z_1 + Z_2}$
Transformer	Figure 6.6 (a)	$Z_{eq} = n^2 Z$
Gyrator	Figure 6.6 (b)	$Z_{eq} = \dfrac{r^2}{Z}$

For a gyrator, the efforts and flows on each side are related through the modulus in the following manner:

$$e_1(s) = r f_2(s) \text{ and } e_2(s) = r f_1(s).$$

By substituting the impedance relation between $e_2(s)$ and $f_2(s)$, the following is derived:

$$e_1(s) = r f_2(s) = r \frac{1}{Z} e_2(s) = r \frac{1}{Z} [r f_1(s)] = \frac{r^2}{Z} f_1(s),$$

which results in the equivalent impedance for a gyrator in tandem with an impedance Z of

$$Z_{eq} = \frac{e_1(s)}{f_1(s)} = \frac{r^2}{Z} \text{ (gyrator).} \tag{6.14}$$

As illustrated in Figures 6.5 and 6.6 and summarized in Table 6.2, the properties of 1-junctions, 0-junctions, transformers, and gyrators can be used to simplify impedance bond graphs through equivalencies.

6.6   Effort and Flow Dividers

Besides summing impedances like resistors in series or resistors in parallel, junctions also serve as effort or flow dividers much like voltage or current dividers in circuits.

A 1-junction like that in Figure 6.7 (a) serves as an effort divider and can be used to relate any of the efforts associated with the attached impedances to the effort on the primary bond. For instance, assume that we wish to determine the transfer function relating the effort on impedance Z_2 labeled e_{out} to the effort on the primary bond labeled e_{in}. Because a 1-junction has common flow, we know

$$e_{out}(s) = Z_2 f(s).$$

We also know from the previous section that e_{in} can be related to the common flow, f, through the equivalent impedance,

$$e_{in}(s) = Z_{eq} f(s) = (Z_1 + Z_2) f(s).$$

The effort divider relation can be derived by dividing the two equations above,

$$\frac{e_{out}(s)}{e_{in}(s)} = \frac{Z_2 f(s)}{(Z_1 + Z_2) f(s)} = \frac{Z_2}{Z_1 + Z_2} \quad \text{(effort-divider)}. \qquad (6.15)$$

What if the 1-junction has more than 2 impedances attached? Can it still be used as an effort divider? Yes it can. To do so, select the bond with the "output" effort you wish to relate to the "input" effort. Using the equivalent impedance relation, combine the other impedances into a single equivalent impedance, and continue as described above. This is illustrated in Figure 6.7 (c).

If a 1-junction serves as an effort divider, it only makes sense then that a 0-junction should serve as a flow divider. A 0-junction can be used to relate any of the flows on the attached impedances to the flow on the primary bond. If, for example, one desires to determine the transfer function relating the flow through impedance Z_2 labeled f_{out} to the flow on the primary bond labeled f_{in}, then the impedance relation for Z_2,

$$\frac{e(s)}{f_{out}(s)} = Z_2 \Rightarrow f_{out}(s) = \frac{e(s)}{Z_2},$$

and the equivalent impedance relation for the 0-junction,

$$\frac{e(s)}{f_{in}(s)} = \frac{Z_1 Z_2}{Z_1 + Z_2} \Rightarrow f_{in}(s) = \frac{Z_1 + Z_2}{Z_1 Z_2} e(s),$$

can be used to derive such a transfer function,

$$\frac{f_{out}(s)}{f_{in}(s)} = \frac{e(s)/Z_2}{\dfrac{Z_1 + Z_2}{Z_1 Z_2} e(s)} = \frac{Z_1}{Z_1 + Z_2} \quad \text{(flow-divider)}. \qquad (6.16)$$

Like the flow divider, it is not necessary that the 0-junction has attached only 2 impedances. Select the bond with the "output" flow you wish to relate to the "input" flow and combine the remaining impedances using the equivalent impedance relation for a 0-junction. This is illustrated in Figure 6.7 (d). The effort- and flow-divider relations are summarized in Table 6.3.

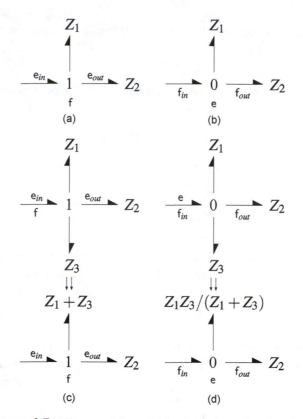

Figure 6.7: *Effort and flow divider impedance bond graphs.*

6.7 Sign Changes

Though most sign changes can be eliminated by choosing an appropriate sign convention, some sign changes are inevitable and must be accounted for when simplifying or condensing impedance bond graphs.

Take for instance the 2-bond 1-junction in Figure 6.8 (a). When accounting for a sign change, one has to be careful to maintain the effort and flow relations on adjacent junctions or elements. The 1-junction has two bonds pointing out of the junction, and therefore

$$e_1 = -e.$$

However, when examining the summation of flows on the adjacent 0-junction, the flow f is positive relative to the 0-junction using a power-in-positive sign convection,

$$f_a - f_b + f = 0.$$

Table 6.3: *Effort and flow dividers.*

Item	Bond Graph	Mathematical Relation
Effort Divider	Figure 6.7 (a)	$\dfrac{e_{out}}{e_{in}} = \dfrac{Z_2}{Z_1 + Z_2}$
Flow Divider	Figure 6.7 (b)	$\dfrac{f_{out}}{f_{in}} = \dfrac{Z_1}{Z_1 + Z_2}$

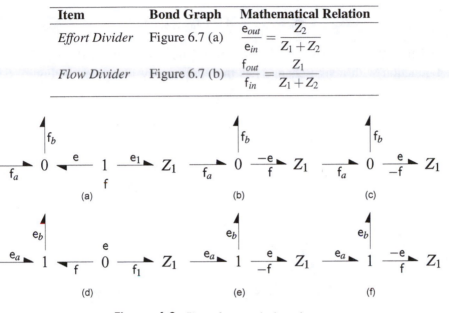

Figure 6.8: *Sign changes in junctions.*

Thus when condensing the 2-bond 1-junction, the resulting equivalent bond graph fragment must maintain not just the relations on the 1-junction but also the relations on the adjacent 0-junction. One cannot simply eliminate the 1-junction and move the impedance and its bond to the 0-junction. Generally, it is preferred to point the bond from the system to the impedance. The impedance relation is

$$Z_1 = \frac{e_1(s)}{f(s)} = \frac{-e(s)}{f(s)} = \frac{e(s)}{-f(s)}.$$

The last part of the above equation is used to maintain the sign of $f(s)$ positive relative to the 0-junction and account for the sign change due to the 1-junction. As illustrated in Figure 6.8 (b), if the left bond attached to the 1-junction is simply eliminated, the effort e would be negated (recall $e_1 = -e$) and the common effort relation at the 0-junction would not be maintained. Additionally, it would change the sign of the flow f relative to the 0-junction. Instead, the effort e should remain positive and the flow f should be negated as shown in Figure 6.8 (c). Because the bond attached to the impedance Z_1 points away from the 0-junction, the associated flow $-f$ is subtracted when the flows are summed at the junction,

$$f_a - f_b - (-f) = f_a - f_b + f,$$

and thus the sign convention at the 0-junction is maintained. The 2-bond 1-junction with a sign change has the equivalency illustrated in Figure 6.9 (a).

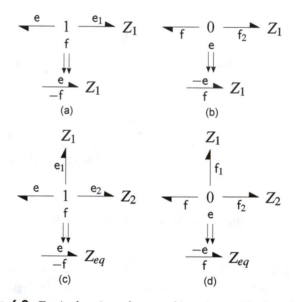

Figure 6.9: *Equivalent impedances of junctions with sign changes.*

The sign convention must be similarly maintained when condensing a 2-bond 0-junction with a sign change (Figure 6.8 (d)). The 0-junction has two bonds pointing out of the junctions, and therefore

$$f_1 = -f.$$

Additionally, the efforts sum at the adjacent 1-junction,

$$e_a - e_b + e = 0.$$

Note that relative to the adjacent 1-junction, the effort e is positive. Hence, the effort e must remain positive relative to the 1-junction when the bond graph is condensed. The impedance relation for the 0-junction is

$$Z_1 = \frac{e(s)}{f_1(s)} = \frac{e(s)}{-f(s)} = \frac{-e(s)}{f(s)}.$$

Again, the last part of the above equation is used to maintain, in this case, the effort $e(s)$ positive relative to the adjacent 1-junction. Figure 6.8 (e) illustrates that if the left bond attached to the 0-junction is simply eliminated, the effort $e(s)$ becomes negative relative to the adjacent 1-junction. Thus, the appropriate equivalent simplification is found in Figure 6.8 (f) because it maintains the impedance relation between $e(s)$ and $f(s)$ and the sign convention at the adjacent 1-junction,

$$e_a - e_b - (-e) = e_a - e_b + e = 0.$$

The 2-bond 0-junction with a sign change has the equivalency illustrated in Figure 6.9 (b).

Using the same basic approach detailed above, it can be shown that the impedance relations for 3-bond junctions with sign changes can be accounted for in much the same manner. That is, for a 3-bond 1-junction with a sign change (refer to Figure 6.9 (c)), the equivalent impedance relation is

$$Z_{eq} = \frac{e(s)}{-f(s)} = Z_1 + Z_2,$$

and for a 3-bond 0-junction with a sign change (refer to Figure 6.9 (d)), the equivalent impedance relation is

$$Z_{eq} = \frac{-e(s)}{f(s)} = \frac{Z_1 Z_2}{Z_1 + Z_2}.$$

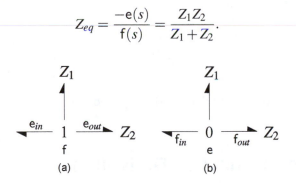

Figure 6.10: *Effort and flow dividers with sign changes.*

That leaves just effort and flow dividers with sign changes (refer to Figure 6.10 (e) and (f)). With effort and flow dividers, the objective is not to determine an equivalent impedance that maintains the sign convention and impedance relations. It is simply to determine the relationship between the effort or flow on one bond relative to the effort or flow on the primary bond. For an effort divider with a sign change at the primary bond, the effort at the bond of interest remains

$$e_{out}(s) = Z_2 f(s).$$

However, the effort at the primary bond as shown above for a 3-bond 1-junction changes sign:

$$e_{in}(s) = -Z_{eq}f(s) = -(Z_1 + Z_2)f(s).$$

The effort-divider relation can be derived by dividing the above two equations,

$$\frac{e_{out}(s)}{e_{in}(s)} = -\frac{Z_2 f(s)}{(Z_1 + Z_2)f(s)} = -\frac{Z_2}{Z_1 + Z_2} \quad \text{(effort-divider).} \qquad (6.17)$$

In a similar manner, it can be shown that the relation for a flow divider with a sign change is also simply negated. The flow on the bond of interest is

$$f_{out}(s) = \frac{e(s)}{Z_2}$$

and on the primary bond is

$$f_{in}(s) = -\frac{e(s)}{Z_{eq}} = -e(s)\frac{Z_1 + Z_2}{Z_1 Z_2}.$$

Dividing f_{out} by f_{in} results in

$$\frac{f_{out}(s)}{f_{in}(s)} = \frac{e(s)/Z_2}{-e(s)\dfrac{Z_1 + Z_2}{Z_1 Z_2}} = -\frac{Z_1}{Z_1 + Z_2}. \qquad (6.18)$$

Table 6.4 summarizes the relations derived above for junctions and dividers with sign changes.

6.8 Transfer Function Derivation

Once the impedance bond graph is synthesized, the derivation of transfer functions results from a few basic steps. The keys to facilitating derivation of transfer functions from impedance bond graphs are (1) identify the outputs and (2) formulate a strategy. The basic procedure is

1. Identify the desired system inputs and outputs to determine the necessary transfer functions,

2. Formulate a strategy to derive outputs by identifying necessary intermediate steps or transfer functions, and

Table 6.4: *Sign changes.*

Item	Bond Graph	Mathematical Relation
2-Bond 1-Junction	Figure 6.9 (a)	$Z_{eq} = \dfrac{e(s)}{-f(s)} = Z_1$
2-Bond 0-Junction	Figure 6.9 (b)	$Z_{eq} = \dfrac{-e(s)}{f(s)} = Z_1$
3-Bond 1-Junction	Figure 6.9 (c)	$Z_{eq} = \dfrac{e(s)}{-f(s)} = Z_1 + Z_2$
3-Bond 0-Junction	Figure 6.9 (d)	$Z_{eq} = \dfrac{-e(s)}{f(s)} = \dfrac{Z_1 Z_2}{Z_1 + Z_2}$
Effort Divider	Figure 6.10 (a)	$\dfrac{e_{out}}{e_{in}} = -\dfrac{Z_2}{Z_1 + Z_2}$
Flow Divider	Figure 6.10 (b)	$\dfrac{f_{out}}{f_{in}} = -\dfrac{Z_1}{Z_1 + Z_2}$

3. Iteratively condense (simplify) the impedance bond graph using the equivalencies while solving for any necessary intermediate transfer functions.

These basic steps are best illustrated through the examples that follow.

Example 6.5

Let us return to examine the simple mass-spring-damper problem from Example 3.1. Derive the transfer function using an impedance bond graph.

Solution. The first step is to identify the desired outputs of interest and the necessary transfer functions. Assume that the outputs of interest are the positions of the first and second cart, $x_1(t)$ and $x_2(t)$, respectively. Thus, to predict the displacement responses of each cart, two transfer functions are necessary: (1) the displacement of the first cart relative to the input force, $x_1(s)/F(s)$, and (2) the displacement of the second cart relative to the input force, $x_2(s)/F(s)$.

The second step is to formulate a strategy. By examining the impedance bond graph in Figure 6.11 (a), it becomes evident that the effort and flow of the bond on the end are the input force, $F(s)$, and the velocity of the second cart which is the derivative of the respective displacement, $\mathcal{L}[\dot{x}_2(t)] = sx_2(s)$. Therefore, if the total equivalent impedance of the entire system is known, then the transfer function relating the displacement of the second cart relative to the input force can be determined:

$$\frac{F(s)}{sx_2(s)} = Z_{total} \quad \rightarrow \quad \frac{x_2(s)}{F(s)} = \frac{1}{sZ_{total}}.$$

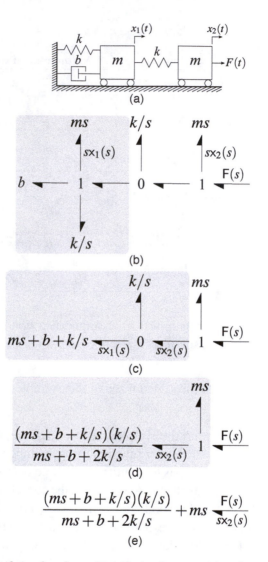

Figure 6.11: *Impedance bond graph analysis of mass-spring-damper problem from Example 3.1.*

Hence, one of the transfer functions can be derived by condensing the entire bond graph and determining the total equivalent impdedance, Z_{total}. Upon closer examination, one should note that the 0-junction in the middle functions as a flow divider that can be used to relate $x_1(s)$ to $x_2(s)$,

$$\frac{sx_1(s)}{sx_2(s)} = \frac{x_1(s)}{x_2(s)},$$

which with $x_2(s)/F(s)$ can be used to derive $x_1(s)/F(s)$,

$$\frac{x_1(s)}{F(s)} = \frac{x_2(s)}{F(s)} \frac{x_1(s)}{x_2(s)}.$$

Thus the strategy is to determine $x_1(s)/x_2(s)$ during an intermediate step and condense the entire bond graph to determine Z_{total}.

The third step is to iteratively condense and simplify. In the first iteration, the impedances highlighted in (b) are summed to arrive at an equivalent impedance for the I-, R-, and C-elements combined off the 1-junction, $ms + b + k/s$. This results in the flow divider highlighted in (c). The flow divider is used to derive the transfer function relating $x_1(s)$ to $x_2(s)$ (refer to Table 6.3 for the flow divider relation),

$$\frac{x_1(s)}{x_2(s)} = \frac{sx_1(s)}{sx_2(s)} = \frac{k/s}{ms + b + 2k/s} = \frac{k}{ms^2 + bs + 2k}.$$

The impedances, highlighted in (c), attached to the 0-junction are then combined to determine their equivalent impedance,

$$\frac{(ms + b + k/s)(k/s)}{ms + b + 2k/s} = \frac{mks^2 + bks + k^2}{s(ms^2 + bs + 2k)}.$$

Finally, the two remaining impedances highlighted in (d) are summed to determine the total equivalent impedance of the system,

$$Z_{total} = \frac{(ms + b + k/s)(k/s)}{ms + b + 2k/s} + ms = \frac{m^2s^4 + mbs^3 + 3mks^2 + bks + k^2}{s(ms^2 + bs + 2k)}.$$

Then $x_2(s)/F(s)$ is derived from the total equivalent impedance,

$$\frac{x_2(s)}{F(s)} = \frac{1}{sZ_{total}} = \frac{ms^2 + bs + 2k}{m^2s^4 + mbs^3 + 3mks^2 + bks + k^2}, \tag{6.19}$$

which is then multiplied by $x_1(s)/x_2(s)$ to determine $x_1(s)/F(s)$,

$$\frac{x_1(s)}{F(s)} = \left(\frac{ms^2 + bs + 2k}{m^2s^4 + mbs^3 + 3mks^2 + bks + k^2} \right) \left(\frac{k}{ms^2 + bs + 2k} \right)$$

$$= \frac{k}{m^2s^4 + mbs^3 + 3mks^2 + bks + k^2}. \tag{6.20}$$

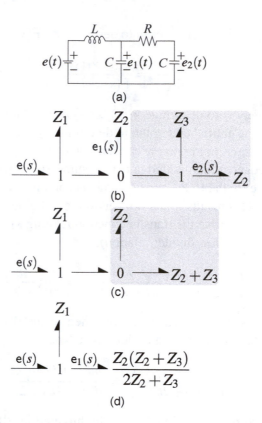

Figure 6.12: *Impedance bond graph analysis of circuit from Example 3.5.*

Example 6.6

Recall the electric circuit from Example 6.4. Derive the transfer function relating the voltage $e_2(t)$ to the input voltage $e(t)$.

Solution. As previously mentioned, the algebraic "bookkeeping" can be facilitated by using symbolic substitutions for the impedances. The circuit impedances are

$$Z_1 = Ls, \; Z_2 = \frac{1}{Cs}, \; \text{and } Z_3 = R.$$

First the inputs and outputs are determined. The input voltage is $e(t)$. The outputs of interest are the voltage drops across the two capacitors, $e_1(t)$ and $e_2(t)$. The necessary transfer functions are then $e_1(s)/e(s)$ and $e_2(s)/e(s)$.

Upon examination of the impedance bond graph in Figure 6.12 (b), it is evident that the two 1-junctions can be used as effort dividers to determine

$e_2(s)/e_1(s)$ and $e_1(s)/e(s)$ and then to calculate $e_2(s)/e(s)$,

$$\frac{e_2(s)}{e(s)} = \frac{e_2(s)}{e_1(s)} \frac{e_1(s)}{e(s)}.$$

The strategy is then to use the 1-junctions as effort dividers to derive $e_1(s)/e(s)$ and the intermediate transfer function $e_2(s)/e_1(s)$, and then solve for $e_2(s)/e(s)$ from these two transfer functions.

As shown in part (b) of the figure, the effort divider on the right end of the impedance bond graph can be used to relate $e_2(s)$ to $e_1(s)$,

$$\frac{e_2(s)}{e_1(s)} = \frac{Z_2}{Z_2+Z_3} = \frac{\dfrac{1}{Cs}}{R+\dfrac{1}{Cs}} = \frac{1}{RCs+1}.$$

As shown in (c), the two impedances are then summed to arrive at the equivalent impedance of the two combined off the 1-junction. The highlighted impedances in (c) can then be combined like resistors in parallel to determine the equivalent impedance of the two impedances off the 0-junction. The resulting bond graph in (d) is an effort divider that can be used to relate $e_1(s)$ to $e(s)$,

$$\frac{e_1(s)}{e(s)} = \frac{\dfrac{Z_2(Z_2+Z_3)}{2Z_2+Z_3}}{Z_1+\dfrac{Z_2(Z_2+Z_3)}{2Z_2+Z_3}} = \frac{Z_2(Z_2+Z_3)}{Z_1(2Z_2+Z_3)+Z_2(Z_2+Z_3)}$$

$$= \frac{\dfrac{1}{Cs}\left(\dfrac{1}{Cs}+R\right)}{Ls\left(\dfrac{2}{Cs}+R\right)+\dfrac{1}{Cs}\left(\dfrac{1}{Cs}+R\right)} = \frac{RCs+1}{LRC^2s^3+2LCs^2+RCs+1},$$

which can be used with the transfer function above to derive $e_2(s)/e(s)$,

$$\frac{e_2(s)}{e(s)} = \left(\frac{Z_2(Z_2+Z_3)}{Z_1(2Z_2+Z_3)+Z_2(Z_2+Z_3)}\right)\left(\frac{Z_2}{Z_2+Z_3}\right)$$

$$= \frac{Z_2}{Z_1(2Z_2+Z_3)+Z_2(Z_2+Z_3)}$$

$$= \left(\frac{RCs+1}{LRC^2s^3+2LCs^2+RCs+1}\right)\left(\frac{1}{RCs+1}\right)$$

$$= \frac{1}{LRC^2s^3+2LCs^2+RCs+1}.$$

Example 6.7

A schematic of a permanent-magnet direct current (PMDC) motor model is provided Figure 6.13. It is desired to determine the response of the output torque, $\tau_2(t)$, relative to the input voltage, $e_{in}(t)$.

Solution. The motor has an incorporated gear reduction where the gear ratio is N_1/N_2. This gear ratio relates the torques, $\tau_1(t)$ and $\tau_2(t)$, and the angular velocities, $\omega_1(t)$ and $\omega_2(t)$:

$$\tau_1 = \frac{N_1}{N_2}\tau_2 \text{ and } \frac{N_1}{N_2}\omega_1 = \omega_2.$$

Thus the transformer modulus, n, for gear reduction is N_1/N_2. The motor constant, k_m, relates the current to the motor torque, τ_m:

$$\tau_m(t) = k_m i(t).$$

The gyrator representing the ideal motor relation therefore has a modulus of k_m.

The output of interest is the output shaft torque, $\tau_2(t)$, and the input is the voltage $e_{in}(t)$. One can identify the intermediate steps needed to relate $T_2(s)$ to $e_{in}(s)$ by examining the impedance bond graph in (b). First, the output shaft torque $T_2(s)$ can be related to the intermediate torque $T_1(s)$ through the gear ratio,

$$T_1(s) = nT_2(s) \;\rightarrow\; \frac{T_2(s)}{T_1(s)} = \frac{1}{n}.$$

Second, the 1-junction on the right can then be used as an effort divider to relate the intermediate torque, $T_1(s)$, to the motor torque, $T_m(s)$ (i.e., to garner $T_1(s)/T_m(s)$). The gyrator is then used to relate the motor torque, $T_m(s)$, to the armature current, $i(s)$:

$$T_m(s) = k_m i(s) \;\rightarrow\; \frac{T_m(s)}{i(s)} = k_m.$$

Finally, the total equivalent impedance can be used to relate the armature current, $i(s)$, to the input voltage, $e_{in}(s)$:

$$\frac{i(s)}{e_{in}(s)} = \frac{1}{Z_{total}}.$$

Hence the overall transfer function is

$$\frac{T_2(s)}{e_{in}(s)} = \frac{i(s)}{e_{in}(s)}\frac{T_m(s)}{i(s)}\frac{T_1(s)}{T_m(s)}\frac{T_2(s)}{T_1(s)} = \frac{1}{Z_{total}}k_m\frac{T_1(s)}{T_m(s)}\frac{1}{n}.$$

The strategy is therefore to condense the impedance bond graph to determine the total equivalent impedance, Z_{total}, and to determine, in an intermediate step, the relation between the intermediate torque, $T_1(s)$, and the motor torque, $T_m(s)$, using the 1-junction on the right as an effort divider.

As shown in (b), the transformer in tandem with the bearing (represented by the impedance B) can be condensed using their equivalent impedance n^2B as shown in (c). This allows for the 1-junction on the right (highlighted in (c)) to then be used as an effort divider:

$$\frac{T_1(s)}{T_m(s)} = \frac{n^2B}{Js+n^2B}.$$

Then, the impedances off the 1-junction can be summed to arrive at their equivalent impedance, $Js+n^2B$, shown in (d). The gyrator and equivalent impedance highlighted in (d) can then be combined using their equivalent impedance, $k_m^2/(Js+n^2B)$. Finally, the overall total equivalent impedance is derived by summing the two remaining impedances,

$$Z_{total} = R + \frac{k_m^2}{Js+n^2B} = \frac{R(Js+n^2B)+k_m^2}{Js+n^2B} = \frac{JRs+n^2BR+k_m^2}{Js+n^2B}.$$

The transfer function of interest is then

$$\frac{T_2(s)}{e_{in}(s)} = \frac{Js+n^2B}{JRs+n^2BR+k_m^2}k_m\frac{n^2B}{Js+n^2B}\frac{1}{m} = \frac{k_mnB}{JRs+n^2BR+k_m^2}.$$

As has been previously mentioned, sometimes sign changes are inevitable. Sign changes usually arise due to the chosen sign convention. In examples from previous chapters, it has been shown that assigning signs to bonds is somewhat arbitrary, and, as long as one is consistent, should always arrive at an equivalent set of equations. However, occasionally one may wish to adapt an accepted or commonly used sign convention. For instance, in many books on Mechanical Vibrations, when calculating relative displacements (velocities) in problems involving a train of masses attached by springs and dampers, it may be customary to subtract the displacement (velocity) of the left mass from the displacement (velocity) of the right mass attached to a particular spring or damper combination. When strictly imposing such a sign convention, this can lead to sign changes in the impedance bond graph. The following example illustrates such a phenomenon and how to deal with it using the sign-change equivalencies previously derived.

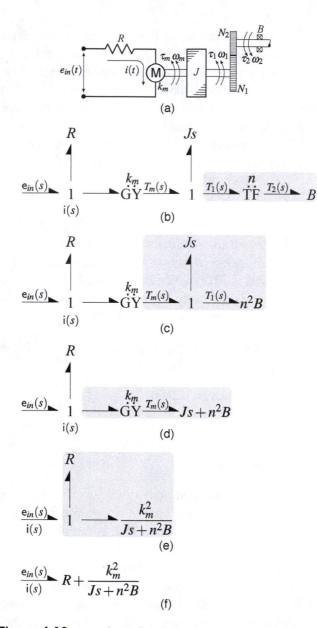

Figure 6.13: *Impedance bond graph of a geared PMDC motor.*

Example 6.8

A mass-spring system is depicted in Figure 6.14. The "train" of masses is excited by a shaker which provides a velocity input $v(t)$ with an as-

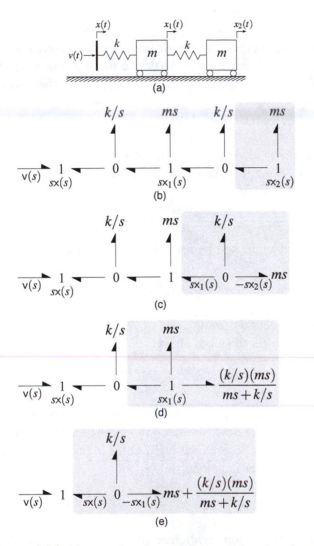

Figure 6.14: *Mass-spring-damper system with a sign change.*

sociated displacement of $x(t)$. Derive the transfer function relating the displacements $x_1(t)$ and $x_2(t)$ to the shaker motion $x(t)$.

Solution. The sign convention used to determine the relative velocities at the 0-junctions is to subtract the velocity of the left node or mass from the right node or mass. That is, at the 0-junctions, the velocity of the element farthest along the x-axis is positive and the other negative. This results in a sign change at the 1-junction attached to the rightmost

I-element.

The desired outputs are the displacements of the respective masses, $x_1(t)$ and $x_2(t)$. Hence, we wish to derive transfer functions relating these two displacements to the shaker motion, $x(t)$. Thus, the transfer functions of interest are

$$\frac{x_1(s)}{x(s)} \text{ and } \frac{x_2(s)}{x(s)}.$$

The strategy is to use the two 0-junctions as flow dividers to relate the displacements. The left 0-junction will be used to derive a transfer function relating $x_1(s)$ to $x(s)$:

$$\frac{sx_1(s)}{sx(s)} = \frac{x_1(s)}{x(s)}.$$

The right 0-junction will be used to derive the transfer function relating $x_2(s)$ to $x_1(s)$:

$$\frac{sx_2(s)}{sx_1(s)} = \frac{x_2(s)}{x_1(s)}.$$

The transfer function relating $x_2(s)$ to $x(s)$ can be calculated by multiplying the two transfer functions above:

$$\frac{x_2(s)}{x(s)} = \frac{x_1(s)}{x(s)}\frac{x_2(s)}{x_1(s)}.$$

An equivalency is used to condense the 1-junction highlighted in (b) as shown in (c). The flow is negated to maintain appropriate sign conventions. The resulting flow divider highlighted in (c) is used to determine the intermediate transfer function $x_2(s)/x_1(s)$:

$$\frac{sx_2(s)}{-sx_1(s)} = -\frac{k/s}{ms+k/s} = -\frac{k}{ms^2+k} \rightarrow \frac{x_2(s)}{x_1(s)} = \frac{k}{ms^2+k}.$$

The resulting 0-junction highlighted in (c) has a sign change. The equivalent impedance is determined, but note that when condensed, the effort on the equivalent impedance is negated and not the flow. For the two impedances off the 0-junction the equivalent impedance is

$$\frac{(k/s)(ms)}{ms+k/s} = \frac{mks}{ms^2+k}.$$

This results in two impedances off a 1-junction as highlighted in (d), which can be simply summed

$$ms + \frac{(k/s)(ms)}{ms+k/s} = ms + \frac{mks}{ms^2+k} = \frac{m^2s^3+2mks}{ms^2+k}.$$

The 1-junction has a sign change, however, and therefore the associated flow, $sx_1(s)$, must be negated as shown in (e). What then remains is the flow divider highlighted in (e) which is used to relate $x_1(s)$ to $x(s)$. Note, though, that this flow divider also has a sign change that must be accounted for. The flow divider relation is

$$\frac{sx_1(s)}{-sx(s)} = -\frac{k/s}{k/s + ms + \dfrac{(k/s)(ms)}{ms + k/s}} \rightarrow \frac{x_1(s)}{x(s)} = \frac{mks^2 + k^2}{m^2 s^4 + 3mks^2 + k^2}.$$

The resulting transfer function can be multiplied by the transfer function for $x_2(s)/x_1(s)$ derived above to calculate $x_1(s)/x(s)$:

$$\frac{x_2(s)}{x(s)} = \frac{mks^2 + k^2}{m^2 s^4 + 3mks^2 + k^2} \frac{k}{ms^2 + k} = \frac{k^2}{m^2 s^4 + 3mks^2 + k^2}.$$

Take special note that the resulting transfer functions are the same that would be derived had the power directions of the inner bonds been assigned in the opposite direction.

Sometimes in circuit diagrams the currents are assigned an assumed flow direction that specifies a distinct sign convention when currents are summed at the nodes (0-junctions). If the bond graph strictly follows the circuit diagram, this can lead to sign changes. However, as in Example 6.8 the sign changes can be readily accounted for using equivalencies. This is further illustrated in the following example.

Example 6.9

Derive the transfer functions relating the loop currents to the input voltage.

Solution. As illustrated in Figure 6.15, when the currents are summed at the top center node, $i_1(t)$ is positive and $i_2(t)$ negative (assuming a current into the node is positive). Thus, for the bond graph to strictly follow the sign convention provided by the circuit schematic in (a), $i_1(t)$ must be positive (i.e., power into the 0-junction) and $i_2(t)$ must be negative (i.e., power out of the 0-junction). The resulting impedance bond graph in (b) has a 1-junction with a sign change.

For this problem, it is desired to determine as outputs the loop currents. Therefore, the necessary transfer functions are

$$\frac{i_1(s)}{e(s)} \text{ and } \frac{i_2(s)}{e(s)}.$$

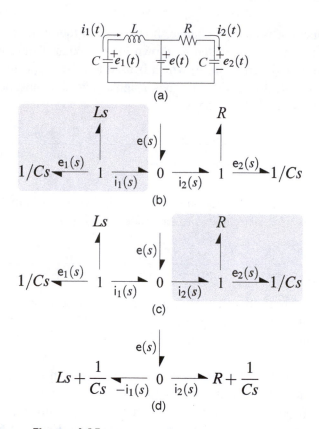

Figure 6.15: *Electric circuit with sign change.*

By inspection, the circuit has three branches: (1) the left branch composed of a capacitor and inductor in series, (2) the center branch which is simply the effort source, and (3) the right branch which is a resistor and capacitor in series. The left branch as highlighted in (b) has a sign change. The equivalent impedances for the right and left branches are easily determined by simply summing the impedances attached off the respective 1-junctions. However, because of the sign change, the equivalent impedance for the left branch relates the input voltage, $e(s)$, to the negative of the current in the left loop, $-i_1(s)$:

$$\frac{e(s)}{-i_1(s)} = Ls + \frac{1}{Cs} = \frac{LCs^2 + 1}{Cs} \rightarrow \frac{i_2(s)}{e(s)} = -\frac{Cs}{LCs^2 + 1}.$$

The other transfer function is determined from the equivalent impedance of the right loop,

$$\frac{e(s)}{i_2(s)} = R + \frac{1}{Cs} = \frac{RCs+1}{Cs} \rightarrow \frac{i_2(s)}{e(s)} = \frac{Cs}{RCs+1}.$$

6.9 Alternate Derivation of a Transfer Function

Occasionally, the impedance bond graph for a system cannot be readily simplified using equivalencies. In these cases, a more generalized approach is necessary. Just as regular bond graphs were used to derive differential equations in the time domain to solve for the necessary states, impedance bond graphs can be used to derive algebraic equations in the s-domain that can be used to solve for the efforts and flows of interest.

As before, each 1-junction leads to an equation summing the efforts on the attached bonds, and each 0-junction leads to an equation summing the flows on the attached bonds. The basic procedure for deriving the transfer functions using the junction summation equations is

1. Label the distinct efforts and flows associated with each junction on the impedance bond graph,

2. Use the impedance relations to determine any remaining unknown efforts or flows in terms of the distinct efforts and flows associated with the junctions in the previous step,

3. Use the impedance bond graph to derive a summation of efforts equation at each 1-junction and summation of flows equation at each 0-junction to solve for any unknown flows and/or efforts in the s-domain,

4. Arrange the resulting equations into a set of linear algebraic equations in terms of your unknown efforts and flows, and

5. Simultaneously solve the set of algebraic equations.

This basic procedure is illustrated with the following example.

Example 6.10
A mass-spring-damper systerm is shown in Figure 6.16. Derive the transfer functions relating the displacements of the masses (i.e., $x_1(t)$, $x_2(t)$, and $x_3(t)$) to the input displacement, $x(t)$.

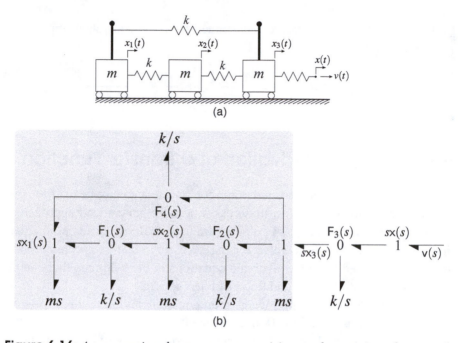

(a)

(b)

Figure 6.16: *A mass-spring-damper system requiring an alternate impedance analysis.*

Solution. Note that the spring attached between the first and third masses creates a "loop" highlighted in (b) in the impedance bond graph that cannot be dealt with using the equivalencies described in the previous sections. However, as mentioned above, the 1- and 0-junctions can be used to derive a set of algebraic equations in the s-domain that can be simultaneously solved to arrive at the necessary transfer functions.

The system is excited by the velocity specified at the right end, $v(t)$. That velocity has an associated displacement, $x(t)$. The outputs of interest are the respective displacements of the three masses, $x_1(t)$, $x_2(t)$, and $x_3(t)$. It is desired to determine the response of these outputs relative to the displacement at the input. Hence, the necessary transfer functions are

$$\frac{x_1(s)}{x(s)}, \frac{x_2(s)}{x(s)}, \text{ and } \frac{x_3(s)}{x(s)}.$$

To determine these transfer functions, though, it is also necessary to know the efforts due to the four springs in the s-domain (i.e., $F_1(s)$, $F_2(s)$, $F_3(s)$, and $F_4(s)$). To derive the necessary algebraic equations, these efforts and the flows $sx_1(s)$, $sx_2(s)$, and $sx_3(s)$ are labeled at the associated junctions in (b) of the figure.

The following equations result when summing the efforts at the 1-junctions associated with $sx_1(s)$, $sx_2(s)$, and $sx_3(s)$:

$$F_1(s) + F_4(s) - ms^2x_1(s) = 0,$$
$$-F_1(s) + F_2(s) - ms^2x_2(s) = 0, \text{ and}$$
$$-F_2(s) + F_3(s) - F_4(s) - ms^2x_3(s) = 0.$$

By summing the flows at the 0-junctions associated with $F_1(s)$, $F_2(s)$, $F_3(s)$, and $F_4(s)$, one can derive the following equations:

$$-sx_1(s) + sx_2(s) - F_1(s)/(k/s) = 0,$$
$$-sx_2(s) + sx_3(s) - F_2(s)/(k/s) = 0,$$
$$-sx_3(s) - F_3(s)/(k/s) = -sx(s), \text{ and}$$
$$-sx_1(s) + sx_3(s) - F_4(s)/(k/s) = 0.$$

Note that the unknown efforts and flows due to the R-, C-, and I-elements shown in the above equations are expressed in terms of the common efforts (F_1 to F_4) and flows ($sx_1(s)$ to $sx_3(s)$) using the impedance relations for the respective elements. The above equations can be arranged into a linear algebraic problem,

$$\begin{bmatrix} 1 & 0 & 0 & 1 & -ms^2 & 0 & 0 \\ -1 & 1 & 0 & 0 & 0 & -ms^2 & 0 \\ 0 & -1 & 1 & -1 & 0 & 0 & -ms^2 \\ -s/k & 0 & 0 & 0 & -s & s & 0 \\ 0 & -s/k & 0 & 0 & 0 & -s & s \\ 0 & 0 & -s/k & 0 & 0 & 0 & -s \\ 0 & 0 & 0 & -s/k & -s & 0 & s \end{bmatrix} \begin{bmatrix} F_1(s) \\ F_2(s) \\ F_3(s) \\ F_4(s) \\ x_1(s) \\ x_2(s) \\ x_3(s) \end{bmatrix} =$$

$$\begin{bmatrix} 0 & 0 & 0 & 0 & 0 & -s & 0 \end{bmatrix}^T x(s).$$

The above linear algebraic problem is in the form

$$\mathbf{AY}(s) = \mathbf{B}x(s)$$

and can be readily solved to determine the forces and displacements,

$$\mathbf{Y}(s) = \mathbf{A}^{-1}\mathbf{B}x(s).$$

The transfer function matrix is then

$$\mathbf{G}(s) = \mathbf{A}^{-1}\mathbf{B}.$$

We can use MATLAB to solve for the transfer functions.

```
>> syms m b k s
>> A = [1 0 0 1 -m*s^2 0 0; ...
-1 1 0 0 0 -m*s^2 0; ...
0 -1 1 -1 0 0 -m*s^2; ...
-s/k 0 0 0 -s s 0; ...
0 -s/k 0 0 0 -s s; ...
0 0 -s/k 0 0 0 -s; ...
0 0 0 -s/k -s 0 s];
>> B = [0; 0; 0; 0; 0; -s; 0];
>> G = inv(A)*B

G =

                                                   0
            k^2*s^2*m/(k^2+4*k*s^2*m+m^2*s^4)
  s^2*m*(3*k+m*s^2)*k/(k^2+4*k*s^2*m+m^2*s^4)
            k^2*s^2*m/(k^2+4*k*s^2*m+m^2*s^4)
                k^2/(k^2+4*k*s^2*m+m^2*s^4)
                k^2/(k^2+4*k*s^2*m+m^2*s^4)
        (k+m*s^2)*k/(k^2+4*k*s^2*m+m^2*s^4)
```

The transfer functions are

$$\frac{x_1(s)}{x(s)} = \frac{x_2(s)}{x(s)} = \frac{k^2}{m^2s^4 + 4mks^2 + k^2}$$

and

$$\frac{x_3(s)}{x(s)} = \frac{mks^2 + k^2}{m^2s^4 + 4mks^2 + k^2}.$$

The alternate approach described above is not limited to problems with "loops." It can generally be used, in place of equivalencies, to derive transfer functions from any impedance bond graph.

6.10 Model Transformations using MATLAB

As we saw in Chapter 4 we can define state-space objects in MATLAB and use them to simulate responses to a variety of inputs and initial conditions.

In Chapter 5 we discovered the `tf` and `zpk` commands which are used to define transfer function objects. The commands we previously used to simulate responses for state-space representations are generally formulated for LTI systems that can be defined in MATLAB in terms of their state-space or transfer-function representations (refer to Table 6.5).

We derived the relationship between the Laplace transform and the state space in Section 6.2. MATLAB includes several functions that facilitate conversion from one form of model to another. The `ss2tf` command converts from the state space to transfer function, and the `tf2ss` converts from the transfer function to the state space. These commands, however, do not operate on symbolic matrices or functions.

Table 6.5: *Functions for LTI model conversion.*

Function	Description
`ss2tf(A,B,C,D)`	converts from state space to transfer function
`tf2ss(num,den)`	converts from transfer function to state space

Example 6.11

Use MATLAB to convert the state-space representation of Example 6.2 assuming that the mass, damping constant, and spring rate are 10 kg, 20 N-m/s, and 60 N/m.

Solution. Recall from Example 6.2 that the state-space representation is

$$
\begin{bmatrix} \dot{x}_1 \\ \dot{p}_1 \\ \dot{\delta} \\ \dot{p}_2 \end{bmatrix} = \begin{bmatrix} 0 & 1/m & 0 & 0 \\ -k & -b/m & k & 0 \\ 0 & -1/m & 0 & 1/m \\ 0 & 0 & -k & 0 \end{bmatrix} \begin{bmatrix} x_1 \\ p_1 \\ \delta \\ p_2 \end{bmatrix} + \begin{bmatrix} 0 \\ 0 \\ 0 \\ 1 \end{bmatrix} F(t)
$$

$$
= \begin{bmatrix} 1 & 0 & 0 & 0 \\ 1 & 0 & 1 & 0 \end{bmatrix} \begin{bmatrix} x_1 \\ p_1 \\ \delta \\ p_2 \end{bmatrix} + \begin{bmatrix} 0 \\ 0 \end{bmatrix} F(t).
$$

Note that this system has two outputs and a single input. Thus, the system can be represented by two transfer functions representing the relations between the displacements and the input force, $x_1(s)/F(s)$ and $x_2(s)/F(s)$. To compute the transfer function, we define the parameters, the matrices, and the state-space representation.

```
>> m = 10; b = 20; k = 60;
>> A = [0 1/m 0 0; -k -b/m k 0; 0 -1/m 0 1/m; 0 0 -k 0];
>> B = [0 0 0 1]';
>> C = [1 0 0 0; 1 0 1 0];
>> D = [0 0]';
>> [num,den] = ss2tf(A,B,C,D)

num =

         0          0          0          0     0.6000
         0          0     0.1000     0.2000     1.2000


den =

    1.0000     2.0000    18.0000    12.0000    36.0000
```

Given the numerical solution, the transfer functions are

$$\frac{x_1(s)}{F(s)} = \frac{0.6}{s^4 + 2s^3 + 18s^2 + 12s + 36}$$

$$= \frac{60}{100s^4 + 200s^3 + 1800s^2 + 1200s + 3600}$$

$$\frac{x_2(s)}{F(s)} = \frac{0.1s^2 + 0.2s + 1.2}{s^4 + 2s^3 + 18s^2 + 12s + 36}$$

$$= \frac{10s^2 + 20s + 120}{100s^4 + 200s^3 + 1800s^2 + 1200s + 3600}.$$

If the parameter values are substituted, the transfer functions match those derived in Examples 6.2 and 6.5. The tf2ss function can be used to convert back to the state-space model.

```
>> [A,B,C,D] = tf2ss(num,den)

A =

   -2.0000   -18.0000   -12.0000   -36.0000
    1.0000          0          0          0
         0     1.0000          0          0
         0          0     1.0000          0


B =

    1
```

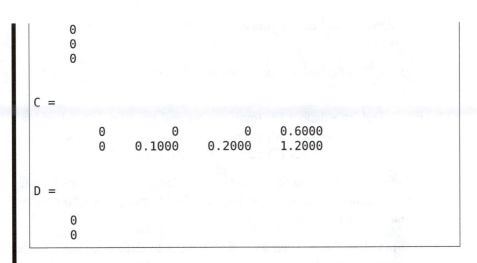

$$C =$$

$$\begin{matrix} 0 & 0 & 0 & 0.6000 \\ 0 & 0.1000 & 0.2000 & 1.2000 \end{matrix}$$

$$D =$$

$$\begin{matrix} 0 \\ 0 \end{matrix}$$

Note, however, that the state-space system that results is generally equivalent and not equal to the original.

6.11 Summary

▷ Laplace transforms can be applied to the differential equations derived from a bond graph to determine transfer functions. This is especially concise if applied to the state-space form of the resulting equations. However, for systems with more than three states, this can be cumbersome to implement analytically and many times must be done so with the aid of symbolic math software.

▷ The concept of impedance used for electric circuits analysis can be adapted to derive equivalent mechanical impedances. The impedance can be derived by applying the transfer function concept to the effort-flow relations on linear I-, R-, and C-elements. The impedance of an I-, R-, or C-element is the ratio of effort to flow after taking the Laplace transform of the constitutive relation and assuming initial conditions are zero.

▷ Impedance bond graphs are synthesized in much the same manner as regular bond graphs with a few exceptions: (1) I-, R-, and C-elements are replaced with their equivalent impedances, (2) causal strokes are unnecessary, and (3) sources are replaced simply by a bond labeled with the Laplace transform of the associated effort or flow.

▷ Impedances attached to 1-junctions sum like resistances in series (Equation 6.11).

▷ Impedances attached to 0-junctions sum like resistances in parallel (Equation 6.12).

▷ Equivalent impedances for transformers and gyrators in tandem with an impedance can be readily determined in terms of the modulus and the attached impedance (Equations 6.13 and 6.14).

▷ A 1-junction can be used as an effort divider to determine the transfer function relating the effort on any attached bond to the effort on the primary bond (Equation 6.15).

▷ A 0-junction can be used as a flow divider to determine the transfer function relating the flow on any attached bond to the flow on the primary bond (Equation 6.16).

▷ Sign changes can be readily accounted for in impedance bond graphs. The basic relations for equivalent impedance remain the same. However, to maintain the sign change the flow on the primary bond of a 1-junction with a sign change is negated, and the effort on the primary bond of a 0-junction with a sign change is negated (Figure 6.8 (c) and (f)). For effort and flow dividers, the overall relations are negated (Equations 6.17 and 6.18).

▷ The basic steps to deriving transfer functions directly from an impedance bond graph are: (1) identify the inputs and outputs to determine the necessary transfer functions, (2) formulate a strategy to determine intermediate steps, and (3) iteratively condense the impedance bond graph using equivalencies while solving for any necessary intermediate transfer functions.

▷ An alternate approach to deriving transfer functions from bond graphs is to use the summation of effort and summation of flow equations resulting from junctions to derive a set of linear algebraic equations in terms of unknown efforts and flows in the s-domain that can be solved simultaneously using Linear Algebra. The basic procedure is: (1) label junctions with distinct efforts and flows, (2) derive remaining unknown efforts and flows using impedance relations, (3) derive summation of effort and summation of flow equations, (4) arrange the resulting equations into set linear algebraic equations in terms of the efforts and flows, and (5) solve the equations simultaneously.

▷ MATLAB includes several commands for converting between state-space and transfer-function representations.

6.12 Review

R6-1 How are the transfer function and state-space representations related?

R6-2 Explain the concept of impedance in the context of bond graphs.

R6-3 List the impedances for basic 1-port elements in each of the energy domains discussed.

R6-4 Are impedances off a 0-junction analogous to resistors connected in series or parallel? What about impedances off a 1-junction?

R6-5 Describe in your own words the 0- and 1-junction impedance equivalencies.

R6-6 What are the impedance relations for a transformer and gyrator in tandem with an impedance?

R6-7 How are electric voltage and current dividers generalized for impedance bond graphs?

R6-8 How do sign changes modify the 0- and 1-junction equivalencies? How do they change the effort and flow divider relations?

R6-9 Describe in your own words how to derive transfer functions directly from the impedance bond graph.

R6-10 Explain an alternate approach for deriving transfer functions from an impedance bond graph when equivalencies will not suffice.

R6-11 List and describe two MATLAB functions for LTI model transformations.

6.13 Problems

For each of the systems, synthesize the impedance bond graph and derive the requested transfer functions.

P6-1 *Mass-Spring-Damper Systems (Figure 1.15).* The input for (a) is the displacement $x(t)$, and the outputs are $x_1(t)$ and $x_2(t)$. Derive the transfer functions relating the outputs to the input (i.e., $x_1(s)/x(s)$ and $x_2(s)/x(s)$). For (b) the input is the forcing function $F(t)$ and the outputs of interest are the displacements of the masses, $x_1(t)$ and $x_2(t)$. Find the associated transfer functions, $x_1(s)/F(s)$ and $x_2(s)/F(s)$. For (c) and (d) the input is the force F. You do not need to consider the effects of gravity. Derive the transfer functions for the vertical displacements of each mass relative to the input force.

P6-2 *Rotational Systems (a) and (b) (Figure 1.16).* The input torque $\tau_m(t)$ is used to generate an angular velocity $\omega_{out}(t)$ at the output shaft. Synthesize the impedance bond graph and use it to determine the transfer function $\Omega_{out}(s)/T_m(s)$.

P6-3 *Electric Circuits (a) and (b) (Figure 1.17).* For both circuits, the input is the voltage e. The outputs of interest are the inductor current i_L and the capacitor voltage v_c. Synthesize the impedance bond graph and derive the transfer functions $i_L(s)/e(s)$ and $e_C(s)/e(s)$ for both circuits.

P6-4 *Hydraulic Circuits (a) and (b) (Figure 1.18).* Both hydraulic systems have an input volumetric flow source $Q(t)$. Synthesize the impedance bond graphs and use each to determine the transfer function that relates the time-varying output volumetric flow rate $Q_{out}(t)$ to the input $Q(t)$.

P6-5 *Permanent Magnet Direct Current (PMDC) Motor with Gear Reduction (Figure 1.3).* The voltage $v_{in}(t)$ is an input used to generate the desired angular velocity ω_2 as an output. Use an impedance bond graph to derive the transfer function $\Omega_2(s)/e_{in}(s)$.

P6-6 *A Simple Wind Turbine Generator (Figure 3.28).* Derive the transfer function by synthesizing the impedance bond graph and relating the output lamp voltage $e_{lamp}(t)$ to the input wind velocity $v(t)$.

Use MATLAB to solve the following problems.

P6-7 Synthesize the impedance bond graph for the mass-spring-damper system depicted in Figure 6.17 (a). Derive the transfer functions relating the two displacements to the input force. Plot the responses $x_1(t)$ and $x_2(t)$ to the input force in Figure 6.17 (b) assuming that the system parameters are $m_1 = 1$ kg, $m_2 = 2$ kg, $k_1 = 20$ N/m, $k_2 = 10$ N/m, and $b = 2$ N-s/m.

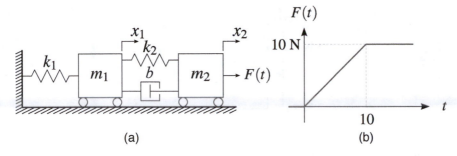

Figure 6.17: *(a) A mass-spring-damper system and (b) piecewise-continuous input.*

P6-8 Derive the state-space representation for the mass-spring-damper system in Figure 6.18. The input is the velocity $v(t)$ and the outputs are the displacements $x_1(t)$ and $x_2(t)$. Use MATLAB to *symbolically* derive the transfer functions relating each of the mass displacements to the input velocity, $v(t)$ (i.e., find $x_1(s)/F(s)$ and $x_2(s)/F(s)$). Refer to Examples 6.1 and 6.2 for assistance.

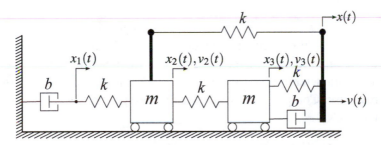

Figure 6.18: *A moderately complex mass-spring-damper system.*

P6-9 Regarding the previous problem, assuming that $m = 10$ kg, $b = 20$ N-s/m, and $k = 60$ N/m, use MATLAB to *numerically* convert the state-space model to transfer functions.

P6-10 Recall the rotational system illustrated in Figure 1.16 (b). Derive the transfer function relating the output angular velocity to the input torque. Assume that the rotational inertias are $J_1 = 0.05$ kg-m^2 and $J_2 = 0.1$ kg-m^2. The bearing constant and shaft rigidity are $\beta = 0.02$ N-m-s/rad and $\kappa = 10$ N-m/rad. The gear ratios are $N_1/N_2 = 5$ and $N_3/N_4 = 2$. Plot the response of the angular velocity of the second rotational inertia, J_2, to the input torque, τ_m, plotted in Figure 6.19.

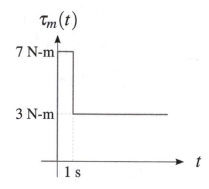

Figure 6.19: *Piecewise continuous motor torque.*

6.14 Challenges

As you did with the practice problems, synthesize the bond graph, assign causality, annotate it, and derive the differential equations. Organize the state equations into state-space form.

C6-1 *A Toy Electric Car.* Synthesize an impedance bond graph for the toy car and use it to derive a transfer function relating the car speed ($v(t)$) to the input motor voltage ($e_{in}(t)$). Use the resulting transfer function and parameters previously provided in Table 4.3 to simulate the step response. Compare your results with those previously attained in Challenge C4-1.

C6-2 *A Quarter-Car Suspension Model.* Synthesize the impedance bond graph for the quarter-car suspension and use it to derive the transfer function between the vertical vehicle displacement ($y(t)$) and the input road displacement ($y_{road}(t)$). Be sure to account for the sprung and unsprung masses and to consider the dynamics due to the tire. Using the transfer functions, plot the impulse responses of the velocities of the sprung and unsprung masses. Use the parameters previously provided in Challenge C4-2 and compare the results attained with the transfer function to those attained with the state-space model.

C6-3 *An Electric Drill.* For the electric drill, generate the impedance bond graph and use it to determine the transfer function between the input voltage e_{in} and the output torque τ_{out}. Use the parameters from Challenge C4-3 and compare with the state-space results.

Chapter 7

Time Domain Analysis

Given what we now know about state-space and transfer-function representations, we can now relate the characteristic roots (i.e., poles) or system eigenvalues to the features of typical responses such as the natural, impulse, step, and ramp responses. The system parameters affect how quickly a response reaches and settles to steady-state. As one or more parameters of the system are varied, the eigenvalues change and so do the response characteristics. Consider the following:

> ▷ How then do the system parameters affect the response?

> ▷ How are the parameters linked to the system poles or eigenvalues?

> ▷ How can Laplace transforms and transfer functions be used to analyze the time domain response of a system?

> ▷ How can Linear Algebra and the state-space model be used to analyze the time domain response?

7.1   Introduction

Linear systems and the concept of superposition were originally introduced in Chapter 2. This was further expanded in Chapter 5 where it was shown that system responses can generally be decomposed into a summation of first- and second-order terms that are derived through partial fraction expansion. Through partial fraction expansion, we decomposed the response in the s-domain into the summation of terms with first- and/or second-order polynomials in the denominator. The inverse Laplace transform was applied to the individual terms to arrive at a summation in the time domain. As will be shown in this chapter, those terms have first- and second-order responses. Hence, the response of a higher-order dynamic system is generally comprised of the summation of lower-order responses. Therefore, by understanding the characteristics of first- and second-order systems, one can also understand higher-order dynamics.

In this chapter we explore responses of first- and second-order systems to a series of typical inputs or initial conditions. In particular we examine natural (or unforced), impulse, step, and ramp responses. As we will discover, systems can be analyzed using both transfer function and state-space representations, and the characteristics of their responses can be linked to the system poles or eigenvalues.

The objectives and outcomes for this chapter are:

▷ *Objectives:*

1. To understand first-order responses,

2. To understand second-order responses,

3. To understand how higher-order responses are composed from first- and second-order terms, and

4. To understand the relation between the system poles or eigenvalues and the overall system response.

▷ *Outcomes:* Upon completion, you will be able to

1. identify the characteristics of first-order responses,

2. identify the characteristics of second-order responses,

3. identify the dominant poles of higher-order systems,

4. use the transfer function or state-space representation to determine a system's characteristic roots, and

5. predict overall response based on pole placement.

7.2 Transient Responses of First-Order Systems

First-order systems are those that have a single dynamic state. They are modeled by a single first-order differential equation. Though the simplest of dynamic systems, there do exist a variety of physical systems that can be modeled by a single, first-order, differential equation. Some examples include a merry-go-round and a compact disk drive.

7.2.1 The Natural Response

The natural response of a system is that which results from a non-zero initial condition. It results when the system is initially not in equilibrium. In vibration systems, this is referred to as the *unforced* or *free vibration*.

Given a prototypical first-order differential equation with an initial condition of $x(0)$,

$$T\frac{dx}{dt} + x = 0, \tag{7.1}$$

the natural response can be solved using Laplace transforms,

$$Ts\mathsf{x}(s) - Tx(0) + \mathsf{x}(s) = 0$$
$$(Ts + 1)\mathsf{x}(s) = Tx(0),$$

which when solved for $\mathsf{x}(s)$ results in

$$\mathsf{x}(s) = Tx(0)\frac{1}{Ts+1} = x(0)\frac{1}{s+1/T}.$$

Applying the inverse Laplace transform results in the time domain response

$$x(t) = x(0)e^{-t/T}, \tag{7.2}$$

where T is referred to as the first-order time constant. The response can be derived any number of ways including using Laplace transforms. The response is plotted in Figure 7.1. Note that the response begins at an initial value of $x(0)$ and decays exponentially. This is characteristic of any first-order system that matches the form of Equation 7.1. Note that the response decays to approximately 36.8% of the initial condition at one time constant, T. At two, three, four, and five times the time constant (i.e., $2T$, $3T$, $4T$, and $5T$) the response decays to approximately 13.5%, 5.0%, 1.8%, and 0.7% of the initial value, respectively. The tangent line evaluated at zero has a slope of $-x(0)/T$ and intersects the time axis at one time constant, T. Similarly, the tangent lines at T, $2T$, $3T$, and $4T$ intersect the time axis at $2T$, $3T$, $4T$, and $5T$, respectively.

7.2.2 The Impulse Response

Let us investigate, instead, the system response when there is a zero initial condition and the input is an impulse function as in the following differential equation:

$$T\frac{dx}{dt} + x = A\tilde{\delta}(t). \tag{7.3}$$

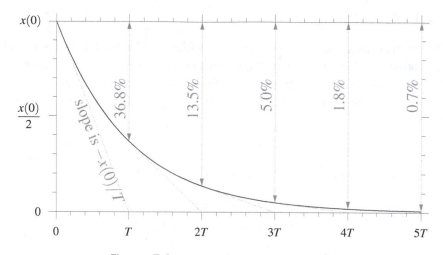

Figure 7.1: *First-order natural response.*

By applying Laplace transforms,

$$Tsx(s) + x(s) = (Ts+1)x(s) = A,$$

one can solve for the $x(s)$,

$$x(s) = A\frac{1}{Ts+1} = \frac{A}{T}\frac{1}{s+1/T}$$

and apply the inverse Laplace transform to solve for $x(t)$,

$$x(t) = \frac{A}{T}e^{-t/T}, \tag{7.4}$$

which when compared with the natural response is exactly the same except for the initial amplitude of the response (i.e., A/T for the impulse response versus $x(0)$ for the natural response). The dynamic characteristics of the impulse response are the same as those depicted in Figure 7.1 for the natural response except that A/T would replace $x(0)$ throughout. For example, at one time constant T the response would reach approximately 36.8% of the initial value A/T.

7.2.3 The Step Response

The step response results from any constant input that is applied at initial time. Given a step input,

$$T\frac{dx}{dt} + x = A\,1(t), \tag{7.5}$$

one can apply the Laplace transform,

$$Tsx(s) + x(s) = (Ts+1)x(s) = \frac{A}{s},$$

to find $x(s)$,

$$x(s) = A\left[\left(\frac{1}{s}\right)\left(\frac{1}{Ts+1}\right)\right] = A\left[\frac{1}{s} - \frac{1}{s+1/T}\right].$$

After applying the inverse Laplace transform, the time domain response is

$$x(t) = A\left(1 - e^{-t/T}\right). \tag{7.6}$$

Graphically, as illustrated in Figure 7.2, the step response is like the mirror image of the impulse or natural response. Instead of decaying to zero, the response asymptotically approaches the steady-state value, A. The time constant continues to be T, and the tangent line intersects the steady-state value A at one time constant. At $2T$, $3T$, $4T$, and $5T$ the response reaches 63.2%, 86.5%, 95.0%, 98.2%, and 99.3% of steady-state, respectively. The lines tangent at T, $2T$, $3T$, and $4T$ intersect steady-state at $2T$, $3T$, $4T$, and $5T$, respectively.

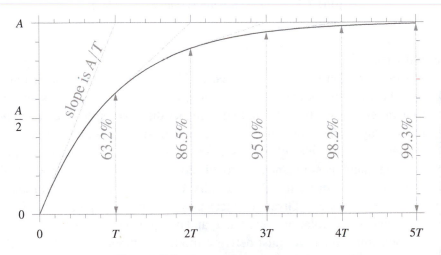

Figure 7.2: *First-order step response.*

Recall from Chapter 4 that the step function is the integral of the impulse function. Consequently, the step response must be the integral of the impulse response,

$$\int_0^t \frac{A}{T} e^{-t/T} = A\left(1 - e^{-t/T}\right),$$

given $x(0) = 0$. Graphically, the area under the impulse response curve, Equation 7.4 and Figure 7.1, is the step response curve, Equation 7.6 and Figure 7.2. It can also be shown that the impulse response is the derivative of the step response,

$$\frac{d}{dt}\left[A\left(1 - e^{-t/T}\right)\right] = \frac{A}{T} e^{-t/T}.$$

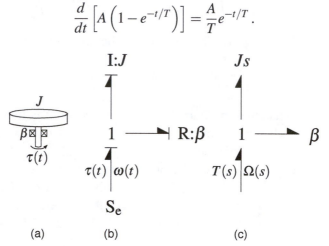

(a)   (b)   (c)

Figure 7.3: *A simple torsion system with damping.*

Example 7.1

A simple torsion system is depicted in Figure 7.3. It is a disk mounted on a bearing and is excited by an input torque. This system is similar to a variety of machines including compact disc drives, hard drives, and record players. The rotational inertia can be readily attained by measuring the mass and diameter of the disk. The bearing damping coefficient, on the other hand, is not something that is commonly advertised or supplied. However, this can be readily determined using experimental data and a model of the system. Given the measured step response plotted in Figure 7.4, what are the rotational inertia, J, and damping coefficient, β?

Solution. First we must derive a model for the system either in the form of a differential equation or a transfer function. Given the impedance bond graph in Figure 7.4 (c) the transfer function is

$$\frac{\Omega(s)}{T(s)} = \frac{1}{Js + \beta} = \frac{1}{J}\left(\frac{1}{s + \beta/J}\right).$$

For a unit step input (i.e., $\tau(t) = 1(t)$), the partial fraction of $\Omega(s)$ is

$$\Omega(s) = \frac{1}{\beta} \left[\frac{1}{s} - \frac{1}{s+\beta/J} \right],$$

which has an inverse Laplace transform of

$$\omega(t) = \frac{1}{\beta} \left[1(t) - e^{-\frac{\beta}{J}t} \right] \text{ rad/s}.$$

By comparing to Equation 7.6 one can surmise that the amplitude (and ergo the steady-state value) must equal to $1/\beta$. It can also be deduced that the time constant is $T = J/\beta$. (Note that $T(s)$ is the Laplace transform of $\tau(t)$ while simply T is the time constant.) As evidenced by the response, the steady-state angular velocity is 50 rad/s. Note that at steady-state the exponential term decays to zero and

$$\omega(t \rightarrow \infty) = \omega_{ss} = \frac{1}{\beta}.$$

Accordingly, the damping coefficient is

$$\beta = \frac{1}{\omega_{ss}} = 0.02 \text{ N-m-s/rad}.$$

Now knowing the damping coefficient, we can determine the rotational inertia. The response reaches 63.2% of ω_{ss}, 50 rad/s (i.e., 31.6 rad/s) at about 0.52 seconds. Thus, the time constant is T is 0.52 seconds. Given this fact, the rotational inertia can be calculated as

$$J = T\beta = (0.52 \text{ seconds})(0.02 \text{ N-m-s/rad}) = 0.0104 \text{ kg-m}^2.$$

7.2.4 The Ramp Response

The ramp response results from a linearly varying input as in the following differential equation:

$$T\frac{dx}{dt} + x = At. \tag{7.7}$$

Laplace transforms can be applied to the differential equation,

$$Tsx(s) + x(s) = \frac{A}{s^2},$$

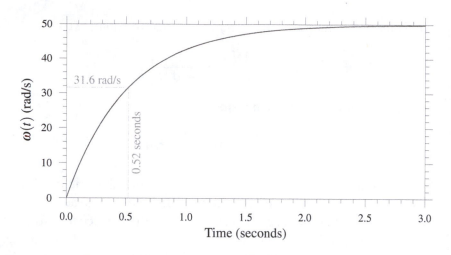

Figure 7.4: *Step response of one-disk torsion system.*

to solve for $x(s)$,

$$x(s) = A\left[\left(\frac{1}{s^2}\right)\left(\frac{1}{Ts+1}\right)\right] = A\left[\frac{1}{s^2} - \frac{T}{s} - \frac{T}{s+1/T}\right],$$

and the inverse Laplace transform then be used to derive the response

$$x(t) = A\left(t - T - Te^{-t/T}\right). \tag{7.8}$$

As depicted in Figure 7.5, the resultant curve at steady-state increases at a constant rate such that there is a constant delay in time of T and error of AT. Thus there exists a constant steady-state error between the input and output. Image that the input represents a desired value for the output. If the system is first-order and the input is a ramp function, there will always be a constant, steady-state error.

Also, recall from Chapter 4 that we determined that the ramp function is the integral of the step function. Thus, the ramp response should be the integral of the step response,

$$\int_0^t A\left(1 - e^{-t/T}\right) = A\left(t - T - Te^{-t/T}\right)$$

given $x(0) = 0$. The area under the step response curve, Equation 7.6 and Figure 7.2, is the ramp response, Equation 7.8 and Figure 7.5. Conversely, the derivative of the ramp response is the step response,

$$\frac{d}{dt}\left[A\left(t-T-Te^{-t/T}\right)\right]=A\left(1-e^{-t/T}\right).$$

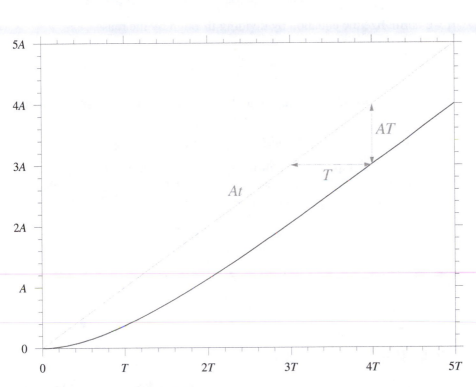

Figure 7.5: *First-order ramp response.*

7.3 Transient Responses of Second-Order Systems

Recall the simple mass-spring-damper system from Example 6.1 depicted in Figure 6.1. Also recall the state-space representation, Equation 6.5, which can be written as two first-order differential equations

$$\dot{x}=\frac{p}{m}\text{ and}$$
$$\dot{p}=-kx-\frac{b}{m}p+F(t).$$

Through a few simple substitutions, $p/m = v = \dot{x}$ and $\dot{p} = m\dot{v} = m\ddot{x}$, we can convert the two first-order differential equations into a single, second-order differential equation,

$$m\ddot{x} + b\dot{x} + kx = F(t).$$

If we normalize the equation by dividing through by the mass,

$$\ddot{x} + \frac{b}{m}\dot{x} + \frac{k}{m}x = \frac{1}{m}F(t), \tag{7.9}$$

the result is a second-order, differential equation like the prototypical second-order system

$$\ddot{x} + 2\zeta\omega_n\dot{x} + \omega_n^2 x = u(t) \tag{7.10}$$

where ζ is the damping ratio, ω_n the undamped natural frequency, and $u(t)$ the input. By comparing Equations 7.9 and 7.10, one can surmise that, for this system, the undamped natural frequency is

$$\omega_n = \sqrt{\frac{k}{m}} \tag{7.11}$$

and the damping ratio is

$$\zeta = \frac{b}{2\sqrt{km}}. \tag{7.12}$$

7.3.1 The Natural Response

The natural response, also referred to as the unforced response in Vibrations, occurs when the input is zero and one or both initial conditions are non-zero:

$$\ddot{x} + 2\zeta\omega_n\dot{x} + \omega_n^2 x = 0. \tag{7.13}$$

In Vibrations, the response is also referred to as free vibration. The *characteristic equation* is

$$s^2 + 2\zeta\omega_n s + \omega_n^2 = 0, \tag{7.14}$$

which has roots

$$s_{1,2} = \frac{-2\zeta\omega_n \pm \sqrt{4(\zeta\omega_n)^2 - 4\omega_n^2}}{2} = -\zeta\omega_n \pm \omega_n\sqrt{\zeta^2 - 1}. \tag{7.15}$$

Herein we will explore the various types of second-order responses based on damping – undamped ($\zeta = 0$), underdamped ($0 < \zeta < 1$), critically damped ($\zeta = 1$), and overdamped ($\zeta > 1$).

Undamped ($\zeta = 0$).   For the mass-spring-damper system discussed earlier, undamped responses occur when there is no dashpot or the damping is zero (i.e., $b = 0$):

$$\ddot{x} + \omega_n^2 x = 0.$$

The characteristic equation of the resultant second-order, differential equation is

$$s^2 + \omega_n^2 = 0,$$

which has imaginary roots

$$s_{1,2} = \pm j\omega. \tag{7.16}$$

With no damping to attenuate the dynamics, we expect the natural response of this system to be oscillatory or sinusoidal. Assuming a non-zero initial condition $x(0)$, the Laplace transform of the differential equation is

$$s^2 x(s) - sx(0) + \omega_n^2 x(s) = 0,$$

which we can use to solve for $x(s)$:

$$x(s) = \frac{sx(0)}{s^2 + \omega_n^2}.$$

The inverse Laplace transform is

$$x(t) = x(0)\cos \omega t. \tag{7.17}$$

Hence, with no damping, the natural response will oscillate forever in the form of a sinusoid with an initial value and amplitude of $x(0)$

Underdamped ($0 < \zeta < 0$).   If the second-order system is underdamped, the roots of the characteristic equation are complex. Using Equation 7.15 we can derive the roots

$$
\begin{aligned}
s_{1,2} &= -\zeta\omega_n \pm \omega_n\sqrt{\zeta^2 - 1} \\
&= -\zeta\omega_n \pm \omega_n\sqrt{-1}\sqrt{1 - \zeta^2} \\
&= -\zeta\omega_n \pm j\omega_n\sqrt{1 - \zeta^2}.
\end{aligned} \tag{7.18}
$$

We know from Chapter 5 that complex roots are associated with completing the square in the partial fraction expansion which results in terms that are the Laplace transforms of exponentials multiplied with sinusoids. We apply Laplace transforms to Equation 7.13,

$$s^2 x(s) - sx(0) + 2\zeta\omega_n[s + x(0)] + \omega_n^2 x(s) = 0$$

and solve for $x(s)$,

$$x(s) = \frac{(s + 2\zeta\omega_n)x(0)}{s^2 + 2\zeta\omega_n s + \omega_n^2}.$$

To complete the square the denominator is manipulated in the following fashion:

$$\begin{aligned}
s^2 + 2\zeta\omega_n s + \omega_n^2 &= [s^2 + 2\zeta\omega_n s + (\zeta\omega_n)^2] + [\omega_n^2 - (\zeta\omega_n)^2] \\
&= (s + \zeta\omega_n)^2 + \omega_n^2(1 - \zeta^2) \\
&= (s + \zeta\omega_n)^2 + (\omega_n\sqrt{1 - \zeta^2})^2.
\end{aligned}$$

Similarly, the numerator can be manipulated,

$$s + 2\zeta\omega_n = (s + \zeta\omega_n) + \zeta\omega_n = (s + \zeta\omega_n) + \frac{\zeta}{\sqrt{1 - \zeta^2}}\omega_n\sqrt{1 - \zeta^2}.$$

Thus partial fraction expansion of $x(s)$ is

$$x(s) = x(0)\left[\frac{s + \zeta\omega_n}{(s + \zeta\omega_n)^2 + (\omega_n\sqrt{1 - \zeta^2})^2} + \frac{\zeta}{\sqrt{1 - \zeta^2}}\frac{\omega_n\sqrt{1 - \zeta^2}}{(s + \zeta\omega_n)^2 + (\omega_n\sqrt{1 - \zeta^2})^2}\right],$$

which has an inverse Laplace transform of

$$\begin{aligned}
x(t) &= x(0)\left[e^{-\zeta\omega_n t}\cos\omega_n\sqrt{1 - \zeta^2}t + \frac{\zeta}{\sqrt{1 - \zeta^2}}e^{-\zeta\omega_n t}\sin\omega_n\sqrt{1 - \zeta^2}t\right] \\
&= x(0)\left[e^{-\zeta\omega_n t}\cos\omega_d t + \frac{\zeta}{\sqrt{1 - \zeta^2}}e^{-\zeta\omega_n t}\sin\omega_d t\right] \\
&= x(0)e^{-\zeta\omega_n t}\left[\cos\omega_d t + \frac{\zeta}{\sqrt{1 - \zeta^2}}\sin\omega_d t\right] \quad\quad (7.19)
\end{aligned}$$

where

$$\omega_d = \omega_n\sqrt{1 - \zeta^2} \quad\quad (7.20)$$

is the damped, natural frequency of oscillation. The overall response is an attenuated sinusoid. The sinusoidal terms generate oscillation, but each is multiplied by a decaying exponential which means that the oscillatory terms will attenuate with an exponential envelope that decays to zero with time.

Critically damped ($\zeta = 1$). When the system is critically damped, the roots of the characteristic equation are repeated, real, and equal to one another:

$$
\begin{aligned}
s_{1,2} &= -\zeta \omega_n \pm \omega_n \sqrt{\zeta^2 - 1} \\
&= -\zeta \omega_n \pm \omega_n \sqrt{1^2 - 1} \\
&= -\zeta \omega_n.
\end{aligned}
\tag{7.21}
$$

This is also illustrated by modifying the characteristic equation using $\zeta = 1$,

$$
s^2 + 2\omega_n s + \omega_n^2 = (s + \omega_n)^2 = 0.
$$

When the system is critically damped, the differential equation is

$$
\ddot{x} + 2\omega_n \dot{x} + \omega_n^2 = 0,
$$

which we apply Laplace transforms to

$$
s^2 x(s) - s x(0) + 2\omega_n \left[s + x(0) \right] + \omega_n^2 x(s) = 0
$$

to derive $x(s)$,

$$
\begin{aligned}
x(s) &= x(0) \frac{s + 2\omega_n}{s^2 + 2\omega_n^2 + \omega_n^2} \\
&= x(0) \frac{s + \omega_n + \omega_n}{(s + \omega_n)^2} \\
&= x(0) \left[\frac{s + \omega_n}{(s + \omega_n)^2} + \frac{\omega_n}{(s + \omega_n)^2} \right] \\
&= x(0) \left[\frac{1}{s + \omega_n} + \frac{\omega_n}{(s + \omega_n)^2} \right].
\end{aligned}
$$

The resultant response is derived by applying the inverse Laplace transform

$$
x(t) = x(0)(1 + \omega_n t) e^{-\omega_n t}.
\tag{7.22}
$$

Note the exponential term. As a result of this term, the response will decay to zero over time.

Overdamped ($\zeta > 1$). When the system is overdamped the roots of the characteristic equation are real, distinct, and negative:

$$
s_{1,2} = -\zeta \omega_n \pm \omega_n \sqrt{\zeta^2 - 1} = -\omega_n (\zeta \mp \sqrt{\zeta^2 - 1}).
\tag{7.23}
$$

Thus the characteristic equation can be factored,

$$s^2 + 2\zeta\omega_n s + \omega_n^2 = \left[s + \omega_n\left(\zeta + \sqrt{\zeta^2 + 1}\right)\right]\left[s + \omega_n\left(\zeta - \sqrt{\zeta^2 + 1}\right)\right].$$

Like the underdamped case, $x(s)$ is

$$x(s) = \frac{(s + 2\zeta\omega_n)x(0)}{s^2 + 2\zeta\omega_n s + \omega_n^2}.$$

However, because the roots of the denominator are real $x(s)$ becomes

$$x(s) = \frac{(s + 2\zeta\omega_n)x(0)}{\left[s + \omega_n\left(\zeta + \sqrt{\zeta^2 + 1}\right)\right]\left[s + \omega_n\left(\zeta - \sqrt{\zeta^2 + 1}\right)\right]},$$

which has the partial fraction expansion

$$x(s) = a\frac{1}{s + \omega_n\left(\zeta + \sqrt{\zeta^2 - 1}\right)} + b\frac{1}{s + \omega_n\left(\zeta - \sqrt{\zeta^2 - 1}\right)}$$

where

$$a = \frac{-\zeta + \sqrt{\zeta^2 - 1}}{2\sqrt{\zeta^2 - 1}}x(0)$$

and

$$b = \frac{\zeta + \sqrt{\zeta^2 - 1}}{2\sqrt{\zeta^2 - 1}}x(0).$$

Therefore, the overdamped response is

$$x(t) = ae^{-\omega_n\left(\zeta + \sqrt{\zeta^2 - 1}\right)t} + be^{-\omega_n\left(\zeta - \sqrt{\zeta^2 - 1}\right)t}. \tag{7.24}$$

The overdamped response like the critically damped decays exponentially to zero with time. As the damping ratio is increased, the oscillation increasingly attenuates until the system is critically damped, at which point there are no oscillations. As the damping ratio is further increased, the response takes longer to reach steady-state.

Figure 7.6 depicts the responses (i.e., Equations 7.17, 7.19, 7.22, and 7.24) derived above. Notice how the undamped responses oscillate with the same period, T_p and amplitude, $x(0)$, continuously. Generally, the *period of oscillation* can be determined from the undamped natural frequency of the system and the damping ratio or simply from the damped natural frequency,

$$T_p = \frac{2\pi}{\omega_n\sqrt{1 - \zeta^2}} = \frac{2\pi}{\omega_d}. \tag{7.25}$$

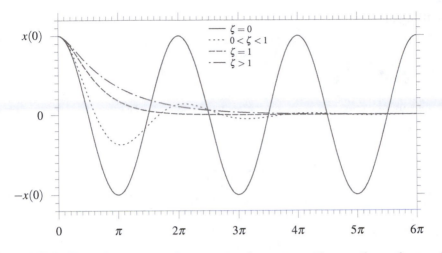

Figure 7.6: *Natural responses of a second-order system with an undamped natural frequency of $\omega_n = 1$ rad/s.*

It should not be confused with the time constant, T, of the exponential decay that bounds the underdamped response. Figure 7.7 illustrates the difference between the period of oscillation, T_p, and the time constant, T.

Using trigonometric identities, Equation 7.19 can be rewritten as (Ogata 2004)

$$x(t) = \frac{x(0)}{\sqrt{1-\zeta^2}} e^{-\zeta \omega_n t} \cos\left(\omega_d t - \tan^{-1} \frac{\zeta}{\sqrt{1-\zeta^2}}\right),$$

which is bonded by the exponentially decaying envelope

$$\frac{x(0)}{\sqrt{1-\zeta^2}} e^{-\zeta \omega_n t}$$

that has a time constant of

$$T = \frac{1}{\zeta \omega_n}. \tag{7.26}$$

Recall from Section 7.2.1 that the first-order natural response decays to 1.8% and 0.7% of the initial value within $4T$ and $5T$, respectively. Though herein the system is second-order and not first-order, the underdamped response is attenuated and bounded by an exponentially decaying function like that found in the first-order natural response. Hence, the underdamped responses will attenuate to within 2% of the final value at $4T$. This is referred to as the settling time,

$$t_s = 4T = \frac{4}{\zeta \omega_n} \ (2\% \ \text{criterion}).$$ (7.27)

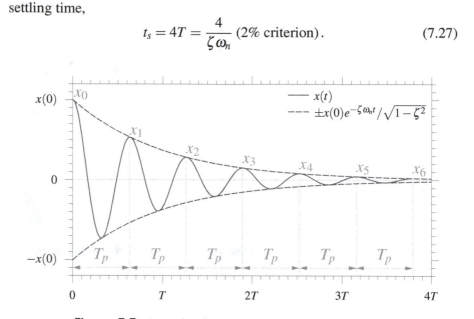

Figure 7.7: *Second-order natural, underdamped response.*

Note that in Figure 7.7 the peeks are labeled x_0 to x_6. Also note that each peak occurs at a time T_p after the previous. In other words, the peaks occur at $T_p, 2T_p, 3T_p$, etc. The peak values are then

$$x_1 = \frac{x(0)}{\sqrt{1 - \zeta^2}} e^{-\zeta \omega_n T_p}$$

$$x_2 = \frac{x(0)}{\sqrt{1 - \zeta^2}} e^{-\zeta \omega_n 2T_p}$$

$$\vdots$$

$$x_n = \frac{x(0)}{\sqrt{1 - \zeta^2}} e^{-\zeta \omega_n n T_p}.$$

The ratio of the first peak to the nth is

$$\frac{x_1}{x_n} = \frac{e^{-\zeta \omega_n T_p}}{e^{-\zeta \omega_n n T_p}} = \frac{1}{e^{-\zeta \omega_n (n-1) T_p}}.$$

Taking the natural logarithm of both sides yields

$$\ln \frac{x_1}{x_n} = \zeta \omega_n (n - 1) T_p,$$

otherwise known as the *logarithmic decrement*. Recall, however, that $T_p = 2\pi/\omega_d$. This can be substituted into the above equation to arrive at

$$\frac{1}{n-1} \ln \frac{x_1}{x_n} = \zeta \omega_n \left(\frac{2\pi}{\omega_d} \right) = \frac{2\pi\zeta}{\sqrt{1-\zeta^2}}$$

which we can solve for ζ,

$$\zeta = \frac{\dfrac{1}{n-1} \ln \dfrac{x_1}{x_n}}{\sqrt{4\pi^2 + \left(\dfrac{1}{n-1} \ln \dfrac{x_1}{x_n} \right)^2}}. \tag{7.28}$$

To utilize the equation, select a peak, use it as the principle peak (i.e., x_1), measure its height, choose a subsequent peak n periods later, and measure its height. That is, x_1 need not be the absolute first peak, but rather the first relative to the nth peak.

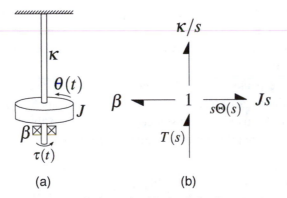

Figure 7.8: *Second-order torsion system.*

Example 7.2

A second-order torsion system is portrayed in Figure 7.8. Assume that the disk used is the same as that from Example 7.1, but the bearing in this system is different. Given the natural response plotted in Figure 7.9, determine the damping coefficient, β, and the shaft rigidity, κ.

Solution. From Example 7.1 we know that the rotational inertia J is 0.0104 kg-m^2. Given the impedance bond graph in Figure 7.8 (b) it can be shown that the transfer function relating the angular displacement, $\theta(t)$, to the input torque, $\tau(t)$, is

$$\frac{\Theta(s)}{T(s)} = \frac{1}{Js^2 + \beta s + \kappa} = \frac{1}{J}\frac{1}{s^2 + \dfrac{\beta}{J}s + \dfrac{\kappa}{J}}.$$

By comparing the denominator to the characteristic equation for a proto-typical second-order system, one can surmise that

$$\omega_n^2 = \frac{\kappa}{J}$$

and

$$2\zeta\omega_n = \frac{\beta}{J}.$$

We can solve the above equations for the unknowns, β and κ, in terms of ζ and ω_n:

$$\kappa = J\omega_n^2$$

and

$$\beta = 2J\zeta\omega_n.$$

The natural frequency and damping ration can be ascertained from the response plot.

Using the peaks provided in the plot, we can calculate the damping ratio from the logarithmic decrement equation (Equation 7.28). Any pair of peaks can be used. For instance, let us use the third peak, $\theta(2T_p) = 0.792$, as the principle and fourth subsequent peak, $\theta(6T_p) = 0.559$ rad/s,

$$\zeta = \frac{\dfrac{1}{4-1}\ln\dfrac{0.792}{0.559}}{\sqrt{4\pi^2 + \left(\dfrac{1}{4-1}\ln\dfrac{0.792}{0.559}\right)^2}} = 0.0185.$$

You can confirm the same answer would be garnered using the third and fourth peaks or any other pair.

Recall that the undamped natural frequency can be calculated in terms of the damped frequency and damping ratio using Equation 7.20. Consequently, we need to determine the damped frequency of oscillation. The damped frequency is calculated using the period of oscillation,

$$\omega_d = \frac{2\pi}{T_p} = \frac{2\pi}{0.2422 \text{ seconds}} = 25.937 \text{ rad/s}.$$

Knowing the damped frequency and damping ratio, the natural frequency
is

$$\omega_n = \frac{\omega_d}{\sqrt{1-\zeta^2}} = \frac{25.937 \text{ rad/s}}{\sqrt{1-0.0185^2}} = 25.942 \text{ rad/s}.$$

Now we can calculate the rigidity and damping constant,

$$\kappa = (0.0104 \text{ kg-m}^2)(25.9 \text{ rad/s})^2 = 7 \text{ N-m/rad}$$

and

$$\beta = 2(0.0104 \text{ kg-m}^2)(0.0185)(25.942 \text{ rad/s}) = 0.01 \text{ N-m-s/rad}.$$

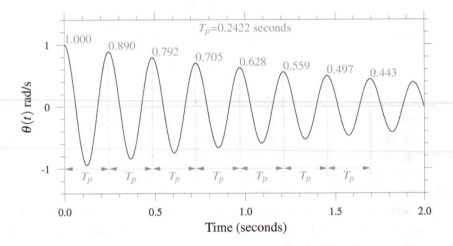

Figure 7.9: *Natural response of torsion system depicted in Figure 7.8.*

7.3.2 The Natural Frequency, Damping Ratio, and Pole Placement

In deriving the natural responses we checked the roots of the characteristic
equation, which is also the polynomial in the denominator of the transfer
function for the system

$$\frac{x(s)}{u(s)} = \frac{1}{s^2 + 2\zeta \omega_n s + \omega_n^2}.$$

Thus the roots of the characteristic equation are the poles of the transfer function. The roots for the four cases previously discussed (recall Equations 7.16, 7.18, 7.21, and 7.23) are of the following forms:

> ▷ Undamped ($\zeta = 0$): purely imaginary conjugate roots

$$s_{1,2} = \pm j\omega$$

> ▷ Underdamped ($0 < \zeta < 1$): complex conjugate roots with negative real parts

$$s_{1,2} = -\zeta\omega_n \pm j\omega_n\sqrt{1-\zeta^2}$$

> ▷ Critically damped ($\zeta = 1$): repeated negative real roots

$$s_{1,2} = -\zeta\omega_n$$

> ▷ Overdamped ($\zeta > 1$): distinct negative real roots

$$s_{1,2} = -\omega_n(\zeta \mp \sqrt{\zeta^2-1})$$

The characteristic roots (or poles) are generally complex and, as discussed in Chapter 5, can be treated like two-dimensional vectors that can be plotted on the real-imaginary plane as illustrated in Figure 7.10. Figure 7.10 (a) illustrates how the poles move in the real-imaginary plane as the damping varies from undamped to overdamped. The poles begin on the imaginary axis when the system is undamped. They travel around a semicircle in the left half of the plane as the damping is increased until they intersect at the real axis where critical damping is reached. They proceed to split and become distinct as they become overdamped.

Figure 7.10 (b) shows the relationships between the pole placement and the damping ratio, undamped natural frequency, and damped natural frequency of the second-order system. The pole placement in the real-imaginary plane is geometrically related to these characteristics. If a vector is drawn from the origin to one of the poles, the length of that vector is equal to the undamped natural frequency,

$$\omega_n = \sqrt{(\zeta\omega_n)^2 + \omega_d^2} = \sqrt{(\zeta\omega_n^2) + \omega_n^2(1-\zeta^2)}. \tag{7.29}$$

The projection of that vector onto the real axis is $\zeta\omega_n$, and the cosine of the angle that vector makes relative to the negative real axis is equal to the damping ratio,

$$\zeta = \cos\theta. \tag{7.30}$$

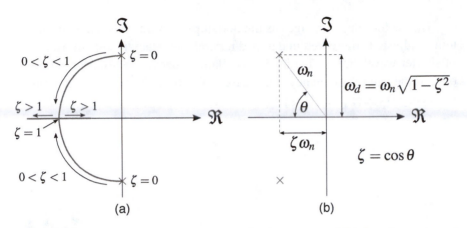

Figure 7.10: *(a) Damping as it relates to pole placement and (b) damping ratio and natural frequency as they relate to pole placement.*

The projection of the vector onto the imaginary axis is the damped natural frequency, ω_d. As the damping ratio is varied from 0 and 1, the vector rotates from vertical to horizontal, and the damped frequency of oscillation, ω_d, decreases resulting in a longer period of oscillation, T_p. Figure 7.11 illustrates how pole placement and resultant response vary as the damping ratio is changed.

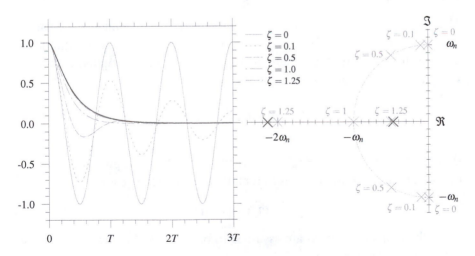

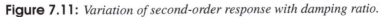

Figure 7.11: *Variation of second-order response with damping ratio.*

The vector length changes as the undamped natural frequency of the system is adjusted. Increased undamped natural frequency results in oscillation of shorter wavelength. That is, the oscillations occur over shorter periods of time as the natural frequency is increased. The oscillations take longer as the frequency is decreased. This phenomenon is demonstrated in Figure 7.12.

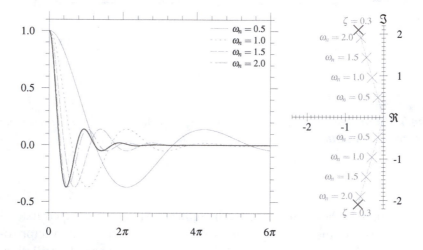

Figure 7.12: *Variation of second-order response with natural frequency.*

7.3.3 The Impulse Response

The impulse responses can be derived in much the same manner as the natural responses. However, unlike the natural responses, the initial conditions are assumed to be zero for the impulse response. To derive the impulse responses, we return to Equation 7.10 and use a generic impulse of magnitude A as the input $u(t)$,

$$\ddot{x} + 2\zeta\omega_n\dot{x} + \omega_n^2 x = A\tilde{\delta}(t). \tag{7.31}$$

Undamped ($\zeta = 0$).   For a generic impulse and a damping ratio of zero, the prototypical second-order differential equation becomes

$$\ddot{x} + \omega_n^2 x = A\tilde{\delta}(t).$$

Since the initial conditions are assumed to be zero, the Laplace transform of the differential equation is

$$s^2 x(s) + \omega_n^2 x(s) = A,$$

which can be solved for $x(s)$,

$$x(s) = \frac{A}{s^2 + \omega_n^2} = \frac{A}{\omega_n} \frac{\omega_n}{s^2 + \omega_n^2}.$$

The inverse Laplace transform of $x(s)$ yields

$$x(t) = \frac{A}{\omega_n} \sin \omega_n t. \tag{7.32}$$

Note that the resultant response is sinusoidal with an amplitude of A/ω_n. Recall that the sine function is zero at initial time as is expected of the initial condition for the impulse response of a second-order system.

Underdamped ($0 < \zeta < 1$). Given a non-zero damping ratio, the differential equation is generally

$$\ddot{x} + 2\zeta \omega_n \dot{x} + \omega_n^2 x = A\tilde{\delta}(t).$$

Applying Laplace transforms,

$$s^2 x(s) + 2\zeta \omega_n x(s) + \omega_n^2 x(s) = A,$$

one can solve for $x(s)$,

$$
\begin{aligned}
x(s) &= \frac{A}{s^2 + 2\zeta \omega_n s + \omega_n^2} \\
&= \frac{A}{\omega_n \sqrt{1 - \zeta^2}} \frac{\omega_n \sqrt{1 - \zeta^2}}{(s + \zeta \omega_n)^2 + (\omega_n \sqrt{1 - \zeta^2})^2} \\
&= \frac{A}{\omega_d} \frac{\omega_d}{(s + \zeta \omega_n)^2 + \omega_d^2},
\end{aligned}
$$

and use the inverse Laplace transform to derive the time domain response,

$$x(t) = \frac{A}{\omega_d} e^{-\zeta \omega_n t} \sin \omega_d t. \tag{7.33}$$

The response is composed of a sinusoid and a decaying exponential. It will oscillate due to the sine term but will decay with time due to the exponential term.

Critically damped ($\zeta = 1$).   When damping ratio is unity, the differential equation simplifies,

$$\ddot{x} + 2\omega_n \dot{x} + \omega_n x = A\delta(t),$$

and results in a Laplace transform,

$$s^2 x(s) + 2\omega_n x(s) + \omega_n^2 x(s) = A,$$

which has a solution of the form

$$x(s) = \frac{A}{s^2 + 2\omega_n s + \omega_n^2}$$

$$= A\frac{1}{(s + \omega_n)^2}.$$

The denominator has repeated real roots, and the resultant response is

$$x(t) = Ate^{-\omega_n t}. \tag{7.34}$$

Notice that the function is the multiplication of time with a decaying exponential. Initially, due to the time term, t, the response will rise, but as time progresses, the exponential term dominates and causes the response to decrease and decay to zero.

Overdamped ($\zeta > 0$).   The transfer function for the overdamped response remains the same as that for the underdamped response, but because there exists two real negative roots for this case, the partial fraction expansion differs from the underdamped or critically damped cases. It can be shown that

$$x(s) = \frac{A}{s^2 + 2\zeta\omega_n s + \omega_n^2}$$

$$= A\frac{1}{\left[s + \omega_n\left(\zeta + \sqrt{\zeta^2 - 1}\right)\right]\left[s + \omega_n\left(\zeta - \sqrt{\zeta^2 - 1}\right)\right]},$$

which has a partial fraction expansion of

$$x(s) = \frac{A}{2\omega_n\sqrt{\zeta^2 - 1}}\left[-\frac{1}{s + \omega_n\left(\zeta + \sqrt{\zeta^2 - 1}\right)} + \frac{1}{s + \omega_n\left(\zeta - \sqrt{\zeta^2 - 1}\right)}\right].$$

Applying the inverse Laplace transform, the response is

$$x(t) = \frac{A}{2\omega_n\sqrt{\zeta^2 - 1}}\left[-e^{-\omega_n \zeta_1 t} + e^{-\omega_n \zeta_2 t}\right] \tag{7.35}$$

where

$$\zeta_1 = \zeta + \sqrt{\zeta^2 - 1} \tag{7.36}$$

and

$$\zeta_2 = \zeta - \sqrt{\zeta^2 - 1}. \tag{7.37}$$

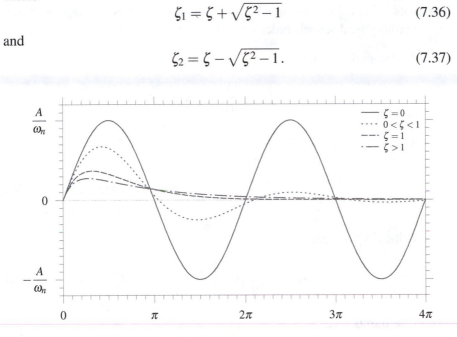

Figure 7.13: *Second-order impulse responses given $\omega_n = 1$ rad/s.*

Figure 7.13 depicts four impulse responses for the prototypical second-order system with an undamped natural frequency of $\omega_n = 1$ rad/s and includes an undamped response, an underdamped response, a critically damped response, and an overdamped response. Unlike the first-order system, the impulse response of the second-order system is not similar to the natural response. Whereas the impulse response of the first-order system jumps instantaneously to some initial value, the second-order impulse response starts at zero. Except for the undamped case, the responses quickly reach their maximum value and then attenuate or decay back to zero. The underdamped response settles to within 2% of zero (the steady-state or final value) beyond four times the time constant, $4T$. As the damping ratio is increased, the peak amplitude decreases. The relations between pole placement and response characteristics including natural frequency and damping ratio still hold true.

7.3.4   The Step Response

It has been previously mentioned that the impulse, step, and ramp responses are related through integration. Therefore, the step responses can be derived by integrating the impulse responses over the range 0 to t. By integrating Equations 7.32, 7.33, 7.34, and 7.35 we obtain the following step responses for the prototypical second-order system:

▷ Undamped ($\zeta = 0$)

$$x(t) = \frac{A}{\omega_n^2}(1 - \cos \omega_n t) \tag{7.38}$$

▷ Underdamped ($0 < \zeta < 1$)

$$x(t) = \frac{A}{\omega_n^2}\left[1 - e^{-\zeta \omega_n t}\left(\frac{\zeta}{\sqrt{1 - \zeta^2}}\sin \omega_d t + \cos \omega_d t\right)\right] \tag{7.39}$$

▷ Critically Damped ($\zeta = 1$)

$$x(t) = \frac{A}{\omega_n^2}\left[1 - e^{-\omega_n t} - \omega_n e^{-\omega_n t}t\right] \tag{7.40}$$

▷ Overdamped ($\zeta > 1$)

$$x(t) = \frac{A}{2\omega_n^2}\left[\frac{\zeta}{\sqrt{\zeta^2 - 1}}\left(e^{-\zeta_1 \omega_n t} - e^{-\zeta_2 \omega_n t}\right) - e^{-\zeta_1 \omega_n t} - e^{-\zeta_2 \omega_n t} + 2\right] \tag{7.41}$$

(ζ_1 and ζ_2 are previously defined in Equations 7.36 and 7.37.)

As Figure 7.14 shows, at steady-state the step response approaches a constant value. Like the natural and impulse responses, the underdamped step response settles to within 2% of the final value beyond four times the time constant, $4T$, the difference being that the steady-state value is not zero. The undamped response oscillates with a mean value equal to the steady-state value of the underdamped, critically damped, and overdamped responses. The underdamped response overshoots the steady-state values and oscillates about that value except that the amplitude is attenuated with each successive oscillation. The critically damped and overdamped responses do not overshoot the final value; rather they asymptotically approach steady-state.

The step response, as we will see in Chapter 9, is often used to assess how well a system responds to a set point situation or control. A set point (or position control) problem is one where it is desired to transition a system from some initial condition to some final, fixed condition at steady-state. Like the step function, the input to a set point problem instantaneously goes

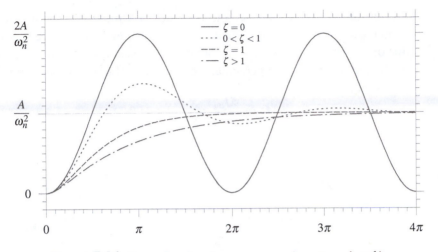

Figure 7.14: *Second-order step responses given $\omega_n = 1$ rad/s.*

from an initial value to a constant final value. Thus, a number of properties portrayed in Figure 7.15 are defined to characterize the performance of the step response. They include the following (Ogata 2004):

1. **Delay time,** t_d: The delay time is the time required to initially reach half the steady-state value.

2. **Rise time,** t_r: For underdamped second-order step responses, the rise time is the time required to rise from 10% to 90% of the steady-state value.

3. **Peak time,** t_p: The peak time is the time it takes to reach the maximum value or peak of the step response.

4. **Settling time,** t_s: As has been previously mentioned, the setting time is the time it takes the step response to settle to within 2% of the steady-state value. In some cases, a 5% criterion rather than the 2% criterion is utilized. Based on the exponentially decaying envelope of the under-damped response, the oscillation attenuates to within 5% of the final value at three times the time constant, $3T$. Therefore, the alternative settling time is

$$t_s = 3T = \frac{3}{\zeta \omega_n} \ (5\% \text{ criteria}). \tag{7.42}$$

5. **Maximum (percent) overshoot, M_p:** The maximum overshoot is the amount by which the initial peak exceeds the steady-state value. It is the difference between the peak value and the steady-state value and is often characterized as a percentage difference. The maximum percent overshoot is

$$M_p = \frac{x(t_p) - x(\infty)}{A/\omega_n^2} \times 100. \tag{7.43}$$

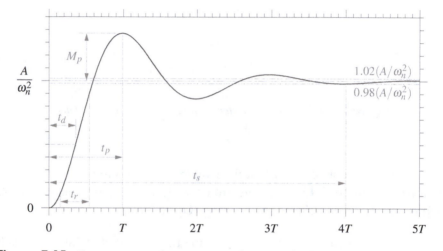

Figure 7.15: *Step response characteristics for an underdamped second-order system.*

7.3.5 The Ramp Response

Integration is again utilized to derive the ramp responses. Integration of Equations 7.38, 7.39, 7.40, and 7.41 yields the following ramp responses:

▷ Undamped ($\zeta = 0$)

$$x(t) = \frac{A}{\omega_n^3} (\omega_n t - \sin \omega_n t) \tag{7.44}$$

▷ Underdamped ($0 < \zeta < 1$)

$$x(t) = \frac{A}{\omega_n^3} \left\{ \frac{e^{-\zeta \omega_n t}}{\omega_n} \left[\frac{\zeta}{\sqrt{1-\zeta^2}} (\omega_d \cos \omega_d t + \zeta \omega_n \sin \omega_d t) \right. \right.$$
$$\left. \left. + \zeta \omega_n \cos \omega_d t - \omega_d \sin \omega_d t \right] + \omega_n t - 2\zeta \right\} \tag{7.45}$$

▷ Critically Damped ($\zeta = 1$)

$$x(t) = \frac{A}{\omega_n^3}\left[\omega_n t + \omega_n e^{-\omega_n t} t + 2e^{-\omega_n t} - 2\right] \tag{7.46}$$

▷ Overdamped ($\zeta > 1$)

$$x(t) = \frac{A}{2\omega_n^3}\left[\frac{\zeta}{\sqrt{\zeta^2-1}}\left(-\zeta_2 e^{-\zeta_1 \omega_n t} + \zeta_1 e^{-\zeta_2 \omega_n t} - 2\sqrt{\zeta^2-1}\right)\right.$$
$$\left.+ \zeta_2 e^{-\zeta_1 \omega_n t} + \zeta_1 e^{-\zeta_2 \omega_n t} - 2\zeta + 2\omega_n t\right] \tag{7.47}$$

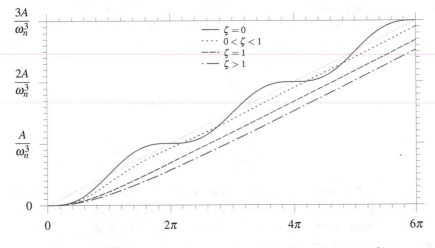

Figure 7.16: *Second-order ramp responses given $\omega_n = 1$ rad/s.*

The ramp responses for the prototypical second-order system are plotted in Figure 7.16. The undamped ($\zeta = 0$) response oscillates about the ramp input displayed in gray. The underdamped response shows some oscillation but settles and follows the input in parallel. The critically damped and over-damped responses are quite similar to the first-order ramp response (Equation 7.8 and Figure 7.5).

7.4   Transient Responses of Higher-Order Systems

Consider an nth-order single input dynamic system represented by a transfer function of the form

$$\frac{\mathsf{x}(s)}{\mathsf{u}(s)} = \frac{\hat{b}_0 s^m + \hat{b}_1 s^{m-1} + \cdots + \hat{b}_{n-1} s + \hat{b}_n}{s^n + \hat{a}_1 s^{n-1} + \cdots + \hat{a}_{n-1} s + \hat{a}_n} \tag{7.48}$$

where $n > m$. It can be shown that the partial fraction expansion of the *step* response of such a system will be of the general form (Ogata 2004)

$$\mathsf{x}(s) = \frac{a}{s} + \sum_{j=1}^{q} \frac{a_j}{s + p_j} + \sum_{k=1}^{r} \frac{b_k(s + \zeta_k \omega_k) + c_k \omega_k \sqrt{1 - \zeta_k^2}}{s^2 + 2\zeta_k \omega_k s + \omega_k^2}. \tag{7.49}$$

Note that generally, the partial fraction expansion is composed of first- and second-order terms like those discussed in the previous sections. That is, a higher-order response can generally be decomposed through partial fraction expansions into the summation of first- and second-order transfer functions. The resultant time domain response can generally contain a step term, decaying exponentials, and decaying sinusoids:

$$
\begin{aligned}
x(t) = a + \sum_{j=1}^{q} a_j e^{-p_j t} &+ \sum_{k=1}^{r} b_k e^{-\zeta_k \omega_k t} \cos \omega_k \sqrt{1 - \zeta_k^2}\, t \\
&+ \sum_{k=1}^{r} c_k e^{-\zeta_k \omega_k t} \sin \omega_k \sqrt{1 - \zeta_k^2}\, t.
\end{aligned}
\tag{7.50}
$$

The step and exponential terms are typical of first-order responses whereas the decaying sinusoids are representative of second-order responses. Hence, by understanding first- and second-order responses, we also garner intuition regarding higher-order responses.

Example 7.3
A two-disk torsion system is depicted in Figure 7.17, and the system parameters are provided in Table 7.1. Determine the lower-order components that contribute to the overall responses $\theta_1(t)$ and $\theta_2(t)$. Plot and compare the individual contributions to the overall responses.

Solution. It can be shown that transfer functions relating the displacements $\theta_1(t)$ and $\theta_2(t)$ to the input torque $\tau(t)$ are

$$\frac{\Theta_1(s)}{T(s)} = \frac{J_2 s^2 + \beta_2 s + \kappa_2 + \kappa_1}{a_0 s^4 + a_1 s^3 + a_2 s^2 + a_3 s + a_4}$$

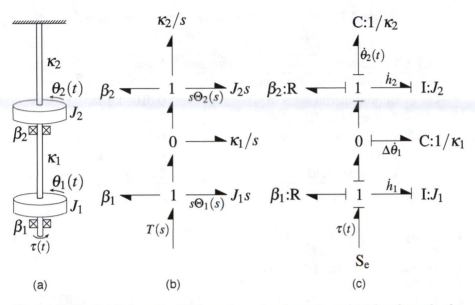

Figure 7.17: (a) Schematic, (b) impedance bond graph, and (c) bond graph of a two-disk torsion system.

and

$$\frac{\Theta_1(s)}{T(s)} = \frac{\kappa_1}{a_0 s^4 + a_1 s^3 + a_2 s^2 + a_3 s + a_4}.$$

where

$$a_0 = J_1 J_2,$$
$$a_1 = J_1 \beta_2 + J_2 \beta_1,$$
$$a_2 = J_1 \kappa_2 + J_1 \kappa_1 + J_2 \kappa_1 + \beta_1 \beta_2,$$
$$a_3 = \beta_1 \kappa_2 + \beta_1 \kappa_1 + \beta_2 \kappa_1, \text{ and}$$
$$a_4 = \kappa_2 \kappa_1.$$

We can use MATLAB to show that the poles of both transfer functions are

$$p_{1,2} = -0.7550 \pm 38.0560i \text{ and}$$
$$p_{3,4} = -0.6872 \pm 9.4265i.$$

```
>> J1 = 0.0104;
>> J2 = J1;
>> K1 = 7;
>> K2 = 2;
```

Table 7.1: *System parameters for two-disk torsion system.*

Parameter	Variable	Value
Bottom Bearing Constant	β_1	0.01 N-m-s/rad
Bottom Inertia	J_1	0.0104 kg-m^2
Bottom Shaft Rigidity	κ_1	7 N-m/rad
Top Bearing Constant	β_2	0.02 N-m-s/rad
Top Inertia	J_2	0.0104 kg-m^2
Top Shaft Rigidity	κ_2	2 N-m/rad

```
>> B1 = 0.01;
>> B2 = 0.02;
>> num1 = [J2 B2 (K1+K2)];
>> num2 = K1;
>> den = [J1*J2 (J1*B2+J2*B1) (J1*(K1+K2)+J2*K1+B1*B2) ...
(B1*(K1+K2)+B2*K1) K1*K2];
>> G1 = tf(num1,den);
>> G2 = tf(num2,den);
>> pole(G1)

ans =

   -0.7550 +38.0560i
   -0.7550 -38.0560i
   -0.6872 + 9.4265i
   -0.6872 - 9.4265i
```

Assume we wish to simulate the impulse response of the system (i.e., $\tau(t) = \delta(t)$ and $\mathcal{L}[\tau(t)] = T(s) = 1$). Because the system poles are two pairs of complex conjugates, the fourth-order system can be decomposed into the sum of two second-order transfer functions,

$$\Theta_1(s) = \frac{-0.03333s + 41.26}{s^2 + 1.51s + 1449} + \frac{0.03333s + 54.88}{s^2 + 1.374s + 89.33} = \Theta_{1a}(s) + \Theta_{1b}(s)$$

and

$$\Theta_2(s) = \frac{0.004746s - 47.6}{s^2 + 1.51s + 1449} + \frac{-0.004746s + 47.6}{s^2 + 1.374s + 89.33} = \Theta_{2a}(s) + \Theta_{2b}(s),$$

each of which will inverse Laplace transform into decaying sinusoids of distinct amplitude and frequency,

$$\theta_1(t) = \theta_{1a}(t) + \theta_{1b}(t)$$
$$= -0.03333\,e^{-0.755t}\,(\cos 38.05t - 32.54\sin 38.05t)$$
$$+0.03333\,e^{-0.687t}\,(\cos 9.427t + 175\sin 9.427t)$$

and

$$\theta_2(t) = \theta_{2a}(t) + \theta_{2b}(t)$$
$$= 0.0047\,e^{-0.755t}\,(\cos 38.05t - 263.6\sin 38.05t)$$
$$-0.0047\,e^{-0.687t}\,(\cos 9.427t - 1064\sin 9.427t)\,.$$

Figure 7.18 shows the components $\theta_{1a}(t)$, $\theta_{1b}(t)$, $\theta_{2a}(t)$, and $\theta_{2b}(t)$ and the overall responses $\theta_1(t)$ and $\theta_2(t)$. Note how the small-amplitude, high-frequency components, $\theta_{1a}(t)$ and $\theta_{2a}(t)$, sum with the large-amplitude, low-frequency pieces, $\theta_{1b}(t)$ and $\theta_{2b}(t)$, to generate the overall response.

7.5   An Introduction to Pole-Zero Analysis

In Chapter 5, we discovered that the transfer functions can be factored to identify the system zeros (roots of the numerator) and poles (roots of the denominator),

$$G(s) = \frac{K(s+z_1)(s+z_2)\ldots(s+z_m)}{(s+p_1)(s+p_2)\ldots(s+p_n)}\,.$$

Also, as highlighted in Section 7.4, a transfer function can generally be decomposed into first- and second-order terms. The second-order terms result from purely imaginary or complex poles. As was shown in Section 7.3.2, the position of the poles in the real-imaginary plane is related to properties of second-order responses such as the damping ratio, undamped natural frequency, and damped natural frequency. This basic concept can be extrapolated to higher-order systems. In higher order systems, however, there are generally more than a few pairs of poles.

Moreover, higher-order systems can also have zeros. Zeros can alter, and even eliminate, the effect of poles that are in close proximity. This can be readily illustrated through a simple example. Imagine a system with two zeros and three poles. Further, imagine that one pole and one zero are of

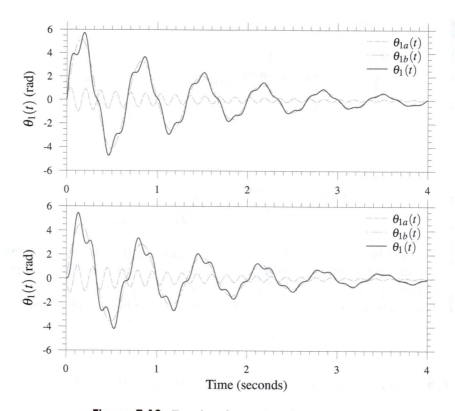

Figure 7.18: *Fourth-order torsion plant response.*

similar value (i.e., $z_2 \approx p_2$); the resultant system is approximately

$$G(s) = \frac{(s+z_1)(s+z_2)}{(s+p_1)(s+p_2)(s+p_3)} \approx \frac{(s+z_1)}{(s+p_1)(s+p_3)}.$$

If the two were exactly equal (i.e., $z_2 = p_2$) the pole and zero would cancel each other out,

$$\frac{s+z_2}{s+p_2} = 1,$$

effectively reducing the overall order of the system. The closer the proximity of the zero to the pole, the more the zero will negate the effects of that pole on the overall response.

Though roots may generally lie in the right-half of the complex plane, for a large class of physical, stable systems, the roots are strictly on or to the left of the imaginary axis. Higher-order systems are composed of first- and second-order parts, each of which have their own poles. When more than one pole exists in the left-half plane, the pole or pole pair nearest the imaginary axis dominates. The extent to which it dominates depends on the placement of the other poles and their proximity to zeros. This will be discussed further in Chapter 9, but for now we introduce the concept and explain why certain poles will dominate.

First-order terms will result in contributions to the time domain response of the form

$$Ae^{-at}.$$

Recall that the roots of the underdamped second-order system are generally

$$s_{1,2} = -\zeta\omega_n \pm j\omega_n\sqrt{1-\zeta^2}$$

and that the underdamped responses (Equations 7.19, 7.33, and 7.39) all have decaying exponential terms of the form

$$e^{-\zeta\omega_n t}.$$

Hence, except for oscillatory or undamped terms, stable responses will attenuate. For first-order terms, the greater the value of a (or the farther to the left the pole) the faster the term decays to zero in the time domain. Underdamped second-order terms attenuate to within 2% of steady-state at $4T = 4/(\zeta\omega_n)$, where $\zeta\omega_n$ is the real part of the generally complex root and the horizontal distance of the pole from the imaginary axis (as previously illustrated in Figure 7.10). Since the time constant, T, is inversely proportional to $\zeta\omega_n$, and $\zeta\omega_n$ is the horizontal distance from the imaginary axis, the closer the pole(s) is (are) to the imaginary axis, the longer it takes for the response to attenuate. The pole placement is affected by the natural frequency, ω_n. Therefore, generally poles with lower natural frequency will have a larger time constant and thus will take longer to dampen. Similarly, for an underdamped pair of complex poles, the poles move farther from the imaginary axis as the damping ratio increases and the response dampens more quickly. The unpaired poles closest to the imaginary axis are referred to as the dominant poles of the system because they take the longest to attenuate. The overall response of the system will tend to have characteristics that are closest to those of the dominant poles.

To summarize, given the affects of zeros on nearby poles, and the decaying nature of real and complex poles on the time domain response, the overall dominant pole(s) will be the pole(s) closest to the imaginary axis and not in close proximity to a zero. Any poles that are five times or more farther from the imaginary axis than the dominant poles will have negligible impact on the overall response (Ogata 2002).

7.6 Pole-Zero Analysis Using MATLAB

MATLAB includes a number of functions that facilitate pole-zero analysis. The functions include a command to determine the natural frequencies and damping ratios associated with the poles of a transfer function.

```
>> damp(sys)
```

The function damp(sys) will calculate the eigenvalues, damping ratios, and natural frequencies of an LTI object sys defined using ss, tf, or zpk.

The pzmap function is utilized to visualize a plot of the poles and zeros in the complex plane. It generates a *pole-zero map*, and the sgrid command can be used to add a grid with lines of constant damping and constant natural frequency.

```
>> pzmap(sys);
>> sgrid
```

Given the matrix **A** of a state-space model as an array A in MATLAB, the characteristic equation (or polynomial) can be calculated with the poly function.

```
>> poly(A)
```

The following example demonstrates some of the functions summarized in Table 7.2.

Table 7.2: *MATLAB commands for pole-zero analysis.*

Function	Description
damp(sys)	computes natural frequencies and damping ratios
pzmap(sys)	plots pole-zero map of dynamic systems
sgrid	generates s-plane grid lines for a pole-zero map
poly(A)	computes characteristic polynomial of matrix

Example 7.4

Recall the mass-spring-damper system from Examples 3.1, 4.2, 4.7, and 6.5 with transfer functions (from Equations 6.19 and 6.20),

$$\frac{x_1(s)}{F(s)} = \frac{k}{m^2 s^4 + mbs^3 + 3mks^2 + bks + k^2}$$

and

$$\frac{x_2(s)}{F(s)} = \frac{ms^2 + bs + 2k}{m^2s^4 + mbs^3 + 3mks^2 + bks + k^2}.$$

The second transfer function, $x_2(s)/F(s)$, has two zeros and two poles. Determine which pole or pole pair is dominant, and explain why.

Solution. Recall that the mass, damping coefficient, and spring constant are 10 kg, 20 N-s/m, and 60 N/m, respectively, and that the zeros are

$$z_{1,2} = -1.0000 \pm 3.3166j$$

and the poles are

$$p_{1,2} = -0.7106 \pm 3.7722j \quad \text{and}$$
$$p_{3,4} = -0.2894 \pm 1.5361j.$$

The poles and zeros can be numerically determined using the MATLAB zero and pole commands and plotted using the pzmap command.

```
>> m = 10; b = 20; k = 60;
>> sys2 = tf([m b 2*k],[m^2 m*b 3*m*k b*k k^2]);
>> Z2 = zero(sys2)

Z2 =

  -1.0000 + 3.3166i
  -1.0000 - 3.3166i

>> P2 = pole(sys2)

P2 =

  -0.7106 + 3.7722i
  -0.7106 - 3.7722i
  -0.2894 + 1.5361i
  -0.2894 - 1.5361i

>> pzmap(sys2); sgrid
```

The pole-zero map and impulse response for this system are plotted in Figure 7.19. The zeros are noticeably closer to poles $p_{1,2}$. Moreover, poles $p_{3,4}$ are closer to the imaginary axis. The natural frequencies associated with each set of poles can be attained from the length of the vector drawn from the origin to the pole. For poles $p_{1,2}$ the natural frequency is

$$\omega_{n1} = \sqrt{(-0.7106)^2 + (3.7722)^2} = 3.8385 \text{ rad/s},$$

and for poles $p_{3,4}$ the frequency is

$$\omega_{n2} = \sqrt{(-0.2894)^2 + (1.5361)^2} = 1.5631 \text{ rad/s}.$$

Recall that the real part of the pole is equal to $-\zeta \omega_n$. Hence, given the natural frequencies, ω_{n1} and ω_{n2}, the damping ratio for poles $p_{1,2}$ is

$$\zeta_1 = \frac{0.7106}{3.8385} = 0.1851$$

and for poles $p_{3,4}$ is

$$\zeta_2 = \frac{0.2894}{1.5631} = 0.1851.$$

The absolute value of the real part, $\zeta \omega_n$, can also be used to determine the settling time. The contribution in the time domain due to poles $p_{1,2}$ should attenuate in

$$t_{s1} = \frac{4}{\zeta_1 \omega_{n1}} = \frac{4}{0.7106 \text{ rad/s}} = 5.6289 \text{ seconds}.$$

For poles $p_{3,4}$, the settling time is

$$t_{s2} = \frac{4}{\zeta_2 \omega_{n2}} = \frac{4}{0.2894 \text{ rad/s}} = 13.8228 \text{ seconds}.$$

From the impulse response, it is apparent that the settling time is indeed closer to t_{s2}. As further evidence of the dominance of poles $p_{3,4}$ consider the period of oscillation denoted in the figure, $T_p = 3.988$ seconds. Based on the damped frequencies of oscillation for each of the pole pairs, the respective periods for the first and second pairs are

$$T_{p1} = \frac{2\pi}{\omega_{d1}} = \frac{2\pi}{\omega_{n1}\sqrt{1 - \zeta_1^2}} = \frac{2\pi}{(3.8385 \text{ rad/s})\sqrt{1 - 0.1851^2}} = 1.6657 \text{ s}$$

and

$$T_{p2} = \frac{2\pi}{\omega_{d2}} = \frac{2\pi}{\omega_{n2}\sqrt{1 - \zeta_2^2}} = \frac{2\pi}{(1.5631 \text{ rad/s})\sqrt{1 - 0.1851^2}} = 4.0904 \text{ s}.$$

The resultant period of oscillation is also closer to that of the second set of poles.

If, in fact, the zeros, $z_{1,2}$ illustrated in Figure 7.19, pair with the poles $p_{1,2}$, poles $p_{3,4}$ will dominate the response. Regardless, poles $p_{3,4}$ are closest to the imaginary axis. Note that despite being a fourth-order system, the impulse response, which is also plotted in Figure 7.19, looks very much like that of a second-order system. Much of this information can be readily garnered using the MATLAB command damp. The impulse response is also readily attained using the impulse commands.

```
>> damp(sys2)

        Eigenvalue              Damping      Frequency

  -2.89e-01 + 1.54e+00i        1.85e-01      1.56e+00
  -2.89e-01 - 1.54e+00i        1.85e-01      1.56e+00
  -7.11e-01 + 3.77e+00i        1.85e-01      3.84e+00
  -7.11e-01 - 3.77e+00i        1.85e-01      3.84e+00

(Frequencies expressed in rad/seconds)

>> pzmap(sys2)
>> sgrid(0:0.1:1,1:4)
>> impulse(sys2)
```

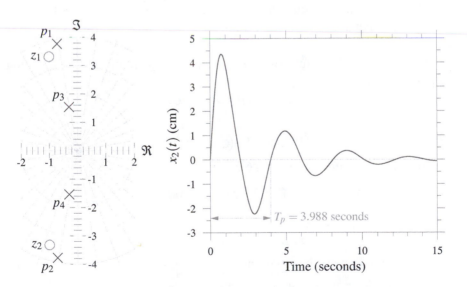

Figure 7.19: *Pole-zero map and impulse response for* $x_2(s)/F(s)$.

7.7 Pole-Zero Analysis in the State Space

In Section 6.2 we discovered that the transfer function could be derived from the state-space representation. It should then come as no surprise that the poles and zeros can be derived from the state-space model directly. Recall Equation 6.4; the single-input-single-output (SISO) version is provided here for convenience:

$$\frac{y(s)}{u(s)} = G(s) = \mathbf{C}(s\mathbf{I} - \mathbf{A})^{-1}\mathbf{B} + D.$$

The poles of the transfer functions $\mathbf{G}(s)$ are the roots of the denominators. That is, they are the values for which the denominator is zero and the resulting transfer function is infinite:

$$G(s) \to \infty \text{ as } s \to p \text{ or } G(p) = \infty.$$

In matrix equations, this condition would be the analog of dividing a scalar by zero. This occurs when

$$\det(s\mathbf{I} - \mathbf{A})|_{s=p} = \det(p\mathbf{I} - \mathbf{A}) = 0. \tag{7.51}$$

The equation above should look familiar. It is the eigenvalue problem (Equation 4.18) previously discussed in Section 4.7. It is also referred to as the characteristic equation. Hence, the values $s = p$ are the eigenvalues of matrix $\mathbf{A}$.

The zeros are the values of s for which the numerator and the resulting output are zero, i.e.,

$$y(z) = G(z)u(z) = 0.$$

Thus when $s = z$

$$\mathbf{C}(z\mathbf{I} - \mathbf{A})^{-1}\mathbf{B}u(z) + Du(z) = 0,$$

which when considering Equation 6.3 yields

$$\mathbf{C}x(z) + Du(z) = 0 \tag{7.52}$$

where

$$x(z) = (z\mathbf{I} - \mathbf{A})^{-1}\mathbf{B}u(z).$$

The above equation can be rewritten

$$(z\mathbf{I} - \mathbf{A})x(z) = (z\mathbf{I} - \mathbf{A})(z\mathbf{I} - \mathbf{A})^{-1}\mathbf{B}u(z)$$
$$= \mathbf{B}u(z)$$
$$(z\mathbf{I} - \mathbf{A})x(z) - \mathbf{B}u(z) = 0$$
$$(\mathbf{A} - z\mathbf{I})x(z) + \mathbf{B}u(z) = 0. \tag{7.53}$$

Equations 7.52 and 7.53 can be formulated as a single linear algebraic equation,

$$\begin{bmatrix} \mathbf{A} - z\mathbf{I} & \mathbf{B} \\ \mathbf{C} & D \end{bmatrix} \begin{bmatrix} \mathsf{x}(z) \\ \mathsf{u}(z) \end{bmatrix} = \begin{bmatrix} 0 \\ 0 \end{bmatrix}.$$

For the above to be true, we require

$$\det \begin{bmatrix} \mathbf{A} - z\mathbf{I} & \mathbf{B} \\ \mathbf{C} & D \end{bmatrix} = 0. \qquad (7.54)$$

The solutions z to Equation 7.54 are the zeros of the system.

Example 7.5

Let us return to the previous example (Example 7.4). Using the state-space model, find the poles and zeros.

   Solution. Matrices **A** and **B** were previously derived in Example 4.2 and were given in Equation 4.7:

$$\mathbf{A} = \begin{bmatrix} 0 & 1/m & 0 & 0 \\ -k & -b/m & k & 0 \\ 0 & -1/m & 0 & 1/m \\ 0 & 0 & -k & 0 \end{bmatrix} \text{ and } \mathbf{B} = \begin{bmatrix} 0 \\ 0 \\ 0 \\ 1 \end{bmatrix}.$$

For this problem, we only require the second displacement, $x_2(t)$, as an output making matrices **C** and D:

$$\mathbf{C} = \begin{bmatrix} 1 & 0 & 1 & 0 \end{bmatrix} \text{ and } D = 0.$$

The poles are determined by solving for the eigenvalues of **A**,

$$\det(p\mathbf{I} - \mathbf{A}) = \det \begin{bmatrix} p & -1/m & 0 & 0 \\ k & p+b/m & -k & 0 \\ 0 & 1/m & p & -1/m \\ 0 & 0 & k & p \end{bmatrix}$$

$$= \frac{m^2 p^4 + mbp^3 + 3mkp^2 + bkp + k^2}{m^2} = 0.$$

Thus the eigenvalues (or poles) are the solutions to

$$m^2 p^4 + mbp^3 + 3mkp^2 + bkp + k^2 = 0.$$

The characteristic polynomial can be computed for both numeric and symbolic matrices using MATLAB.

```
>> syms m b k
>> A = [0 1/m 0 0; -k -b/m k 0; 0 -1/m 0 1/m; 0 0 -k 0];
>> poly(A)

ans =

x^4 + k^2/m^2 + (b*x^3)/m + (3*k*x^2)/m + (b*k*x)/m^2
```

By default, the polynomial is reported in terms of x. Note that the resultant polynomial is the same as the denominator of the transfer function $x_2(s)/F(s)$ except that p has been substituted for s. One can further confirm that the eigenvalues are the same as the poles using MATLAB.

```
>> m = 10; b = 20; k = 60;
>> A = [0 1/m 0 0; -k -b/m k 0; 0 -1/m 0 1/m; 0 0 -k 0];
>> eig(A)

ans =

  -0.7106 + 3.7722i
  -0.7106 - 3.7722i
  -0.2894 + 1.5361i
  -0.2894 - 1.5361i
```

The zeros are derived using Equation 7.54,

$$\det \begin{bmatrix} \mathbf{A} - z\mathbf{I} & \mathbf{B} \\ \mathbf{C} & \mathbf{D} \end{bmatrix} = \det \left[\begin{array}{cccc|c} -z & 1/m & 0 & 0 & 0 \\ -k & -b/m-z & k & 0 & 0 \\ 0 & -1/m & -z & 1/m & 0 \\ 0 & 0 & -k & -z & 1 \\ \hline 1 & 0 & 1 & 0 & 0 \end{array} \right]$$

$$= \frac{mz^2 + bz + 2k}{m^2} = 0.$$

The zeros are therefore solved by finding the roots of

$$mz^2 + bz + 2k = 0.$$

You should recognize the polynomial as being the numerator of $x_2(s)/F(s)$. The determinants can also be confirmed using the symbolic toolbox in MATLAB.

```
>> syms p z m b k
>> A = [0 1/m 0 0; -k -b/m k 0; 0 -1/m 0 1/m; 0 0 -k 0];
```

```
>> B = [0; 0; 0; 1];
>> C = [1 0 1 0];
>> D = 0;
>> det(p*eye(4)-A)

ans =

(k^2 + 3*k*m*p^2 + b*k*p + m^2*p^4 + b*m*p^3)/m^2

>> det([A-z*eye(4) B; C D])

ans =

(m*z^2 + b*z + 2*k)/m^2
```

7.8 Steady-State Space Analysis

Thus far the analyses discussed in this chapter have focused on transient responses. Sometime, however, we seek information regarding the steady-state response. For natural, impulse, and step responses, it has been noted that the responses attenuate to final constant values. For these types of excitations, the time rate of change of the state vector, $\dot{\mathbf{x}}(t)$, at steady-state is $\mathbf{0}$ because the state vector, $\mathbf{x}(t)$, remains constant, i.e.,

$$\dot{\mathbf{x}}(t \to \infty) = \dot{\mathbf{x}}_{ss} = \mathbf{0}.$$

This means we can readily determine the steady-state condition. For the natural and impulse responses this is trivial. The states should all go to zero at steady-state for the natural response,

$$\mathbf{A}\mathbf{x}_{ss} = \mathbf{0} \quad \Rightarrow \quad \mathbf{x}_{ss} = \mathbf{0}.$$

For the impulse and step responses

$$\mathbf{A}\mathbf{x}_{ss} + \mathbf{B}u_{ss} = \mathbf{0} \tag{7.55}$$

where $u_{ss} = u(t \to \infty)$. Impulse inputs are zero at infinity which means, like the natural response, the steady-state condition is $\mathbf{x}_{ss} = \mathbf{0}$. For step responses, the input is constant. Consequently, using Equation 7.55 we can show

$$\mathbf{A}\mathbf{x}_{ss} = -\mathbf{B}u_{ss}$$
$$\mathbf{x}_{ss} = -\mathbf{A}^{-1}\mathbf{B}u_{ss}. \tag{7.56}$$

Knowing the steady-state input and states, one can also find the steady-state outputs:

$$
\begin{aligned}
\mathbf{y}_{ss} &= \mathbf{C}\mathbf{x}_{ss} + \mathbf{D}u_{ss} \\
&= -\mathbf{C}\mathbf{A}^{-1}\mathbf{B}u_{ss} + \mathbf{D}u_{ss} \\
&= \left[-\mathbf{C}\mathbf{A}^{-1}\mathbf{B} + \mathbf{D} \right] u_{ss}.
\end{aligned}
\tag{7.57}
$$

The above solution can be generalized for multiple input systems with inputs that are constant at steady-state (e.g., $\mathbf{x}_{ss} = -\mathbf{A}^{-1}\mathbf{B}\mathbf{u}_{ss}$).

Example 7.6

Given the two-disk torsion system from Example 7.3, find the angular displacements at steady-state for a constant input torque.

Solution. First, we must derive the state-space model. Figure 7.17 (c) provides the bond graph. One can show that the state equations are

$$
\dot{h}_1 = \tau(t) - \frac{\beta_1}{J_1} h_1 - \kappa_1 \Delta\theta_1,
$$

$$
\Delta\dot{\theta}_1 = \frac{h_1}{J_1} - \frac{h_2}{J_2},
$$

$$
\dot{h}_2 = \kappa_1 \Delta\theta_1 - \frac{\beta_2}{J_2} h_2 - \kappa_2 \theta_2, \text{ and}
$$

$$
\dot{\theta}_2 = \frac{h_2}{J_2}.
$$

The outputs of interest are $\theta_1(t) = \Delta\theta_1 + \theta_2$ and $\theta_2(t)$. The state-space representation is therefore

$$
\begin{bmatrix} \dot{h}_1 \\ \Delta\dot{\theta}_1 \\ \dot{h}_2 \\ \dot{\theta}_2 \end{bmatrix} =
\begin{bmatrix}
-\beta_1/J_1 & -\kappa_1 & 0 & 0 \\
1/J_1 & 0 & -1/J_2 & 0 \\
0 & \kappa_1 & -\beta_2/J_2 & -\kappa_2 \\
0 & 0 & 1/J_2 & 0
\end{bmatrix}
\begin{bmatrix} h_1 \\ \Delta\theta_1 \\ h_2 \\ \theta_2 \end{bmatrix} +
\begin{bmatrix} 1 \\ 0 \\ 0 \\ 0 \end{bmatrix} \tau(t)
$$

and

$$
\begin{bmatrix} \theta_1 \\ \theta_2 \end{bmatrix} =
\begin{bmatrix} 0 & 1 & 0 & 1 \\ 0 & 0 & 0 & 1 \end{bmatrix}
\begin{bmatrix} h_1 \\ \Delta\theta_1 \\ h_2 \\ \theta_2 \end{bmatrix} +
\begin{bmatrix} 0 \\ 0 \end{bmatrix} \tau(t).
$$

The steady-state angular deflections are solved using Equation 7.57:

$$
\begin{bmatrix} (\theta_1)_{ss} \\ (\theta_2)_{ss} \end{bmatrix} = \left\{ \begin{bmatrix} 0 & 1 & 0 & 1 \\ 0 & 0 & 0 & 1 \end{bmatrix} \begin{bmatrix} -\beta_1/J_1 & -\kappa_1 & 0 & 0 \\ 1/J_1 & 0 & -1/J_2 & 0 \\ 0 & \kappa_1 & -\beta_2/J_2 & -\kappa_2 \\ 0 & 0 & 1/J_2 & 0 \end{bmatrix}^{-1} \begin{bmatrix} 1 \\ 0 \\ 0 \\ 0 \end{bmatrix} \right\} \tau_{ss}
$$

$$
= \begin{bmatrix} 2/\kappa_1 \\ 1/\kappa_1 \end{bmatrix} \tau_{ss}.
$$

7.9 Summary

▷ The natural response (also known as the unforced response) is the dynamic response that results when there are no inputs but the initial conditions are generally not zero.

▷ Natural and impulse responses (Equations 7.2 and 7.4 and Figure 7.1) for first-order systems exponentially decay. The time constant, T, is the amount of time required for the response to decay to 36.8% of the initial value. The responses decay to under 2% in $4T$.

▷ The first-order step response (Equation 7.6 and Figure 7.2) asymptotically approaches the steady-state value. The time constant, T, is the time required to reach 63.2% of the steady-state value. Within $4T$ the step response is within 98% of the final value.

▷ The first-order ramp response (Equation 7.8 and Figure 7.5) increases at a constant rate that runs parallel to the input. At steady-state, there is a constant time delay between the input and response equal to the time constant, T. Also, at steady-state, the difference between the response and the input is AT, where A is the rate.

▷ Second-order systems can be categorized based on the amount of damping:

  • Undamped, $\zeta = 0$ (Equations 7.17, 7.32, 7.38, and 7.44)
  • Underdamped, $0 < \zeta < 1$ (Equations 7.19, 7.33, 7.39, and 7.45)
  • Critically damped, $\zeta = 1$ (Equations 7.22, 7.34, 7.40, and 7.46)
  • Overdamped, $\zeta > 1$ (7.24, 7.35, 7.41, and 7.47)

Undamped responses oscillate forever. Underdamped responses will oscillate, but attenuate to the steady-state value. Critically damped and overdamped responses do not oscillate, but rather asymptotically approach steady-state.

▷ The roots of the characteristic equation are referred to as the poles. Poles are generally complex. The placement of second-order poles in the real-complex plane is correlated with the damping ratio (ζ), undamped natural frequency (ω_n), and the damped frequency of oscillation (ω_d):

- Undamped, purely imaginary roots (Equation 7.16)

- Underdamped, complex conjugate roots (Equation 7.18)

- Critically damped, real repeated negative roots (Equation 7.21)

- Overdamped, real distinct negative roots (Equation 7.23)

▷ The second-order natural and impulse responses oscillate about, attenuate to, or asymptotically approach zero.

▷ The unit step responses for second-order systems oscillate about, attenuate to, or asymptotically approach one.

▷ The ramp responses of second-order systems oscillate about the ramp or settle to a steady-state response that runs parallel to the ramp with a constant error.

▷ Higher-order responses are just combinations of first- and second-order responses.

▷ Transfer functions can be factored into first- and second-order terms. The second-order terms have complex poles that have associated damping ratios and natural frequencies. Generally, the poles closest to the imaginary axis have dynamics that take longer to settle. If not negated by the presence of zeros, the poles closest to the imaginary axis will tend to dominate the overall response.

▷ MATLAB includes functions that facilitate transient analysis including commands to determine eigenvalues, damping ratios, and natural frequencies. It also includes functions for generating and annotating pole-zero maps.

▷ Though poles and zeros are commonly determined by factoring the numerator and denominator of the transfer function representation, they can also be readily determined using the state-space model. The poles, for example, are the eigenvalues of the **A** matrix in the state-space representation.

▷ For systems that reach a static condition at steady-state, the rate of change of the state vector, $\dot{\mathbf{x}}$, is **0**. Thus, the steady-state condition, $\mathbf{x}_{ss}$, can be determined algebraically from the steady-state input, $\mathbf{u}_{ss}$.

7.10 Review

R7-1 Explain the difference between the "forced" and "unforced" or natural response.

R7-2 Describe the characteristics of the natural response of a first-order system.

R7-3 What are the similarities between the natural and impulse responses of a first-order system?

R7-4 Given the step response of a first-order system, how does one determine the time constant of the system?

R7-5 Describe the characteristics of the first-order step and ramp responses.

R7-6 What are the damping ratio ranges for undamped, underdamped, critically damped, and overdamped second-order systems?

R7-7 Sketch and describe the undamped, underdamped, critically damped, and overdamped second-order step responses.

R7-8 Given an underdamped, second-order natural response, how does one determine the damping ratio?

R7-9 What is the difference between the second-order time constant and the period of oscillation?

R7-10 What is the settling time?

R7-11 What is the natural frequency?

R7-12 For a second-order system, how is the pole placement related to the natural frequency and damping ratio?

R7-13 List and describe the characteristics used to describe and assess second-order, underdamped, step responses.

R7-14 How are higher-order system responses related to lower-order responses?

R7-15 What is pole-zero analysis, and how can it be used to determine the characteristics of a system?

R7-16 Are the transfer-function and state-space representations associated? If so, how does one determine poles and zeros from the state-space model?

R7-17 How can one determine the system conditions at steady-state?

7.11   Problems

P7-1 *An Inductor-Resistor Circuit.* Figure 7.20 depicts a circuit with an inductor and resistor connected in series. The inductance is $L = 10$ mH and the resistance is 20 Ω. The input voltage is $e(t) = 5$ V. Derive the step response of the voltage drop across the resistor and determine, analytically, how long it takes for the voltage to reach steady-state.

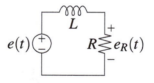

Figure 7.20: *A simple inductor-resistor circuit.*

P7-2 *A Simple Hydraulic Circuit.* A simple hydraulic circuit is illustrated in Figure 7.21. Derive the impulse response of the output flow, $Q_{out}(t)$. Determine how much time it takes the impulse response to settle to less than 1% of its initial value. Do so in terms of the valve resistance, R_f, and the accumulator capacitance, C_f.

P7-3 *A Mass-Spring System.* For the simple mass-spring system shown in Figure 7.22, use an impedance bond graph to derive the transfer function relating the input force, $F(t)$, to the output displacement, $x(t)$. Use Laplace transforms, partial fraction expansions, and Table 5.1 to determine the response to a ramp input force of the form $F(t) = At$.

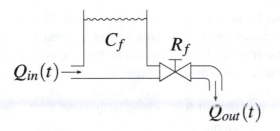

Figure 7.21: *A simple hydraulic circuit.*

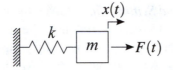

Figure 7.22: *Simple mass-spring system.*

P7-4 *A Mass-Spring-Samper System.* A mass-spring-damper system is depicted in Figure 7.23. The transfer function relating the displacement, $x(t)$, to the forcing function, $F(t)$, is

$$\frac{x(s)}{F(s)} = \frac{1}{ms^2 + bs + k}.$$

The mass, damping constant, and spring rate are 3 kg, 6 N-s/m, and 15 N/m, respectively. Derive the unit step response. Calculate the expected settling time for the 2% criteria and plot the step response.

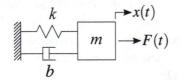

Figure 7.23: *Mass-spring-damper system.*

P7-5 *Logarithmic Decrement.* Assume the damping for the mass-spring-damper system in the previous problem is unknown. If the amplitude decreases by 80% every five oscillations, what is the viscous damping coefficient?

P7-6 *Dominant Poles.* For the system depicted in Figure 1.15 (a), assume that the masses and spring rates are 100 kg and 50 N/m, respectively. Using the transfer function relating the displacement of the second mass relative to the input displacement (i.e. $x_2(s)/x(s)$), find the poles of the system using the transfer function of the . What are the damping ratios and natural frequencies associated with each? Which pole pair, if any, is dominant? Explain using evidence such as a pole-zero plot. What will the approximate settling time be based on the dominant poles?

P7-7 *Pole Placement and Settling Time.* Given the mass-spring-damper system from Figure 1.15 (d), describe how the settling time of the second mass can be reduced by 50% or more by adjusting one or more of the system parameters. Assume that $m_1 = 10$ kg, $m_2 = 20$ kg, $k_1 = 50$ N/m, $k_2 = 200$ N/m, $b_1 = 25$ N-s/m, and $b_2 = 75$ N-s/m.

P7-8 *Steady-State Analysis of a Mass-Spring-Damper System.* For the mass-spring-damper system in Figure 1.15 (d), use the state-space representation to determine the steady-state displacements of the masses for a unit step input. Use the same system parameters provided in P7-7.

P7-9 *State-Space Pole-Zero Analysis of a Torsion Plant.* Derive the state-space representation for the two-disk torsion plant in Figure 7.17. Assume that the only output is the displacement of the bottom mass, $\theta_1(t)$. Use the state-space representation to find the poles and zeros of the system. Remember that the system parameters are provided in Table 7.1.

P7-10 *Steady-State Analysis of a Wind Turbine Generator.* Determine the steady-state voltage supplied to the lamp of the simple wind turbine generator (Figure 3.28) if the wind speed remains relatively constant. Assume that the Pittman motor from Example 4.5 is used as the generator. The motor parameters were provided in Table 4.2. The constant wind speed is 5 m/s and the fan modulus is $k_{fan} = 2$ N-s. The resistance of the lamp is 20 Ω.

P7-11 *Pole-Zero Analysis of an Electric Circuit.* Recall the electric circuit in Figure 3.27. Assume the system parameters are $L = 10$ mH, $C_1 = 60\ \mu$F, $C_2 = 20\ \mu$F, $R_1 = 30\ \Omega$, $R_2 = 20\Omega$, and $R_3 = 5\ \Omega$. Use the state-space representation to determine the poles and zeros if the output is the voltage drop across the capacitor. Identify the dominant poles, and based on the characteristics of the dominant poles, predict the settling time of the first capacitor voltage response to a step input voltage.

7.12   Challenges

C7-1 *A Toy Electric Car.* If the inductance is neglected, the toy car model is first-order. Determine the time constant and the time required to reach more than 99% of the steady-state step response. When the inductance is not neglected, the toy electric car is represented by a second-order model. Determine whether the system is underdamped or overdamped and find the system's damping ratio. Analytically determine the steady-state response v_{ss} for both models.

C7-2 *A Quarter-Car Suspension Model.* The quarter-car suspension model is a fourth-order system. Identify the dominant poles, and based on those poles, estimate the settling time for the step response. Confirm whether the actual step response has dynamics that approximately match the characteristics of the dominant poles.

C7-3 *An Electric Drill.* What order is the electric drill model? Does the system have a dominant pole or set of poles? What are the associated dynamics? What is the expected steady-state output angular velocity?

Chapter 8

Frequency Domain Analysis

Thus far we have concentrated analyses primarily on the transient dynamics of the system. The analyses in the prior chapter focused on characterizing the response over short periods where the system transitioned from one state of equilibrium to another. Some systems, however, operate for extended periods under dynamic load or input (e.g., cyclic loads) and do not settle to a fixed or constant equilibrium condition. For these systems, the transient response makes up a relatively small fraction of the pertinent dynamics we wish to analyze. An alternative formulation is necessary to examine such dynamics. Ask yourself the following:

▷ What is the steady-state response of a linear system excited by a cyclic or oscillatory input?

▷ How does one characterize the response at steady-state when the system is exposed to a consistent oscillatory input?

▷ Is the time domain still appropriate for conducting our analyses of such systems?

▷ What tools are useful for examining such dynamics?

8.1   Introduction

Systems that operate for extended periods cyclically due to sinusoidal or cyclic inputs that maintain the system in a dynamic state cannot be readily examined using the tools from Chapter 7. For such systems, the transient dynamics are not of as great interest as the steady-state, but, unlike the systems in the previous chapter, the steady-state response is not a constant static equi-librium condition. Rather, it is a consistent oscillatory state. Some types of systems that operate under such conditions include alternating current (AC) circuits, rotating machines, and vibrating structures. Take for example me-chanical vibration. Though the vibration may occur over an abbreviated pe-riod, many repeated oscillations occur over the short time span. Over this time span, the vibrations effectively may not attenuate significantly. Thus, characteristics like settling time and overshoot have very little meaning or provide little to no insight. What is more interesting is the number of oc-currences of cycles over a unit of time. This is referred to as the *frequency*. Systems with cyclic responses at steady-state are more aptly analyzed in what is referred to as the *frequency domain*.

Frequency domain analysis involves characterizing and examining steady-state oscillatory dynamics. The input to such systems is often sinusoidal. Lin-ear systems excited by sinusoidal inputs transition quickly from equilibrium through the transient to a cyclic condition at steady-state. As we will see, the steady-state response will also be sinusoidal with the same frequency as the input but with, generally, a different amplitude and a time delay or shift.

In this chapter we will learn about the sinusoidal transfer function and its use in analyzing mechanical vibration and AC circuits. In this chapter, you will learn about vibration attenuation and will study how to use some of the tools previously introduced to determine natural frequencies and modes of vibration. You will also learn how the sinusoidal transfer function provides the premise for phasor analysis used in the study of AC circuits. Furthermore, we will learn how to visualize frequency domain plots and discover how the plots illustrate the impact on the response of varying system parameters.

For this chapter, the outcomes and objectives are as follows:

▷ *Objectives:*

1. To analyze mechanical vibration systems including transmission and modal analysis,

2. To be able to analyze basic AC circuits, and

3. To conduct frequency response analysis.

▷ *Outcomes:* You will be able to

1. determine the steady-state response of a linear time-invariant sys-tem to a sinusoidal input,

2. calculate the force or motion transmitted by a vibration isolation system,

3. conduct basic modal analysis of free vibration systems,

4. conduct basic analyses of AC circuits,

5. identify the characteristics for frequency responses of first- and second-order systems, and

6. compose Bode plots that visualize the frequency response of an oscillatory system.

8.2 Properties of Sinusoids

Sinusoid generally refers to a mathematical function of the form

$$g(t) = A\sin(\omega t + \phi)$$

where A is the amplitude, ω is the frequency of oscillation in rad/s, and ϕ is the phase angle measured in radians (where one cycle is 2π radians). The sinusoid oscillates about zero, and the amplitude is the distance from the peaks and valleys to the time axis. Figure 8.1 depicts two sinusoids of the same frequency but differing amplitude.

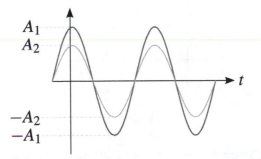

Figure 8.1: *Sinusoids with differing amplitude.*

As has been discussed in the previous chapter, the period of oscillation, T_p, is the time required for a single cycle. The period of oscillation can be measured from peak-to-peak, valley-to-valley, or start-to-finish as illustrated in Figure 8.2. Basically, it can be measured from any point along the cycle to the next time when the specific point in the cycle is repeated.

The *phase angle* measures how much a sinusoid is shifted along the time access. A phase angle of $\pi/2$ would indicate that the sinusoid is shifted to the left a quarter of a cycle along the time axis. Hence, phase angles of $\pi/4$ and π would be time shifts of one eighth and one half cycles, respectively. A

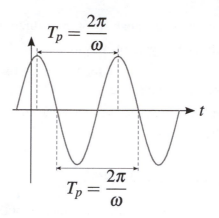

Figure 8.2: *Period of oscillation of a sinusoid measured two ways.*

negative phase angle would mean the sinusoid is shifted to the right instead. Figure 8.3 illustrates two sinusoids with the same amplitude and frequency but different phase angle.

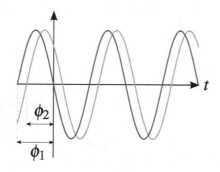

Figure 8.3: *Sinusoids with different phase angle.*

8.3   The Sinusoidal Transfer Function

Linear time-invariant systems will respond with a sinusoid at steady-state when excited by a sinusoid. Moreover, the output response will be a sinusoid with the same frequency but generally different amplitude and phase angle. It can be shown that for a SISO, LTI system with transfer function

$$\frac{y(s)}{u(s)} = G(s) = \frac{K(s+z_1)(s+z_2)\cdots(s+z_m)}{(s+p_1)(s+p_2)\cdots(s+p_n)}$$

the steady-state response to a sinusoidal response will generally be

$$y(t) = |G(j\omega)|A\sin(\omega t + \phi)$$

where $|G(j\omega)|$ is the magnitude of the sinusoidal transfer function and ϕ is generally different from the input phase angle.

The partial fraction expansion of the output transform, $y(s)$, given a sinusoidal input of the form

$$u(t) = A\sin\omega t$$

will be

$$y(s) = G(s)\frac{A\omega}{s^2 + \omega^2}$$
$$= \frac{a}{s + j\omega} + \frac{\bar{a}}{s - j\omega} + \frac{b_1}{s + p_1} + \frac{b_2}{s + p_2} + \cdots + \frac{b_n}{s + p_n}$$

where p_k ($k = 1, \ldots, n$) can generally be complex. Assuming the output response is stable, the terms

$$\frac{b_1}{s + p_1} + \frac{b_2}{s + p_2} + \cdots + \frac{b_n}{s + p_n}$$

will inverse Laplace transform into either decaying exponentials (e.g., $b_k e^{-p_k t}$) if the poles are real and/or attenuating sinusoids (e.g., $A_k e^{-p_k t}\sin\omega_k t$) if the poles are complex. Such functions go to zero at steady-state leaving only the inverse Laplace transforms of the first two terms. Consequently, the response at steady-state will be

$$y_{ss}(t) = ae^{-j\omega t} + \bar{a}e^{j\omega t}.$$

The complex coefficients a and $\bar{a}$ are constants and will be the same regardless at what value of s one evaluates the transform. Therefore, to determine a it is convenient to chose $s = -j\omega$,

$$G(s)\frac{A\omega}{s^2 + \omega^2} = \frac{a}{s + j\omega} + \frac{\bar{a}}{s - j\omega}$$

$$G(s)\frac{A\omega}{(s + j\omega)(s - j\omega)} = \frac{a}{s + j\omega} + \frac{\bar{a}}{s - j\omega}$$

$$a = \left[G(s)\frac{A\omega}{(s + j\omega)(s - j\omega)} - \frac{\bar{a}}{s - j\omega}\right](s + j\omega)$$

$$= \left[G(s)\frac{A\omega}{s - j\omega} - \frac{\bar{a}}{s - j\omega}(s + j\omega)\right]_{s=-j\omega}$$

$$= G(-j\omega)\frac{A\omega}{-j\omega - j\omega} = G(j\omega)\frac{A\omega}{-2j\omega}$$

$$a = -\frac{A}{2j}G(-j\omega),$$

and to calculate $\bar{a}$ we equate s to $j\omega$,

$$\bar{a} = \left[G(s)\frac{A\omega}{(s + j\omega)(s - j\omega)} - \frac{a}{s + j\omega}\right](s - j\omega)$$

$$= \left[G(s)\frac{A\omega}{s + j\omega} - \frac{a}{s + j\omega}(s - j\omega)\right]_{s=j\omega}$$

$$= G(j\omega)\frac{A\omega}{j\omega + j\omega} = G(j\omega)\frac{A\omega}{2j\omega}$$

$$\bar{a} = \frac{A}{2j}G(j\omega).$$

The sinusoidal transfer function is generally complex and can be decomposed into its real and imaginary parts as shown in Figure 8.4. As we saw in Chapter 5, complex numbers can also be written in polar form. This is also true for complex functions,

$$\begin{aligned} G(j\omega) &= \Re[G(j\omega)] + j\,\Im[G(j\omega)] \\ &= |G(j\omega)|\cos\phi + j\,|G(j\omega)|\sin\phi \\ &= |G(j\omega)|(\cos\phi + j\,\sin\phi) \\ &= |G(j\omega)|e^{j\phi} \end{aligned}$$

where $\Re[G(j\omega)]$ and $\Im[G(j\omega)]$ are the real and imaginary parts of $G(j\omega)$. It can be similarly shown that

$$G(-j\omega) = |G(-j\omega)|e^{-j\phi} = |G(j\omega)|e^{-j\phi}$$

because, as illustrated in Figure 8.4 $G(-j\omega)$ and its conjugate $G(-j\omega)$ have the same magnitude. That makes the complex coefficients

$$a = -\frac{A}{2j}|G(j\omega)|e^{-j\phi}$$

and

$$\bar{a} = \frac{A}{2j}|G(j\omega)|e^{j\phi}.$$

Ergo, the steady-state response is

$$
\begin{aligned}
y_{ss}(t) &= ae^{-j\omega t} + \bar{a}e^{j\omega t} \\
&= -\frac{A}{2j}|G(j\omega)|e^{-j\phi}e^{-j\omega t} + \frac{A}{2j}|G(j\omega)|e^{j\phi}e^{j\omega t} \\
&= |G(j\omega)|A\frac{e^{j(\omega t+\phi)} - e^{-j(\omega t+\phi)}}{2j}.
\end{aligned}
$$

Recall Euler's Theorem from Chapter 5. In Chapter 5, we reviewed how sinusoids can be represented in terms of complex exponentials using Euler's Theorem. The complex exponential representation of sine is

$$\sin\theta = \frac{e^{j\theta} - e^{-j\theta}}{2j}.$$

Applying this identity to the prior equation yields the steady-state response

$$
\begin{aligned}
y_{ss}(t) &= |G(j\omega)|A\sin(\omega t + \phi) \\
&= Y\sin(\omega t + \phi)
\end{aligned}
$$

where $Y = |G(j\omega)|A$ and $\phi = \angle G(j\omega)$. The output response will have the same frequency as the input but generally distinct amplitude and phase angle. The output amplitude is amplified by a factor equal to the magnitude of the sinusoidal transfer function. Based on the preceding analysis, we can construe that for sinusoidal inputs

$$|G(j\omega)| = \left|\frac{y(j\omega)}{u(j\omega)}\right| = \text{output to input amplitude ratio} \qquad (8.1)$$

and

$$
\begin{aligned}
\angle G(j\omega) &= \angle\frac{y(j\omega)}{u(j\omega)} = \tan^{-1}\left\{\frac{\Im[G(j\omega)]}{\Re[G(j\omega)]}\right\} \\
&= \tan^{-1}\left\{\frac{\Im[y(j\omega)]}{\Re[y(j\omega)]}\right\} - \tan^{-1}\left\{\frac{\Im[u(j\omega)]}{\Re[u(j\omega)]}\right\} \\
&= \text{phase shift of output with respect to input.} \qquad (8.2)
\end{aligned}
$$

The *phase shift* is the difference between the input and output phase angles. Thus, the characteristics of the steady-state output response can be found directly from $G(j\omega)$.

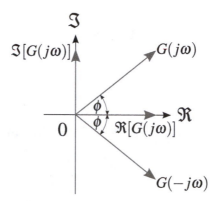

Figure 8.4: *Complex function and its conjugate.*

Example 8.1

In Example 5.8 we derived the impulse response of the quarter car suspension model assuming the system parameters were $m = 500$ kg, $b = 8000$ N-s/m, and $k = 34,000$ N/m, respectively. The resulting response turned out to be underdamped as evidenced by the fact that the system had complex conjugate poles, and we had to complete the square to find the response. For this problem, find the steady-state response to a sinusoidal input displacement of the form

$$y_{road}(t) = (A \sin \omega t) \text{ m/s}$$

where $A = 0.03$ m and $\omega = 10$ rad/s.

Solution. Remember from Example 5.8 that the system transfer function is

$$\frac{y(s)}{y_{road}(s)} = G(s) = \frac{bs + k}{ms^2 + bs + k}.$$

To find the steady-state response, we need to determine the magnitude, $|G(j\omega)|$, and phase angle, $\angle G(j\omega)$, of the sinusoidal transfer function. The output amplitude will be $|G(j\omega)|A = 0.03|G(j\omega)|$ m. The amplitude ratio of the output sinusoid to the input sinusoid is

$$|G(j\omega)| = \left| \frac{y(j\omega)}{y_{road}(j\omega)} \right| = \left| \frac{bs+k}{ms^2+bs+k} \right|_{s=j\omega}$$

$$= \left| \frac{b(j\omega)+k}{m(j\omega)^2+b(j\omega)+k} \right| = \left| \frac{k+jb\omega}{(k-m\omega^2)+jb\omega} \right|$$

$$= \left| \left[\frac{k+jb\omega}{(k-m\omega^2)+jb\omega} \right] \left[\frac{(k-m\omega^2)-jb\omega}{(k-m\omega^2)-jb\omega} \right] \right|$$

$$= \left| \frac{k^2-km\omega^2-jkb\omega+jkb\omega-jbm\omega^3+b^2\omega^2}{(k-m\omega^2)^2+b^2\omega^2} \right|$$

$$= \left| \frac{(k^2-km\omega^2+b^2\omega^2)-jbm\omega^3}{(k-m\omega^2)^2+b^2\omega^2} \right|$$

$$= \frac{\sqrt{(k^2-km\omega^2+b^2\omega^2)^2+(bm\omega^3)^2}}{(k-m\omega^2)^2+b^2\omega^2},$$

and the phase angle is

$$\angle G(j\omega) = \tan^{-1} \left[\frac{-bm\omega^3}{k^2-km\omega^2+b^2\omega^2} \right].$$

Substituting for the system parameters and input sinusoid frequency garners the amplitude ratio,

$$|G(j\omega)| = \Big\{ \big[(34,000 \text{ N/m})^2 - (34,000 \text{ N/m})(500 \text{ kg})(10 \text{ rad/s})^2$$

$$+ (8,000 \text{ N-s/m})^2 (10 \text{ rad/s})^2 \big]^2$$

$$+ \big[(8,000 \text{ N-s/m})(500 \text{ kg})(10 \text{ rad/s})^3 \big]^2 \Big\}^{1/2}$$

$$\div \Big\{ \big[(34,000 \text{ N/m}) - (500 \text{ kg})(10 \text{ rad/s})^2 \big]^2$$

$$+ (8,000 \text{ N-s/m})^2 (10 \text{ rad/s})^2 \Big\}$$

$$= 1.0655.$$

To determine the phase angle, we first calculate

$$\frac{-bm\omega^3}{k^2-km\omega^2+b^2\omega^2} = \big[-(8,000 \text{ N-s/m})(500 \text{ kg})(10 \text{ rad/s})^3 \big]$$

$$\div \big[(34,000 \text{ N/m})^2 - (34,000 \text{ N/m})(500 \text{ kg})(10 \text{ rad/s})^2$$

$$+ (8,000 \text{ N-s/m})^2 (10 \text{ rad/s})^2 \big] = -0.6831.$$

The phase angle is then

$$\underline{/G(j\omega)} = \tan^{-1}[-0.6831]$$
$$= -0.5993 \text{ rad}.$$

Thus, the steady-state sinusoidal response is

$$y_{ss}(t) = 1.0655(0.03)\sin(10t - 0.5993) \text{ m/s}$$
$$= 0.031965\sin(10t - 0.5993) \text{ m/s}.$$

8.4   Complex Operations Using MATLAB

MATLAB provides a several functions that facilitate calculating the magnitude and phase angle of the sinusoidal transfer function amongst other things. When evaluated in MATLAB, the sinusoidal transfer function is just a complex number. MATLAB includes various commands for operations on complex numbers, variables, and functions. Those functions are listed in Table 8.1.

The abs and angle functions are used to determine the magnitude and phase angle of complex functions like a transfer function.

```
>> abs(X);
>> angle(X);
```

In the above command-line scripts, the X can be an array of complex numbers. The conj command calculates the complex conjugate of the array X.

```
>> conj(X);
```

The real and imaginary parts can be numerically determined using the real and imag commands.

```
>> real(X);
>> imag(X);
```

Use of these commands is demonstrated in the following example.

Example 8.2
Solve the previous example using MATLAB. Use least two different approaches and five commands provided in Table 8.1.

Table 8.1: *MATLAB commands for complex operations.*

Function	Description
abs(X)	returns magnitude(s) of complex element(s) in X
angle(X)	returns phase angle(s) of complex element(s) in X
conj(X)	returns complex conjugate(s) of complex element(s) in X
imag(X)	returns imaginary part(s) of complex element(s) in X
real(X)	returns real part(s) of complex element(s) in X

Solution. The response can be solved with the aid of MATLAB a number of ways. The simplest is to use the commands for finding the magnitude and phase angle of $G(j\omega)$:

$$G(j\omega) = \frac{k + jb\omega}{(k - m\omega^2)^2 + jb\omega}.$$

In MATLAB we specify the parameters, define the sinusoidal transfer function as a complex function, and use the provided functions to find the magnitude and phase angle.

```
>> m = 500; b = 8000; k = 34e3; w = 10; A = 0.03;
>> G = (k+j*b*w)/((k-m*w^2)+j*b*w)

G =

   0.8798 - 0.6010i

>> abs(G)

ans =

   1.0655

>> angle(G)

ans =

  -0.5993
```

Alternatively, we could define the numerator and denominator as separate complex functions and then determine the magnitude and angle step-by-step with the aid of the identifies for complex division (Equations 5.2

and 5.3) previously reviewed in Chapter 5. Using Equation 5.2 provide the following magnitude and phase.

```
>> num = k+j*b*w;
>> den = (k-m*w^2)+j*b*w;
>> G = num*conj(den)/(den*conj(den))

G =

   0.8798 - 0.6010i

>> magnitude = abs(num)/abs(den)

magnitude =

   1.0655

>> phase = angle(num)-angle(den)

phase =

  -0.5993
```

Using Equation 5.3 instead provides the same result.

```
>> magitude = sqrt(real(G)^2+imag(G)^2)

magitude =

   1.0655

>> phase = atan(imag(G)/real(G))

phase =

  -0.5993
```

All three approaches yield the same result. Given the magnitude and angle, we can substitute into the steady-state response equation

$$y_{ss}(t) = 1.0655(0.03)\sin(10t - 0.5993) \text{ m/s}$$
$$= 0.0320\sin(10t - 0.5993) \text{ m/s}.$$

8.5 Mechanical Vibration

Mass-spring-damper systems are often utilized to attenuate undesirable dynamics in machines, vehicles, and structures. Masses and springs can absorb or store energy and dampers will dissipate the energy. They can be combined in a number of manners to isolate a machine, vehicle, or structure from an input force or displacement. Instruments used for precision machining and manufacturing like those used to manufacture semiconductors must be isolated to ensure quality and products within specified tolerances. It would obviously be undesirable to drive automobiles without a suspension system; the ride would be uncomfortable and would likely result in undesirable noise. Vibration isolation is used in structures to attenuate excessive vibrations that can be induced by wind drag and earthquakes.

8.5.1 Transmissibility

As depicted in Figure 8.5, a vehicle's suspension is devised to absorb energy that would result in detrimental stresses on the fasteners, cabin, and passengers. The purpose of such isolation is to reduce the relative amplitude of the displacement transmitted from the road to the vehicle mass. Other similar systems are used to attenuate force input. For instance, industrial-sized machines in a manufacturing facility can produce large-amplitude, cyclic forces that can be transferred through the foundation to an adjacent machine where a precision process takes place. Isolation pads beneath the machines are used to isolate force transmission between the first machine and the foundation and displacement transmission between the foundation and the second machine.

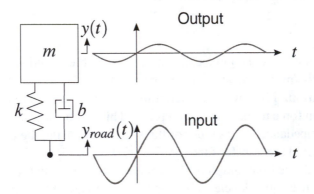

Figure 8.5: *Vehicle suspension vibration isolation.*

The primary purpose for analyzing vibration isolation is to determine the *transmissibility* of the excitation force or displacement. The transmissibility is the amplitude ratio of the transmitted force (displacement) to the excitation force (displacement). The isolation minimizes the amplitude of the transmitted force or displacement as portrayed by Figure 8.6. In terms of input and output forces (i.e., $F_{in}(t)$ and $F_{out}(t)$, respectively), the transmissibility is

$$TR = \left| \frac{F_{out}(j\omega)}{F_{in}(j\omega)} \right|.$$
(8.3)

The following example illustrates how to determine *force transmissibility*.

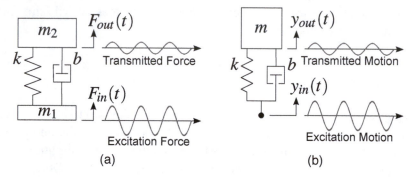

(a) (b)

Figure 8.6: *Vibration isolation of (a) force excitation and (b) motion excitation.*

Example 8.3

For the system depicted in Figure 8.6 (a), find the transmissibility if the foundation is forced by an excitation

$$F_{in}(t) = (5 \sin 2t) \text{ N}$$

and the first mass, damping constant, spring stiffness, and second mass are 1 kg, 2 N-s/m, 5 N/m, and 2 kg, respectively.

Solution. First, we must determine the transfer function relating the excitation force to the transmitted force. This can be derived by synthesizing an impedance bond graph for the system and using the resulting bond graph to find the transfer function $F_{out}(s)/F_{in}(s)$. Figure 8.7 (b) illustrates the impedance bond graph of the system. To derive the transfer function, we must use equivalencies to collapse the top part of the graph. Then we can employ the bottom 1-junction as an effort divider to find the desired transfer function.

To collapse the top part of the impedance bond graph and find the equivalent impedance, Z_{eq}, shown in Figure 8.7 (c), we first combine b and k/s at the top 1-junction. The equivalent impedance of those two off the 1-junction is

$$b + \frac{k}{s} = \frac{bs + k}{s}.$$

Then we can combine this impedance with $m_s s$ using the 0-junction equivalency,

$$Z_{eq} = \frac{m_2 s \left(\dfrac{bs + k}{s}\right)}{m_2 s + \left(\dfrac{bs + k}{s}\right)} = \frac{m_2 s(bs + k)}{m_2 s^2 + bs + k}.$$

Applying the effort-divider relation to the lower 1-junction yields

$$\begin{aligned}
\frac{F_{out}(s)}{F_{in}(s)} &= \frac{\dfrac{m_2 s(bs + k)}{m_2 s^2 + bs + k}}{m_1 s + \dfrac{m_2 s(bs + k)}{m_2 s^2 + bs + k}} \\[2ex]
&= \frac{m_s s(bs + k)}{m_1 s(m_2 s^2 + bs + k) + m_2 s(bs + k)} \\[2ex]
&= \frac{m_2 bs + m_2 k}{m_1 m_2 s^2 + (m_1 + m_2)bs + (m_1 + m_2)k}.
\end{aligned}$$

Now, we need to convert the resulting transfer function to a sinusoidal transfer function,

$$\begin{aligned}
\frac{F_{out}(j\omega)}{F_{in}(j\omega)} &= \frac{m_2 b\, j\omega + m_2 k}{m_1 m_2 (j\omega)^2 + (m_1 + m_2)b\, j\omega + (m_1 + m_2)k} \\[2ex]
&= \frac{m_2 k + j m_2 b\omega}{[(m_1 + m_2)k - m_1 m_2 \omega^2] + j(m_1 + m_2)b\omega}.
\end{aligned}$$

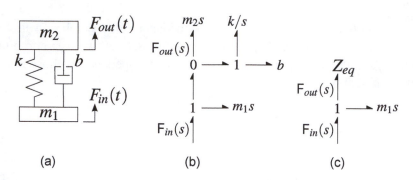

Figure 8.7: *The (a) schematic, (b) impedance bond graph, and (c) simplified impedance bond graph of a force isolation system.*

The transmissibility is then

$$TR = \frac{|F_{out}(j\omega)|}{|F_{in}(j\omega)|}$$

$$= \frac{\sqrt{(m_2 k)^2 + (m_2 b\omega)^2}}{\sqrt{[(m_1 + m_2)k - m_1 m_2 \omega^2]^2 + [(m_1 + m_2)b\omega]^2}}$$

$$= \left\{ (1 \text{ kg})^2 (5 \text{ N/m})^2 + (1 \text{ kg})^2 (2 \text{ N-s/m})^2 (2 \text{ rad/s})^2 \right\}^{1/2}$$

$$\div \left\{ [(10 \text{ kg} + 1 \text{ kg})(5 \text{ N/m}) - (1 \text{ kg})(2 \text{ kg})(2 \text{ rad/s})]^2 \right.$$

$$\left. + [(10 \text{ kg} + 1 \text{ kg})(2 \text{ N-s/m})(2 \text{ rad/s})]^2 \right\}^{1/2}$$

$$= 0.1377 .$$

Motion transmissibility is determined in much the same manner. However, to determine motion transmissibility the sinusoidal transfer function must relate the output and input displacements as opposed to the forces:

$$TR = \frac{|F_{out}(j\omega)|}{|F_{in}(j\omega)|} . \tag{8.4}$$

In Example 8.1 we derived the steady-state response to a sinusoidal input. The input was

$$y_{road}(t) = (0.03 \sin 10t) \text{ m/s}$$

and the output response was

$$y_{ss}(t) = 0.0320 \sin(10t - 0.5993) \text{ m/s} .$$

The amplitude ratio of the output to the input was

$$|G(j\omega)| = \frac{0.031965}{0.03} = 1.0655,$$

which also turns out to be the motion transmissibility

$$
\begin{aligned}
TR &= \frac{|y(j\omega)|}{|y_{road}(j\omega)|} \\
&= \frac{\sqrt{k^2 + (b\omega)^2}}{\sqrt{(k - m\omega^2)^2 + (b\omega)^2}} \\
&= \frac{\sqrt{(34,000 \text{ N/m})^2 + (8,000 \text{ N-s/m})^2 (10 \text{ rad/s})^2}}{\sqrt{[(34,000 \text{ N/m}) - (500 \text{ kg})(10 \text{ rad/s})^2] + (8,000 \text{ N-s/m})^2 (10 \text{ rad/s})^2}} \\
&= 1.0655 \\
&= |G(j\omega)|.
\end{aligned}
$$

Hence, the magnitude of the sinusoidal transfer function may also be used to determine force or motion transmissibility if it relates the output force or displacement to the input force or displacement.

Recall from the prototypical second-order system that the natural frequency and damping ratio of a mass-spring-damper system are (refer to Equations 7.11 and 7.12)

$$\omega_n = \sqrt{\frac{k}{m}} \quad \text{and} \quad \zeta = \frac{b}{2\sqrt{km}}.$$

Note that $k/m = \omega_n^2$ and $b/m = 2\zeta\omega_n$. Also note that the normalized frequency is the input sinusoidal frequency divided by the natural frequency, ω/ω_n. *Resonanace* is a phenomenon that occurs in linear systems exposed to sinusoidal inputs. It is the pronounced maximum or minimum that occurs in and around the *resonant frequency*. For $0 \le \zeta \le 1$, the resonant frequency for a prototypical second-order system is defined as (Ogata 2002)

$$\omega_r = \omega_n \sqrt{1 - 2\zeta^2}. \tag{8.5}$$

For lightly damped systems, the natural and resonant frequencies are approximately the same (i.e., $\omega_r \approx \omega_n$). The motion transmissibility of the quarter-car suspension model can be rewritten in terms of the damping ratio and normalized frequency,

$$TR = \frac{\sqrt{k^2 + (b\omega)^2}}{\sqrt{(k - m\omega^2)^2 + (b\omega)^2}} = \frac{\sqrt{\dfrac{k^2}{m^2} + \dfrac{(b\omega)^2}{m^2}}}{\sqrt{\dfrac{(k - m\omega^2)^2}{m^2} + \dfrac{(b\omega)^2}{m^2}}}$$

$$= \frac{\sqrt{\omega_n^4 + (2\zeta\omega_n\omega)^2}}{\sqrt{(\omega_n^2 - \omega^2)^2 + (2\zeta\omega\omega_n)^2}} = \frac{\sqrt{1 + \left(2\zeta\dfrac{\omega}{\omega_n}\right)^2}}{\sqrt{\left[1 - \left(\dfrac{\omega}{\omega_n}\right)\right]^2 + \left(2\zeta\dfrac{\omega}{\omega_n}\right)^2}}.$$

The normalized transmissibility is plotted in Figure 8.8 for various damping ratios.

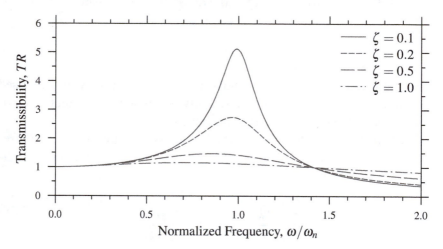

Figure 8.8: *Quarter-car suspension model transmissibility versus normalized frequency curves at various damping ratios.*

Note how the transmissibility peaks at or near the natural frequency when the damping is low. For $\zeta = 0.1$ and $\zeta = 0.2$, the natural frequency is approximately equal to the resonant frequency. The resonant frequency is where the transmissibility reaches its peak. When lightly damped, if excited by an input with a frequency equal to the natural frequency, the system will tend to resonate and amplify the magnitude of the output relative to the input. The resonant frequency appears to shift to the left as the damping increases. At relatively low frequency (i.e., $\omega/\omega_n < 0.5$), the output amplitude is approximately equal to the input amplitude regardless of damping. At relatively high frequency (i.e., ω/ω_n), the output amplitude becomes lower with less damping. At resonance, less damping results in a pronounced amplification of the output motion relative to the input. Because the amplification at and near resonance is so pronounced, one must be careful to select an appropriate damping if the suspension is expected to regularly operate near the natural frequency. The damping ratio for the suspension in Examples 8.1 and 8.2 is

$$\zeta = \frac{b}{2\sqrt{km}} = \frac{8000 \text{ N-s/m}}{2\sqrt{(34,000 \text{ N/m})(500 \text{ kg})}} \approx 0.97$$

and the input frequency is 10 rad/s, which is near the natural frequency of

$$\omega_n = \sqrt{\frac{k}{m}} = \sqrt{\frac{34,000 \text{ N/m}}{500 \text{ kg}}} \approx 8.25 \text{ rad/s}.$$

Consequently, this is why the transmissibility is greater than one (remember that $TR = 1.0655$).

8.5.2 Modal Analysis of Free Vibration

Modal analysis is the study of structures subject to mechanical vibrations. Structures are often modeled by a series of mass-spring elements that approximate the distribution of structural mass and rigidity. In its basic form, modal analysis involves examining multi-degree-of-freedom mass-spring systems like that depicted in Figure 8.9. The system equations derived from the bond graph are of the form

$$\dot{x}_1 = \frac{p_1}{m_1}$$

$$\dot{p}_1 = -k_1 x_1 + k_2 \delta_2$$

$$\dot{\delta}_2 = -\frac{p_1}{m_1} + \frac{p_2}{m_2}$$

$$\dot{p}_2 = -k_2 \delta_2 + k_3 \delta_3$$

$$\vdots$$

$$\dot{\delta}_{n-1} = -\frac{p_{n-2}}{m_{n-2}} + \frac{p_{n-1}}{m_{n-1}}$$

$$\dot{p}_{n-1} = -k_{n-1} \delta_{n-1} + k_n \delta_n$$

$$\dot{\delta}_n = -\frac{p_{n-1}}{m_{n-1}} + \frac{p_n}{m_n}$$

$$\dot{p}_n = -k_n \delta_n.$$

This section is just a cursory introduction to modal analysis and its relation to tools learned in System Dynamics.

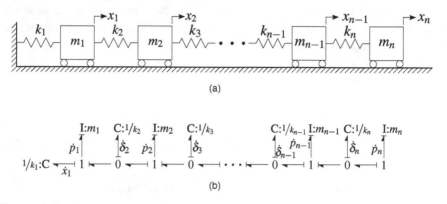

(a)

(b)

Figure 8.9: *Distributed-parameter approximation of a beam or column with axial vibration.*

In Vibrations, systems are represented often using sets of second-order differential equations as opposed to first-order differential equations. Moreover, instead of vectors and matrices, Vibrations literature tends to use a set and tensor notation where, for example, instead of the vector

$$\mathbf{a} = \begin{bmatrix} a_1 \\ a_2 \\ \vdots \\ a_n \end{bmatrix}$$

we use the set

$$\{a\} = \begin{Bmatrix} a_1 \\ a_2 \\ \vdots \\ a_n \end{Bmatrix}$$

and the matrix

$$\mathbf{A} = \begin{bmatrix} a_{11} & \cdots & a_{1n} \\ \vdots & \ddots & \vdots \\ a_{n1} & \cdots & a_{nn} \end{bmatrix}$$

is alternatively denoted $[A]$. Thus, the system of second-order differential equations used to conduct free vibration modal analysis is generally denoted

$$[M]\{\ddot{x}\} + [K]\{x\} = \{0\} \tag{8.6}$$

where $[M]$ and $[K]$ are the mass and stiffness tensors and $\{x\}$ and $\{\ddot{x}\}$ are the displacement and acceleration sets.

Therefore, if we are to use the set of first-order differential equations derived using a bond graph approach, we need a way to transform the equations into a set of second-order differential equations. Traditional vibration models typically use mass displacements and velocities as opposed to spring deflections and mass momenta. However, the spring deflections are just relative or differential displacements. In the distributed-parameter system depicted in Figure 8.9 the spring deflection δ_k is the difference between two displacements,

$$\delta_k = x_k - x_{k-1}, \tag{8.7}$$

where $k = 1, \ldots, n$. Also, recall that the momenta are

$$p_k = mv_k = m\dot{x}_k \tag{8.8}$$

making $\dot{p}_k = m\ddot{x}_k$. By substituting for the deflections and momenta, the set of first-order equations can be converted to an equivalent set of second-order equations in terms of the displacements and velocities,

$$m_1\ddot{x}_1 = -k_1 x_1 + k_2(x_2 - x_1)$$
$$m_2\ddot{x}_2 = -k_2(x_2 - x_1) + k_3(x_3 - x_2)$$
$$\vdots$$
$$m_{n-1}\ddot{x}_{n-1} = -k_{n-1}(x_{n-1} - x_{n-2}) + k_n(x_n - x_{n-1})$$
$$m_n\ddot{x}_n = -k_n(x_n - x_{n-1})$$

or

$$m_1\ddot{x}_1 + (k_1 + k_2)x_1 - k_2 x_2 = 0$$
$$m_2\ddot{x}_2 - k_2 x_1 + (k_2 + k_3)x_2 - k_3 x_3 = 0$$
$$\vdots$$
$$m_{n-1}\ddot{x}_{n-1} - k_{n-1}x_{n-2} + (k_{n-1} + k_n)x_{n-1} - k_n x_n = 0$$
$$m_n\ddot{x}_n - k_n x_{n-1} + k_n x_n = 0.$$

In set and tensor form, the set of second-order equation is

$$[M]\{\ddot{x}\} + [K]\{x\} = \begin{bmatrix} m_1 & 0 & \cdots & 0 & 0 \\ 0 & m_2 & \cdots & 0 & 0 \\ \vdots & \vdots & \ddots & \vdots & \vdots \\ 0 & 0 & \cdots & m_{n-1} & 0 \\ 0 & 0 & \cdots & 0 & m_n \end{bmatrix} \begin{Bmatrix} \ddot{x}_1 \\ \ddot{x}_2 \\ \vdots \\ \ddot{x}_{n-1} \\ \ddot{x}_n \end{Bmatrix} +$$

$$\begin{bmatrix} k_1 + k_2 & -k_2 & 0 & 0 & \cdots & 0 \\ -k_2 & k_2 + k_3 & -k_3 & 0 & \cdots & 0 \\ 0 & -k_3 & k_3 + k_4 & -k_4 & \cdots & 0 \\ \vdots & \vdots & \ddots & \ddots & \ddots & \vdots \\ 0 & 0 & \cdots & -k_{n-1} & k_{n-1} + k_n & -k_n \\ 0 & 0 & \cdots & 0 & -k_n & k_n \end{bmatrix} \begin{Bmatrix} x_1 \\ x_2 \\ x_3 \\ \vdots \\ x_{n-1} \\ x_n \end{Bmatrix} = \begin{Bmatrix} 0 \\ 0 \\ 0 \\ \vdots \\ 0 \\ 0 \end{Bmatrix}.$$

Mechanical vibration responses are similar to steady-state sinusoidal responses of general LTI systems. For free vibration, we can assume that the displacements will be of the form

$$x_k = A_k \sin \omega t.$$

Consequently, the respective accelerations will be

$$\ddot{x} = -A_k \omega^2 \sin \omega t.$$

Substituting into the modal analysis equation (Equation 8.6) yields

$$[M]\{A_k\}(-\omega^2 \sin \omega t) + [K]\{A_k\} \sin \omega t = \{0\}$$
$$\left[-\omega^2 [M] + [K]\right] \{A_k\} \sin \omega t = \{0\}$$
$$\left[-\omega^2 [M]^{-1}[M] + [M]^{-1}[K]\right] \{A_k\} \sin \omega t = \{0\}$$
$$\left[-\omega^2 [I] + [M]^{-1}[K]\right] \{A_k\} \sin \omega t = \{0\}$$
$$\left[\omega^2 [I] - [M]^{-1}[K]\right] \{A_k\} \sin \omega t = \{0\}.$$

Note that the terms in square brackets were left multiplied by the $[M]^{-1}$. For the equation to hold true for all time, we require

$$\left[\omega^2 [I] - [M]^{-1}[K]\right] \{A_k\} = 0$$
$$\left[\lambda [I] - [M]^{-1}[K]\right] \{A_k\} = 0$$
$$\left[\lambda [I] - [A]\right] \{A_k\} = 0 \tag{8.9}$$

where $[A] = [M]^{-1}[K]$. Equation 8.9 is the modal form of the eigenvalue problem where the eigenvalues, λ, are the squares of the system natural frequencies (i.e., $\omega_1^2, \omega_2^2, \ldots, \omega_n^2$) and the corresponding eigenvectors $\{A_k\}$ are the respective *mode shapes*. Though mathematically

$$\omega = \pm\sqrt{\lambda}$$

only positive roots have physical meaning.

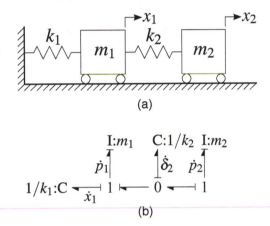

(a)

(b)

Figure 8.10: *Mechanical system with two degrees of freedom.*

Example 8.4

Find the natural frequencies of vibration and respective mode shapes for the two-degree-of-freedom structure model of an axially vibrating beam depicted in Figure 8.10 assuming that $k_1 = 2k_2$ and $m_1 = 2m_2$.

Solution. For simplicity and to drop the subscripts, let $k_2 = k$ and $m_2 = m$. Given the model and respective bond graph, the state equations are

$$\dot{x}_1 = \frac{p_1}{m_1} = \frac{p_1}{2m},$$

$$\dot{p}_1 = -k_1 x_1 + k_2 \delta_2 = -2k x_1 + k\delta_2,$$

$$\dot{\delta}_2 = -\frac{p_1}{m_1} + \frac{p_2}{m_2} = -\frac{p_1}{2m} + \frac{p_2}{m}, \text{ and}$$

$$\dot{p}_2 = -k_2 \delta_2 = -k\delta_2.$$

Recall from Equations 8.7 and 8.8 that $\delta = x_2 - x_1$ and that $\dot{p}_1$ and $\dot{p}_2$ are $m_1\ddot{x}_1 = 2m\ddot{x}_1$ and $m\ddot{x}_2$. As a result, we can reformulate the four first-order differential equations into an equivalent set of two second-order differential equations,

$$2m\ddot{x}_1 = -2kx_1 + k(x_2 - x_1)$$
$$m\ddot{x}_2 = -k(x_2 - x_1)$$

or

$$2m\ddot{x}_1 + 3kx_1 - kx_2 = 0$$
$$m\ddot{x}_2 - kx_1 + kx_2 = 0$$

In set and tensor form, the two second-order differential equations become

$$\begin{bmatrix} 2m & 0 \\ 0 & m \end{bmatrix} \begin{Bmatrix} \ddot{x}_1 \\ \ddot{x}_2 \end{Bmatrix} + \begin{bmatrix} 3k & -k \\ -k & k \end{bmatrix} \begin{Bmatrix} x_1 \\ x_2 \end{Bmatrix} = \begin{Bmatrix} 0 \\ 0 \end{Bmatrix}$$

where

$$[M] = \begin{bmatrix} 2m & 0 \\ 0 & m \end{bmatrix} \quad \text{and} \quad [K] = \begin{bmatrix} 3k & -k \\ -k & k \end{bmatrix}.$$

To find the natural frequencies we need to solve for the values that satisfy

$$\det\left[\omega^2[I] - [M]^{-1}[K] \right] = 0.$$

Substituting for $[M]$ and $[K]$ results in

$$\det\left\{ \begin{bmatrix} \omega^2 & 0 \\ 0 & \omega^2 \end{bmatrix} - \begin{bmatrix} 2m & 0 \\ 0 & m \end{bmatrix}^{-1} \begin{bmatrix} 3k & -k \\ -k & k \end{bmatrix} \right\} = 0$$

$$\det\left\{ \begin{bmatrix} \omega^2 & 0 \\ 0 & \omega^2 \end{bmatrix} - \frac{1}{2m} \begin{bmatrix} 1 & 0 \\ 0 & 2 \end{bmatrix} k \begin{bmatrix} 3 & -1 \\ -1 & 1 \end{bmatrix} \right\} = 0$$

$$\det\left\{ \begin{bmatrix} \omega^2 & 0 \\ 0 & \omega^2 \end{bmatrix} - \frac{k}{2m} \begin{bmatrix} 3 & -1 \\ -2 & 2 \end{bmatrix} \right\} = 0$$

$$\det\begin{bmatrix} \omega^2 - \dfrac{3k}{2m} & \dfrac{k}{2m} \\ \dfrac{k}{m} & \omega^2 - \dfrac{k}{m} \end{bmatrix} = 0$$

$$\left(\omega^2 - \frac{3k}{2m} \right)\left(\omega^2 - \frac{k}{m} \right) - \frac{k^2}{m^2} = 0$$

$$\omega^4 - \frac{5k}{2m}\omega^2 + \frac{k^2}{2m^2} = 0$$

$$\left(\omega^2 - \frac{2k}{m} \right)\left(\omega^2 - \frac{k}{2m} \right) = 0,$$

which has solutions

$$\omega_1^2 = \frac{2k}{m} \quad \text{or} \quad \omega_1 = \sqrt{\frac{2k}{m}}$$

and

$$\omega_2^2 = \frac{k}{2m} \quad \text{or} \quad \omega_2 = \sqrt{\frac{k}{2m}}.$$

The solutions can be derived by substituting $\lambda = \omega^2$ and applying the quadratic equation.

To determine the respective eigenvectors, we need to substitute each eigenvalue back into the eigenvalue problem

$$\left[\omega^2 [I] - [M]^{-1}[K] \right] \{A_k\} = 0$$

$$\begin{bmatrix} \omega_k^2 - \dfrac{3k}{2m} & \dfrac{k}{2m} \\ \dfrac{k}{m} & \omega_k^2 - \dfrac{k}{m} \end{bmatrix} \begin{Bmatrix} A_1 \\ A_2 \end{Bmatrix} = 0.$$

The first eigenvector is found by substituting ω_1 into the eigenvalue problem,

$$\begin{bmatrix} \dfrac{2k}{m} - \dfrac{3k}{2m} & \dfrac{k}{2m} \\ \dfrac{k}{m} & \dfrac{2k}{m} - \dfrac{k}{m} \end{bmatrix} \begin{Bmatrix} A_1 \\ A_2 \end{Bmatrix} = 0$$

$$\frac{k}{2m} \begin{bmatrix} 1 & 1 \\ 2 & 2 \end{bmatrix} \begin{Bmatrix} A_1 \\ A_2 \end{Bmatrix} = 0.$$

The resulting eigenvector is

$$\begin{Bmatrix} 1 \\ -1 \end{Bmatrix}.$$

Similarly, substituting the second eigenvalue, ω_2, yields

$$\begin{bmatrix} \dfrac{k}{2m} - \dfrac{3k}{2m} & \dfrac{k}{2m} \\ \dfrac{k}{m} & \dfrac{k}{2m} - \dfrac{k}{m} \end{bmatrix} \begin{Bmatrix} A_1 \\ A_2 \end{Bmatrix} = 0$$

$$\frac{k}{2m} \begin{bmatrix} -2 & 1 \\ 2 & -1 \end{bmatrix} \begin{Bmatrix} A_1 \\ A_2 \end{Bmatrix} = 0$$

making the eigenvector

$$\begin{Bmatrix} 1 \\ 2 \end{Bmatrix}.$$

The analysis indicates that the masses will naturally oscillate at a frequency of $\omega_1 = \sqrt{2k/m}$ in opposing directions with equal amplitude. Alternatively, the masses can oscillate at the second natural frequency $\omega_2 = \sqrt{k/2m}$ in the same direction, but the second mass will move with twice the amplitude of the first.

8.6   AC Circuits

Alternating current (AC) circuits, like the simple circuit illustrated in Figure 8.11, are excited by sinusoidal inputs and have sinusoidal voltages and currents. They are analyzed in much the same manner as the mass-spring-damper systems in the previous section. Steady-state analysis of AC circuits is called *phasor analysis* and dates back to the nineteenth century when alternating current was first introduced (Schwarz and Oldham 1993). AC circuits are ubiquitous. They are used in every household device and appliance connected to an electric socket. As with mechanical vibration, this is only a brief introduction to AC circuits and phasor analysis.

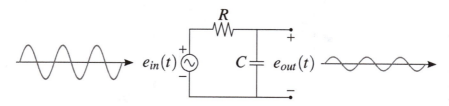

Figure 8.11: *A basic AC circuit.*

In electrical circuit problems, phasor analysis is a classical tool for evaluating AC circuits (Schwarz and Oldham 1993). As depicted in Figure 8.12, *phasors* are vector-like representations in the complex plane of the sinusoidal signals in AC circuits. The phasor magnitude is the amplitude of the sinusoid and the phasor angle varies around the circle as the sinusoid progresses in time through its cycles.

The concept of impedance introduced in Chapter 6 is a generalization of phasor analysis. However, because the inputs are assumed to be sinusoidal, the AC circuit impedances are complex and are determined by simply setting s to $j\omega$ (i.e., $s = j\omega$).

The impedance of a circuit or network, in phasor analysis, is defined as the ratio of the sinusoidal voltage phasor across it to the sinusoidal current

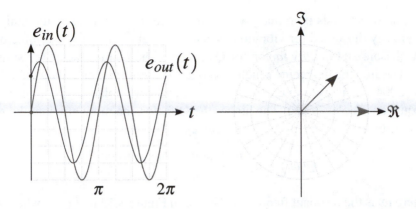

Figure 8.12: *Relation between phasor and sinusoid.*

phasor through it (Schwarz and Oldham 1993),

$$Z = \frac{e(j\omega)}{i(j\omega)}.$$

Recall the impedances for a resistor, capacitor, and inductor listed in Table 6.1. The phasor equivalents are

$$Z_R = \frac{e_R(j\omega)}{i_R(j\omega)} = R, \tag{8.10}$$

$$Z_C = \frac{e_C(j\omega)}{i_C(j\omega)} = \frac{1}{j\omega C} = \frac{1}{j\omega C}\left(\frac{j}{j}\right) = -\frac{j}{\omega C}, \text{ and} \tag{8.11}$$

$$Z_L = \frac{e_L(j\omega)}{i_L(j\omega)} = j\omega L. \tag{8.12}$$

Notice that the impedance of the resistive element is real while those of the capacitive and inductive elements are imaginary. It is common to denote the real and imaginary parts of the impedance as the *resistive* and *reactive* components. The general impedance is

$$Z(j\omega) = \Re[Z(j\omega)] + j\Im[Z(j\omega)]$$

where the real part, $\Re[Z(j\omega)]$, is the resistive component and the imaginary part, $\Im[Z(j\omega)]$, is the reactive. In passive circuits, the resistive part of the impedance is always positive while the reactive part can be either positive or negative.

Phasor analysis is in many ways similar to the transmissibility analysis previously discussed for vibration systems. It can be used to determine output phasors and *quality factor* (or Q factor). The quality factor (or simply Q) measures the degree to which a resonating system is underdamped (Harlow 2004). It is used to characterize the bandwidth relative to the resonant frequency (Tooley 2006). The higher the Q factor, the more efficient the resonator. In AC circuits, the quality factor is

$$Q(\omega_r) = \omega_r \frac{\text{maximum energy stored}}{\text{power loss}}$$

where ω_r is the resonant frequency. Notice in Figure 8.13 that ω_r is where the peak magnitude occurs. The quality factor can also be determined in terms of the resonant frequency and bandwidth,

$$Q = \frac{\omega_r}{\Delta\omega} \tag{8.13}$$

where the bandwidth, $\Delta\omega$, as illustrated in Figure 8.13, is the frequency range over which the magnitude of the resonator impedance, $|Z|$, exceeds $1/\sqrt{2}$ of the maximum. This is a resonator specific definition of bandwidth. As we shall see in Section 8.7, a more general definition exists in terms of the magnitude of the sinusoidal transfer function.

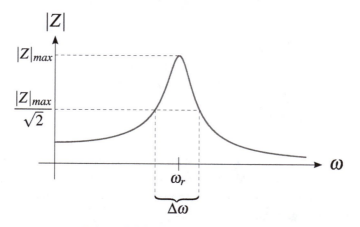

Figure 8.13: *Resonator bandwidth.*

When excited by a sinusoidal input, resonators with a high quality factor will oscillate at a higher amplitude near the resonant frequency. The higher the quality factor, the higher the amplitude. However, increasing the quality factor narrows the bandwidth as illustrated by Figure 8.14. Because of their narrow bandwidth, circuits with a high quality factor can be used to selectively filter a radio frequency, for example. Though most commonly used in the analysis of AC circuits, the concept of quality factor can be generally related to other forms of resonators.

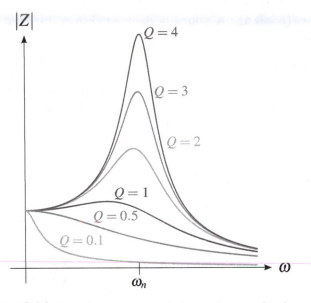

Figure 8.14: *Impedance magnitude for various quality factors.*

Physically, the quality factor is the ratio of energy stored to energy dissipated per cycle. It is a dimensionless quantity that compares the natural frequency of oscillation to the rate at which the resonator dissipates energy (Crowell 2010). The damping ratio and quality factor are related:

$$\zeta = \frac{1}{2Q}. \tag{8.14}$$

Recall the characteristic polynomial for the prototypical second-order system (Equation 7.14). Using Equation 8.14 to substitute for the damping ratio in Equation 7.14 provides an alternative form,

$$s^2 + \frac{\omega_n}{Q}s + \omega_n^2 = 0. \tag{8.15}$$

In general terms, the quality factor is an alternative to damping ratio as a means of quantifying the damped behavior of a simple oscillator:

▷ An overdamped system ($\zeta > 1$) is said to have *low quality factor* ($Q <$ 0.5). When disturbed, it does not oscillate but rather exponentially decays.

▷ A critically damped system ($\zeta = 1$) is said to have *intermediate quality factor* ($Q = 0.5$). A critically damped system does not oscillate nor overshoot when perturbed by a step input.

▷ An underdamped system ($\zeta < 0.5$) is said to have *high quality factor* ($Q > 0.5$). Underdamped systems have an attenuated oscillatory response. When lightly damped, they have higher quality and will oscillate many times before damping out.

Example 8.5

Given the circuit in Figure 8.15, determine the resistance and reactance of the overall impedance.

Solution. Using the bond graph in Figure 8.15 (b) we can find the overall circuit impedance

$$
Z_{total}(s) = \frac{(Ls+R)\left(\dfrac{1}{Cs}\right)}{Ls+R+\dfrac{1}{Cs}} = \frac{(Ls+R)\left(\dfrac{1}{Cs}\right)}{Ls+R+\dfrac{1}{Cs}}\left(\frac{Cs}{Cs}\right)
$$

$$
= \frac{Ls+R}{LCs^2+RCs+1}.
$$

The phasor equivalent impedance is thus

$$
Z_{total}(j\omega) = \frac{R+j\omega L}{(1-\omega^2 LC)+j\omega RC} \tag{8.16}
$$

$$
= \frac{R+j\omega L}{(1-\omega^2 LC)+j\omega RC}\left[\frac{(1-\omega^2 LC)-j\omega RC}{(1-\omega^2 LC)-j\omega RC}\right]
$$

$$
= \frac{(R-\omega^2 LRC+\omega^2 LRC)+j\omega(L-\omega^2 L^2 C-R^2 C)}{(1-\omega^2 LC)^2+(\omega RC)^2}
$$

$$
= \frac{R}{(1-\omega^2 LC)^2+(\omega RC)^2} + j\,\frac{\omega(L-\omega^2 L^2 C-R^2 C)}{(1-\omega^2 LC)^2+(\omega RC)^2}. \tag{8.17}
$$

Note that to isolate the real and imaginary parts of the total impedance, we multiplied the top and bottom by the complex conjugate of the denominator of the impedance. The resistive and reactive parts of the impedance in Equation 8.17 are the real and imaginary parts, respectively. The resistance and reactance are

$$\Re[Z(j\omega)] = \frac{R}{(1-\omega^2 LC)^2 + (\omega RC)^2}$$

and

$$\Im[Z(j\omega)] = \frac{\omega(L-\omega^2 L^2 C - R^2 C)}{(1-\omega^2 LC)^2 + (\omega RC)^2}.$$

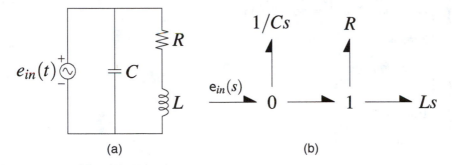

Figure 8.15: *A sample circuit for phasor analysis.*

Example 8.6

What are the resonant frequency, bandwidth, and quality factor of the circuit in Figure 8.15 if $L = 20$ mH, $R = 10\,\Omega$, and $C = 40\,\mu$F.

Solution. This problem is easiest solved numerically by plotting the magnitude frequency response and identifying the resonant frequency and bandwidth and then using Equation 8.13 to calculate the quality factor.

Let us plot the magnitude of the impedance and determine its characteristics. This can be readily done using MATLAB. To do so, we need to determine the magnitude of the impedance. Given the impedance in Equation 8.16, the magnitude of the impedance is

$$|Z_{total}| = \frac{\sqrt{R^2 + (\omega L)^2}}{\sqrt{(1-\omega^2 LC)^2 + (\omega RC)^2}}.$$

The function can be plotted in MATLAB and the data cursor used to determine the maximum impedance magnitude. Additionally, the plot can be used to find the upper and lower bounds of the bandwidth which is evaluated at $1/\sqrt{2}$ times the maximum impedance magnitude. Alternatively, instead of using the data cursor to find the frequency at the maximum magnitude, the max command can be used to find this value in the array Z and its corresponding index (i.e., its place in the array). Knowing the index, we can then find the correlating frequency which should be ω_r.

```
>> Z = ...
@(w) sqrt((R^2+(w*L).^2)./((1-w.^2*L*C).^2+(w*R*C).^2));
>> w = 1:1:2500;
>> plot(w,Z(w))
>> xlabel('\omega (rad/s)')
>> ylabel('|Z(j\omega)|')
>> [Zmax,Imax] = max(Z(w))

Zmax =

    44.7785


Imax =

    987

>> wr = w(Imax)

wr =

    987
```

The lower and upper bounds are 756 and 1,276 rad/s, respectively Thus the bandwidth is 500 rad/s. Though the data cursor can be used to pinpoint the bounds of the bandwidth, yet another MATLAB function can be employed to determine this numerically. The MATLAB find command is used to find elements in an array that meet a specific criteria. The command also has options for finding either the first and final values in the array that meet the criteria if there exists more than one. The first input is the condition, the second is the number of values to find that meet the condition, and the third specifies whether to seek the first or the last values that meet the criteria. In the following, we seek the first, single value and the last, single value that is less than or equal to the bandwidth magnitude. Note that there are a whole range of values that meet either criteria, but we only seek one in each case – the lower bound and the upper bound.

```
>> Zbw = Zmax/sqrt(2);
>> First = find((Z(w)>=Zbw),1,'first');
>> Last = find((Z(w)>=Zbw),1,'last');
```

```
>> w1 = w(First)

w1 =

    756

>> w2 = w(Last)

w2 =

        1256

>> BW = w2-w1

BW =

    500
```

Figure 8.16 shows the magnitude of the impedance plotted and identifies the bandwidth. Note that the peak is centered at the resonant frequency.

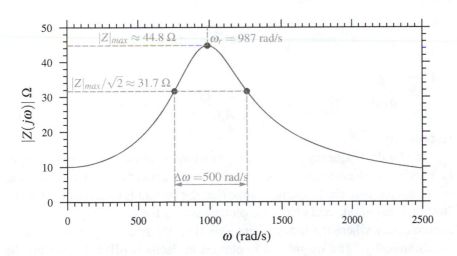

Figure 8.16: *Frequency-varying impedance magnitude for the circuit in Figure 8.15.*

8.7 The Bode Diagram

Thus far, we have used several frequency-varying, magnitude plots to examine transmissibility of vibrating systems and quality factor of AC circuits. The most commonly used frequency response plots are Bode diagrams. Developed by the engineer Hendrik Wade Bode (1905-1982) while working at Bell Labs in the 1930s (Ogata 2004), a Bode plot or diagram is a semilogarithmic plot of both the transfer function magnitude and phase angle.

As has been previously discussed and as is depicted in Figure 8.17, if the input to the LTI system is $u(t) = A_u \sin \omega t$ we expect the output to take the form $y(t) = A_y \sin(\omega t + \phi)$. Furthermore, a system with a transfer function $G(s)$ with a sinusoidal input can be represented as $G(j\omega)$. Therefore, for sinusoidal inputs, the input-output relationship for the system transfer function is a function of the frequency of the input, ω,

$$\frac{y(j\omega)}{u(j\omega)} = G(j\omega) = \Re[G(j\omega)] + j\Im[G(j\omega)].$$

Recall that the magnitude (Equation 8.1) and phase angle (Equation 8.2) of the system transfer function are

$$|G(j\omega)| = \sqrt{\Re[G(j\omega)]^2 + \Im[G(j\omega)]^2} = \frac{A_y}{A_u}$$

and

$$\phi(\omega) = \underline{/G(j\omega)} = \tan^{-1}\left\{\frac{\Im[G(j\omega)]}{\Re[G(j\omega)]}\right\} = \phi_y(\omega) - \phi_u(\omega)$$

respectively.

To plot the frequency response, we must measure two things – the amplitude ratio and phase angle – as functions of the frequency. The two quantities are measured over the frequency range that the system is expected to operate. Both the magnitude and phase are plotted versus the frequency on semilogarithmic axes where the independent axis (i.e., the frequency axis) is varied logarithmically. The magnitude is plotted in decibels (dB). To convert the amplitude ratio, $|G(j\omega)|$, to decibels, its base 10 logarithm is taken and multiplied by 20,

$$M(\omega) = 20\log|G(\omega)|.$$

Experimentally, Bode plot data can be generated by conducting a *sine sweep test*. As Figure 8.17 (a) implies, this is accomplished by exciting the linear time invariant system with a sinusoidal input of known magnitude and

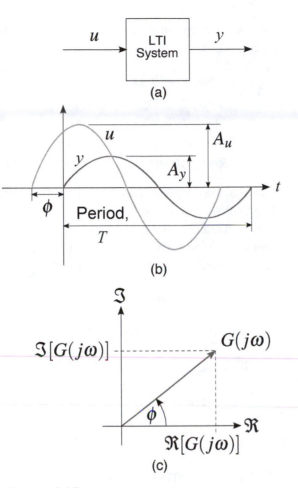

Figure 8.17: *Measuring the frequency response.*

frequency. The steady-state response of the system is measured. In particular, the magnitude and relative phase shift depicted in Figure 8.17 (b) are logged. The frequency of the input sinusoid is varied in logarithmic increments over the frequency range of interest. For each discrete frequency, the measurements of output amplitude and phase shift are repeated. The resultant sinusoidal transfer function will have generally distinct magnitude and phase shift for each frequency.

The frequency response of first-, second-, and higher-order systems are derived in the following sections. To plot the magnitude we must apply a logarithm. To facilitate the discussions that follow recall some basic logarithmic properties. The most common properties of logarithms are the product, quotient, and power rules or

$$\log(xy) = \log x + \log y, \tag{8.18}$$

$$\log(x/y) = \log x - \log y, \text{ and} \tag{8.19}$$

$$\log(x^n) = n \log x. \tag{8.20}$$

8.7.1   Frequency Responses of First-Order Factors

In transfer functions, first-order factors are generally of the form s, $1/s$, $s+a$, or $1/(s+a)$. To generate a Bode plot, we study how the magnitude and phase shift vary with frequency. To garner intuition, we examine how they vary at the extremes (low and high frequency) and determine if there exist asymptotes. Take for example $G(s) = s$. The sinusoidal transfer function is

$$G(j\omega) = j\omega.$$

The real and imaginary parts are 0 and ω, respectively. The phasor for $G(j\omega) = j\omega$ is depicted in Figure 8.18. Applying Equations 8.1 and 8.2, the magnitude and phase shift are

$$|G(j\omega)| = \sqrt{0^2 + \omega^2} = \omega$$

and

$$\underline{/G(j\omega)} = 90°.$$

The magnitude in decibels is $20\log\omega$. The sinusoidal frequency response $G(j\omega) = j\omega$ rises logarithmically at a constant rate of 20 dB/decade.* The Bode plot for $G(j\omega) = j\omega$ is illustrated in Figure 8.19.

A Bode plot can be generated with the aid of MATLAB. To generate a Bode plot in MATLAB we must define an system object using either the ss, tf, or zpk commands. In MATLAB, the command Bode is used to plot the frequency response. The response of $G(s) = s+a$ can be plotted using the following commands assuming the value for a is already specified or defined in the MATLAB workspace.

```
>> G = tf([1 a],1);
>> Bode(G)
```

---

*. A decade is a factor of 10. For instance, a decade would be the range from 1 to 10, 0.1 to 1, 10 to 100, etc.

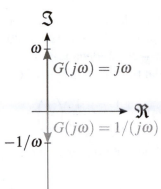

Figure 8.18: *Phasors for $G(j\omega) = j\omega$ and $G(j\omega) = 1/(j\omega)$.*

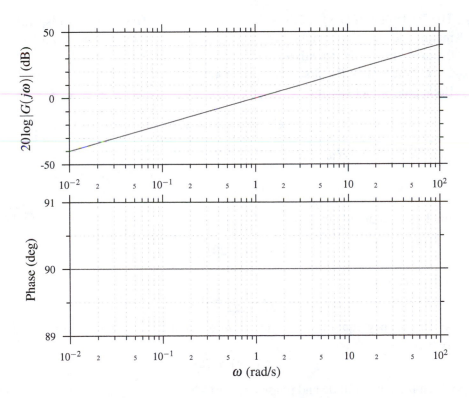

Figure 8.19: *Frequency response of $G(s) = s$.*

Conversely, as the Bode plot in Figure 8.20 portrays, the sinusoidal frequency response of $G(s) = 1/s$ falls at a rate of 20 db/decade (i.e., the slope on a logarithmic scale is -20 db/decade). The phasor is also depicted in Figure 8.18. The sinusoidal transfer function of $G(s) = 1/s$ is

$$G(j\omega) = \frac{1}{j\omega} = \frac{1}{j\omega}\left(\frac{j}{j}\right) = -j\frac{1}{\omega}.$$

The corresponding magnitude and phase shift are

$$|G(\omega)| = \sqrt{0^2 + \frac{1}{\omega^2}} = \frac{1}{\omega}$$

and

$$\underline{/G(j\omega)} = -90°.$$

The Bode plot of $G(s) = 1/s$ is the mirror image of $G(s) = s$. That is, if the magnitude and phase angle plots of $G(s) = s$ were reflected about 0 dB and $0°$, respectively, the resulting Bode plot would be that of $G(s) = 1/s$.

Consider now the first-order polynomial $G(s) = s + a$. To generate the Bode plot, we let $s = j\omega$,

$$G(j\omega) = j\omega + a.$$

At low frequencies ($\omega \ll a$),

$$G(j\omega) \approx a$$

and has a magnitude in decibels of

$$M(\omega) = 20\log a$$

and a phase angle of

$$\underline{/G(j\omega)} = 0°.$$

At a frequency equal to a, $G(j\omega) = ja + a$ and the angle is $45°$. For high frequencies ($\omega \gg a$),

$$G(j\omega) \approx j\omega$$

which has a magnitude and phase angle of

$$M(\omega) = 20\log \omega$$

and

$$\underline{/G(j\omega)} = 90°,$$

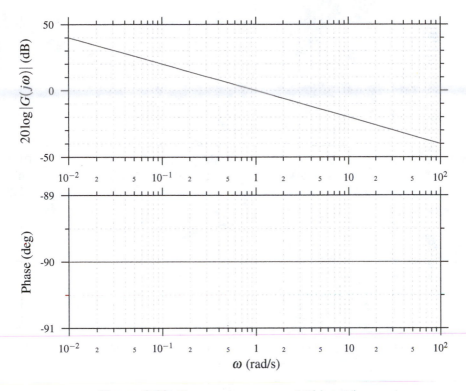

Figure 8.20: *Frequency response of $G(s) = 1/s$.*

respectively. Thus, at high frequency, the magnitude rises logarithmically at a rate of 20 dB/decade. The Bode plot in Figure 8.21 is normalized; the frequency response for $G(j\omega)/a$ is plotted. This is done to illustrate that the magnitude in dB will remain relatively close to the low frequency value, a, until it approaches the *corner frequency*, $\omega = a$, at which point it transitions to a rise at a rate of 20 dB/decade. The corner frequency, also referred to as the *break frequency*, is the point where the low and high frequency asymptotes meet. At the corner frequency, the magnitude for this first-order factor is 3 dB. The phase plot goes from $0°$ at low frequency ($\omega/a \ll 1$), to $45°$ at $\omega = a$ (or $\omega/a = 1$), to $90°$ at high frequency ($\omega/a \gg 1$). The phase has low and high frequency asymptotes of $0°$ and $90°$, respectively.

Now consider the inverse of the previous function, $G(s) = 1/(s+a)$. The

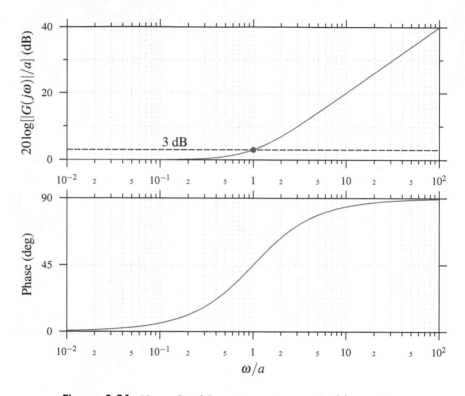

Figure 8.21: *Normalized frequency response of* $G(s) = s + a$.

sinusoidal transfer function is

$$G(j\omega) = \frac{1}{j\omega + a}.$$

At low frequencies,

$$G(j\omega) \approx 1/a$$

which has a magnitude in dB of

$$M(\omega) = 20\log(1/a) = -20\log a$$

and a phase angle of

$$\underline{/G(j\omega)} = 0°.$$

At $\omega = a$, $G(j\omega) = 1/(ja + a)$ and the angle is $-45°$. Approaching high frequencies

$$G(j\omega) \approx j\omega$$

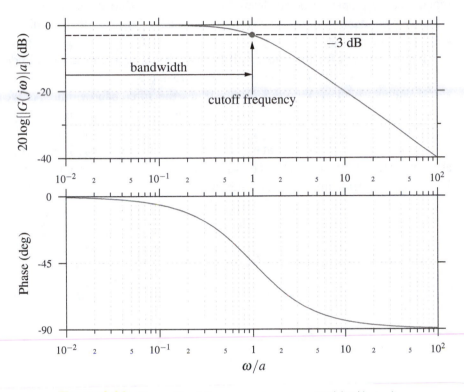

Figure 8.22: *Normalized frequency response of $G(s)1/(s+a)$.*

resulting in a magnitude and phase angle of

$$M(\omega) = 20\log(1/\omega) = -20\log\omega$$

and

$$\underline{/G(j\omega)} = -90°,$$

respectively.

The normalized sinusoidal transfer function is $G(j\omega)a$, and the corresponding Bode plot is provided in Figure 8.22. At low frequency, the magnitude is approximately $1/a$ resulting in a normalized gain of $a/a = 1$ or 0 dB. As the response approaches the break frequency, $\omega = a$ or $\omega/a = 1$, the magnitude transitions from being relatively constant to falling at a rate of 20 dB/decade. For a first-order factor of this form, the break frequency is also referred to as the *cutoff frequency* because beyond this the frequency response attenuates rapidly. More generally, the cutoff frequency is where the magnitude frequency response drops 3 dB below the value at zero frequency. Because we have normalized the sinusoidal transfer function, the magnitude is 0 dB or unity at low frequency.

The bandwidth for the filter is all frequencies for which the magnitude is greater than -3 dB, or, more generally, greater than 3 dB less than the zero-frequency value. Figure 8.22 illustrates the filter bandwidth. As we will see in Example 8.7, this property is used in electric circuits to filter high frequency dynamics. The phase transitions from $0°$ at low frequency, to $-45°$ at $\omega = a$ (or $\omega/a = 1$), to $-90°$ at high frequency. The low and high frequency asymptotes are $0°$ and $-90°$. The Bode plots for $G(s) = s + a$ and $G(s) = 1/(s+a)$ are mirror images of one another.

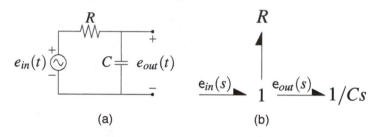

(a)   (b)

Figure 8.23: *An RC circuit.*

Example 8.7

A low-pass filter is an electronic circuit that attenuates frequencies above the cutoff frequency. Take, for example, a simple RC circuit like that depicted in Figure 8.23 if the resistance is 10 Ω and the capacitance is 60 μF. Determine the cutoff frequency and show that the circuit filters sinusoidal components of frequencies higher than the cutoff.

Solution. The cutoff frequency can be found graphically by plotting the frequency response and determining where the response crosses -3 dB. To plot the frequency response, we must first determine the transfer function. Given the bond graph in Figure 8.23 (b), the transfer function relating the output voltage, $e_{out}(t)$, to the input voltage, $e_{in}(t)$, is

$$\frac{e_{out}(s)}{e_{in}(s)} = \frac{\dfrac{1}{Cs}}{R + \dfrac{1}{Cs}} = \frac{1}{RCs + 1} = \left(\frac{1}{RC}\right)\frac{1}{s + \dfrac{1}{RC}}.$$

Given the transfer function, the break frequency is

$$\omega = \frac{1}{RC} = \frac{1}{(10\,\Omega)(100 \times 10^{-6}\,\text{F})} = 1,000 \text{ rad/s}.$$

We can confirm this by plotting the response using MATLAB.

```
>> R = 10;
>> C = 100e-6;
>> G = tf(1,[R*C 1])

G =

        1
   -------------
   0.001 s + 1

Continuous-time transfer function.

>> Bode(G)
>> grid
```

The plot in Figure 8.24 can be used to confirm that the cutoff frequency is indeed 1,000 rad/s. To show that the circuit can filter frequencies above the cutoff, we can excite the circuit with an input with multiple sinusoidal components at frequencies below and above the cutoff. Take, for example,

$$u(t) = u_1(t) + u_2(t) = \sin 100t - \cos 5000t,$$

which is composed of two sinusoids ($u_1(t)$ and $u_2(t)$), one at a frequency of 100 rad/s and another at a frequency of 5,000 rad/s. The first is below the cutoff and the second above. If the circuit functions as a filter, it should attenuate the component at 5,000 rad/s while passing the component at 100 rad/s. This can be confirmed by plotting the time domain response.

```
>> w1 = 100; w2 = 5000;
>> t = (0:1e-5:0.05)';;
>> u1 = sin(w1*t); u2 = -cos(w2*t); u = u1+u2;
>> [Y,T] = lsim(G,u,t);
>> subplot(2,1,1); plot(1000*t,u1,t,u2);
>> ylabel('Input (Volts)')
>> legend('sin(100t)','-cos(5000t)');
>> subplot(2,1,2); plot(1000*t,u,t,Y);
>> ylabel('Response'); xlabel('Time (ms)')
>> legend('u(t)','y(t)')
```

The time domain response is plotted in Figure 8.25. The top subplot shows the components of the input function, $u_1(t)$ and $u_2(t)$. The bottom subplot provides the input, $u(t)$, and output, $y(t)$. Though both components of the input are of the same amplitude, in the output the higher frequency component is attenuated.

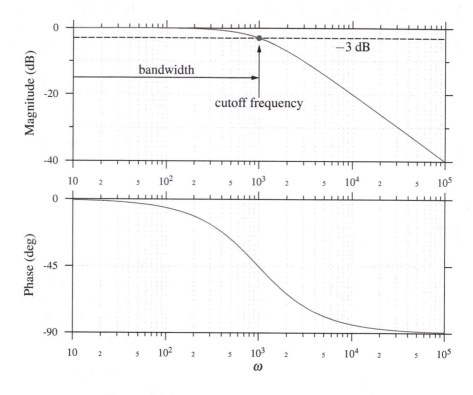

Figure 8.24: *Frequency response of RC circuit.*

8.7.2   Frequency Responses of Second-Order Factors

The second-order transfer function factors, $s^2 + 2\zeta\omega_n s + \omega_n^2$ and $1/(s^2 + 2\zeta\omega_n s + \omega_n^2)$, can be examined in much the same manner to derive their frequency responses. Consider the function $G(s) = s^2 + 2\zeta\omega_n s + \omega_n^2$. Its sinusoidal transfer function is

$$G(j\omega) = (j\omega)^2 + 2\zeta\omega_n j\omega + \omega_n^2$$
$$= (\omega_n^2 - \omega^2) + j2\zeta\omega_n\omega \tag{8.21}$$
$$= \omega_n^2\left\{\left[1 - \left(\frac{\omega}{\omega_n}\right)^2\right] + j2\zeta\frac{\omega}{\omega_n}\right\}. \tag{8.22}$$

At low frequencies ($\omega \ll \omega_n$),

$$G(j\omega) \approx \omega_n^2$$

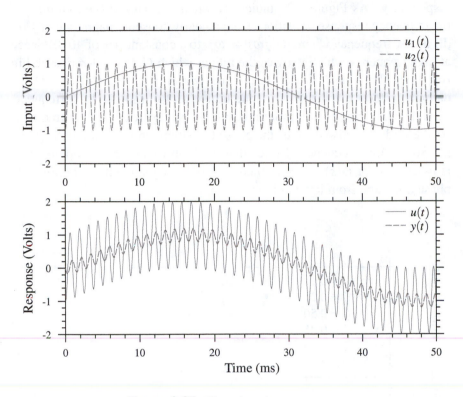

Figure 8.25: *Time domain response.*

which has a magnitude in decibels of

$$M(\omega) = 20 \log \omega_n^2 = 40 \log \omega_n$$

and a phase angle of

$$\underline{/G(j\omega)} = 0°.$$

At the break frequency, $\omega = \omega_n$, $G(j\omega) = j2\zeta\omega_n^2$ and the angle is 90°. For high frequencies ($\omega \gg \omega_n$),

$$G(j\omega) \approx -\omega^2$$

which has a magnitude and phase angle of

$$M(\omega) = 20 \log \omega^2 = 40 \log \omega$$

and

$$\underline{/G(j\omega)} = 180°,$$

respectively. As Figure 8.26 indicates, when normalized (i.e., $G(j\omega)/\omega_n^2$) the frequency response transitions from a nearly constant magnitude of 1 or 0 dB at low frequency where $|G(j\omega)| \approx \omega_n^2$, to a constant rise of 40 dB/decade at high frequency which is twice the rate at which $G(s) = s + a$ rises at high frequency. The transition is centered at the corner frequency ($\omega/\omega_n = 1$). The damping affects the sharpness of the transition from low to high frequency. Less damping results in a more pronounced valley at the corner frequency. As the Bode plot also depicts, the phase transitions from $0°$ at low frequency to $180°$ at high frequency. Notice that the high frequency asymptote, $180°$, is twice that for $G(s) = s + a$. Increased damping smooths or prolongs the phase transition from low to high frequency.

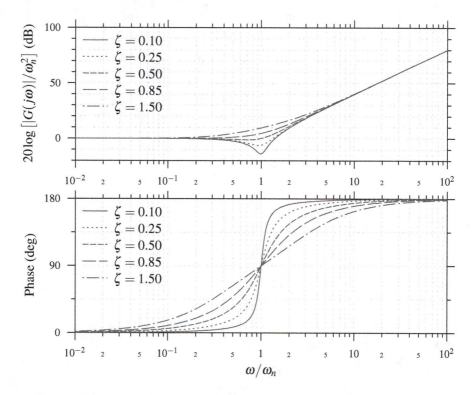

Figure 8.26: *Normalized frequency response of $G(s) = s^2 + 2\zeta\omega_n s + \omega_n^2$.*

Now consider the function $G(s) = 1/(s^2 + 2\zeta\omega_n s + \omega_n^2)$. The sinusoidal transfer function is

$$G(j\omega) = \frac{1}{(j\omega)^2 + 2\zeta\omega_n j\omega + \omega_n^2}$$

$$= \frac{1}{(\omega_n^2 - \omega^2) + j2\zeta\omega_n\omega}$$

$$= \frac{1}{\omega_n^2\left\{\left[1 - \left(\frac{\omega}{\omega_n}\right)^2\right] + j2\zeta\frac{\omega}{\omega_n}\right\}}. \tag{8.23}$$

At low frequencies,

$$G(j\omega) \approx 1/\omega_n^2$$

which has a magnitude in dB of

$$M(\omega) = 20\log 1/\omega_n^2 = -40\log\omega_n$$

and a phase angle of

$$\underline{/G(j\omega)} = 0°.$$

At the corner frequency, $\omega = \omega_n$, $G(j\omega) = 1/j2\zeta\omega_n^2$ and the angle is $-90°$. Approaching high frequencies,

$$G(j\omega) \approx 1/(-\omega^2)$$

resulting in a magnitude and phase of

$$M(\omega) = 20\log(1/\omega^2) = -40\log\omega$$

and

$$\underline{/G(j\omega)} = -180°,$$

respectively. The Bode plot for the normalized transfer function, $G(j\omega)\omega_n^2$, is provided in Figure 8.27. The frequency response mirrors that of $G(s) = s^2 + 2\zeta\omega_n s + \omega_n^2$. The break frequency occurs at $\omega = \omega_n$ and the magnitude has a slope of -40 db/decade at high frequencies. The low and high frequency asymptotes for the phase angle are $0°$ and $-180°$, respectively. The transition is center at $-90°$ and the break frequency, $\omega/\omega_n = 1$. As with $G(s) = s^2 + 2\zeta\omega_n s + \omega_n^2$, the transition is sharper as the damping is decreased. The peak is larger for smaller damping values.

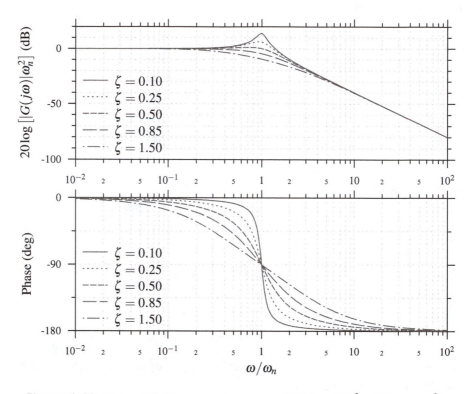

Figure 8.27: *Normalized frequency response of* $G(s) = 1/(s^2 + 2\zeta\omega_n s + \omega_n^2)$.

8.7.3 Frequency Responses of Higher-Order Systems

As has been illustrated, higher-order systems are linear combinations of first- and second-order terms. As was the case with the time-domain response, the frequency-domain response can also be generated through combinations of first- and second-order frequency responses.

The transfer function of any LTI system can be written in the form

$$G(s) = \frac{K(s+z_1)(s+z_2)\ldots(s+z_m)}{s^k(s+p_{k+1})(s+p_{k+2})\ldots(s+p_n)}.$$

The zeros and poles may be real or complex. This means that the numerator and denominator of $G(s)$ can be generally written as the multiplication of a series of first- and/or second-order polynomials. The magnitude of the sinusoidal transfer function is

$$|G(j\omega)| = \frac{K|s+z_1||s+z_2|\ldots|s+z_m|}{|s^k||s+p_{k+1}||s+p_{k+2}|\ldots|s+p_n|}\bigg|_{s\to j\omega}.$$

The product and quotient rules can be used to derive the magnitude in decibels as a summation where the zero terms are added and the pole terms subtracted:

$$\begin{aligned} M(\omega) &= 20\log|G(j\omega)| \\ &= \left[20\log K + 20\log|s+z_1| + \cdots + 20\log|s+z_k|\right. \\ &\quad \left. - 20\log|s^k| - 20\log|s+z_{k+1}| - \cdots - 20\log|s+z_n|\right]_{s\to j\omega}. \end{aligned}$$

In other words, the magnitude of the frequency response for each component of $G(j\omega)$ can be summed graphically to compose the overall magnitude, $M(\omega)$, plot. Similarly, according to Equations 5.1 and 5.2, the phase angles of the individual factors that compose $G(j\omega)$ can be summed – zero terms added and pole terms subtracted – to generate the overall phase angle, $\angle G(j\omega)$, plot. This will be illustrated in the example that follows.

Example 8.8

A vehicle suspension functions much like a filter. It "filters" vibrations such that certain frequencies are not transmitted to the vehicle or its occupants. Plot the frequency response of the quarter-car suspension model (Examples 8.1 and 8.2) and determine which frequencies are attenuated.

Solution. Recall from previous examples that the quarter-car suspension transfer function is

$$\frac{y(s)}{y_{road}(s)} = G(s) = \frac{bs+k}{ms^2+bs+k}.$$

Also recall that the mass, damping, and stiffness are 500 kg, 8,000 N-s/m, and 34,000 N/m. The frequency response is readily generated with MATLAB.

```
>> m = 500; b = 8000; k = 34000;
>> G = tf([b k],[m b k])

G =

       8000 s + 34000
   ------------------------
   500 s^2 + 8000 s + 34000
```

```
Continuous-time transfer function.

>> w = logspace(0,3,100);
>> Bode(G,w); grid
```

The response is plotted in Figure 8.28. The data cursor can be used to confirm that the cutoff frequency is just under 21 rad/s. Thus, any frequency components that exceeds 21 rad/s will be significantly attenuated. It can be shown that like the low-pass filter from Example 8.7, the suspension will pass low frequencies or those significantly less than 21 rad/s.

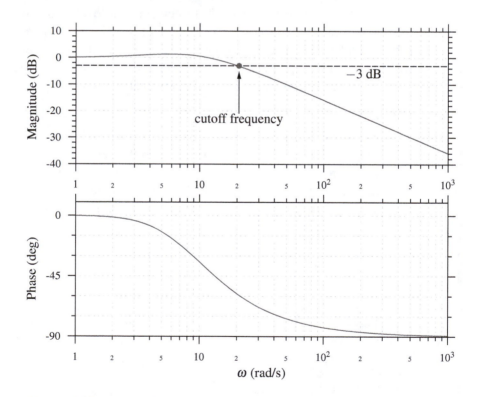

Figure 8.28: *Magnitude frequency response of quarter-car suspension model.*

Example 8.9

Notice that the though the quarter-car suspension system in the previous example is second-order, its transfer function is composed of a first-order factor in the numerator and a second-order factor in the denominator. Notice that the high-frequency asymptote of the magnitude in Figure 8.28 falls at a rate of 20 dB/decade like the first-order factor $G(s) = 1/(s+a)$. That is because the frequency responses of the factors of the transfer function combine at high frequency to generate the equivalent of a first-order frequency response. Show that this is the case by decomposing the frequency response of the quarter-car suspension.

Solution. The quarter-car suspension transfer function is composed of the factors

$$G_1(s) = bs + k = b\left(s + \frac{k}{b}\right)$$

and

$$G_2 = \frac{1}{ms^2 + bs + k} = \left(\frac{1}{m}\right)\frac{1}{s^2 + \dfrac{b}{m}s + \dfrac{k}{m}}$$

where

$$G(s) = G_1 G_2.$$

If

$$G_1(j\omega) = k + j\omega b,$$

at low frequency

$$G_1(j\omega) \approx k.$$

Thus, the magnitude of the first factor will have a low frequency $(\omega \ll k/b)$ magnitude of $20\log k = 20\log(34{,}000) \approx 90.6$ dB. The cutoff frequency occurs at

$$\omega = \frac{k}{b} = \frac{34{,}000 \text{ N/m}}{8{,}000 \text{ N-s/m}} = 4.25 \text{ rad/s}.$$

At high frequency

$$G(j\omega) \approx j\omega b,$$

and the magnitude will be $20\log \omega b = 20\log \omega + 20\log b$. Therefore, at high frequency $(\omega \gg k/b)$ the first component will rise at a rate of 20 dB/decade. Because this is a first-order factor, the phase angle for this factor goes from $0°$ at low frequency, to $45°$ at the $\omega = k/b$ to $90°$, at high frequency.

The second factor is second-order. The sinusoidal form of the second factor is

$$G_2(j\omega) = \left(\frac{1}{m}\right) \frac{1}{-\omega^2 + j\omega\dfrac{b}{m}s + \dfrac{k}{m}} = \left(\frac{1}{m}\right) \frac{1}{\left(\dfrac{k}{m} - \omega^2\right) + j\omega\dfrac{b}{m}}.$$

At low frequency,

$$G_2(j\omega) \approx 1/k$$

making the asymptote $20\log(1/k) = -20\log k = -20\log(34{,}000) \approx -90.6$ dB. The cutoff frequency is

$$\omega_n = \sqrt{\frac{k}{m}} = \sqrt{\frac{34{,}000 \text{ N/m}}{500 \text{ kg}}} \approx 8.25 \text{ rad/s}.$$

The factor at high frequency is

$$G(j\omega) \approx \frac{1}{m\omega^2}$$

which will have a magnitude in dB of $20\log[1/(m\omega^2)] = -20\log m - 40\log\omega$. Hence, at high frequency, the second factor will fall at a rate of 40 dB/decade. The phase angle will transition from $0°$ at low frequency, to $-90°$ at the cutoff frequency ($\omega = \sqrt{k/m} \approx 8.25$ rad/s), to $-180°$ at high frequency.

As depicted in Figure 8.29, by summing the contributions of the two factors, we can arrive at the overall Bode plot. The factors have low frequency magnitudes that are equal in absolute value but opposite in sign. They cancel each other out at frequencies below the break frequency of the first factor. Because the first factor has a lower corner frequency, it begins to rise before the second factor begins to fall. This is evidenced by the slight hump apparent between the $\omega = k/b$ and $\omega = \sqrt{k/m}$ in the magnitude plot provided in Figure 8.28. Eventually, the second factor begins to fall at a rate twice that of which the first factor rises. This results in an overall magnitude response at high frequency that falls at a rate of 20 db/decade. Notice that net sum of a rise of 20 dB/decade and a fall of 40 dB/decade is a fall of 20 dB/decade. Since both factors approach $0°$ phase angle at low frequency, the overall phase angle starts at a low frequency value of $0°$. The first factor rises to $90°$ and the second falls to $-180°$. The overall high frequency phase angle approaches $90° - 180° = -90°$. Moreover, the phase angle transition is center somewhere between the break frequency of the first factor and the cutoff frequency of the second.

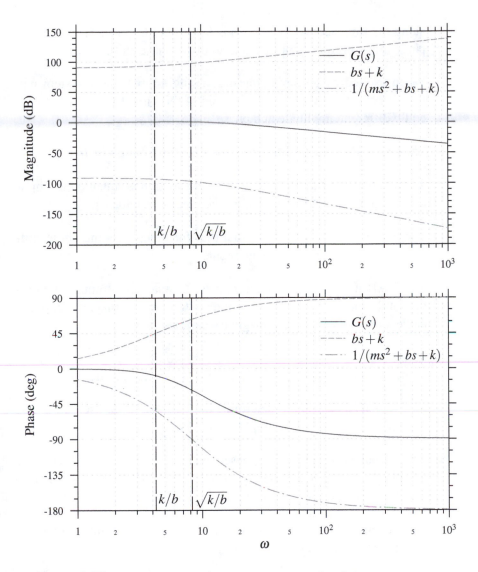

Figure 8.29: *Decomposition of quarter-car suspension frequency response.*

8.8   Summary

▷ The period of oscillation is the time required for a single oscillation cycle.

▷ The phase angle is a measurement of how much a sinusoid is shifted along the time access. One cycle is equivalent to 360°. Hence, a $\pi/2$ phase angle is equivalent to a quarter cycle phase shift.

▷ When excited by an input sinusoid, a single-input linear system will respond at steady-state with a sinusoid of the same frequency but generally different amplitude and phase angle.

▷ The sinusoidal transfer function is attained by letting $s = j\omega$.

▷ The magnitude of the sinusoidal transfer function is equal to the amplitude ratio of the output sinusoid to the input sinusoid.

▷ The phase angle for the sinusoidal transfer function is the phase shift of the output with respect to the input.

▷ MATLAB includes several commands for analysis of complex numbers and functions including commands to compute magnitude, phase angle, complex conjugate, imaginary part, and real part.

▷ In mechanical vibration systems, transmissibility is defined as the ratio of the output displacement (force) to the input displacement (force).

▷ Modal analysis is used to determine the frequencies of oscillation and the relative-displacement, amplitude ratios.

▷ Phasor analysis is a frequency-based method for analyzing AC circuits. The phasor magnitude is equal to the amplitude of the sinusoidal signal. The phasor angle varies as the sinusoidal signal cycles.

▷ The impedance, in phasor analysis, is a sinusoidal transfer function relating effort and flow.

▷ The resonator bandwidth is defined as the frequency range over which the magnitude of the impedance exceeds $1/\sqrt{2}$ of the maximum value.

▷ For a second-order AC circuit, the magnitude of the impedance peaks near the resonant frequency, and the peak decreases with increased damping ratio.

▷ The Bode diagram or plot is a semilogarithmic plot of the transfer function magnitude and phase angle as they vary with frequency. The magnitude is plotted in decibels.

▷ First-order terms of the form s have a magnitude in decibels that *increases* at a rate of 20 dB/decade and a constant phase angle of 90°. The magnitude for terms of the form $1/s$ *decreases* at a rate of 20 dB/decade. Such terms have a constant $-90°$ phase angle.

▷ The corner or break frequency is the frequency at which the magnitude of a term of the form $s+a$ crosses 3 dB and transitions to the high frequency response.

▷ First-order terms of the form $s+a$ have a low frequency (i.e., $\omega \ll a$) magnitude of 0 dB and transition to a high frequency (i.e., $\omega \gg a$) magnitude that increases at a rate of 20 dB/decade as the frequency approaches and crosses the corner or break frequency, $\omega = a$. The phase angle transitions from 0° at low frequency to 90° at high frequency and crosses 45° at the corner frequency.

▷ The cutoff frequency is the frequency at which a frequency response crosses -3 dB. The system tends to attenuate frequencies where the magnitude is less than the cutoff.

▷ At the cutoff frequency, the magnitude of the term $1/(s+1)$ transitions from 0 dB at low frequency to a constant drop of 20 dB/decade. The phase angle goes from 0° at low frequency to $-90°$ at high frequency and crosses $-45°$ at $\omega = a$.

▷ The second-order factor $s^2 + 2\zeta\omega_n 2 + \omega_n^2$ has a magnitude that shifts from 0 dB at low frequency (i.e., $\omega \ll \omega_n$) to a rise of 40 dB/decade at high frequency (i.e., $\omega \gg \omega_n$) as the natural frequency is crossed. The phase angle goes from 0° to 180° and crosses 90° at the natural frequency.

▷ The magnitude for the factor $1/(s^2 + 2\zeta\omega_n 2 + \omega_n^2)$ goes from 0 db to a fall of 40 dB/decade, and the phase angle goes from 0° to $-180°$.

▷ When it comes to higher-order terms, the Bode plot can be composed by summing the plots of the individual first- and second-order factors.

8.9 Review

R8-1 When and why would one prefer to use frequency analysis as opposed to time domain analysis?

R8-2 Explain in your words the concept of phase shift.

R8-3 How can one derive the sinusoidal transfer function?

R8-4 Explain, mathematically, the magnitude and phase angle of the sinusoidal transfer function.

R8-5 List MATLAB functions for analysis of complex numbers and functions.

R8-6 Define the term transmissibility as used in the analysis of mass-spring-damper systems.

R8-7 Describe modal analysis in your own words. In particular, explain the key features or characteristics that are determined using modal analysis.

R8-8 What is the relation between phasors and sinusoids?

R8-9 With regards to second-order, AC circuits, what are the bandwidth and quality factor, and how are they determined?

R8-10 What is the resonant frequency?

R8-11 Describe Bode plots. In particular, detail what is plotted, scales used, and why.

R8-12 How does one compose Bode plots for higher-order systems?

8.10   Problems

P8-1 *Motion Transmissibility.* Determine the transmissibility of the force transferred to the ground relative to the input force, $F(t)$, for the mass-spring-damper system in Figure 1.15 (d).

P8-2 *Steady-State Sinusoidal Response.* Consider the mass-spring-damper systems in Figure 1.15 (d). Obtain the steady-state responses – $y_1(t)$ and $y_2(t)$ – if the force is $F(t) = 10\sin 2t$. Use the system parameters provided in P7-7.

P8-3 *Modal Analysis of Two-Degree-of-Freedom Systems.* For the mass-spring-damper systems in Figure 8.30 (a) and (c), use modal analysis to determine the modes of vibration including the frequencies and amplitude ratios. For (c) assume that $m_2 = 6m_1$ and $k_2 = 4k_1$.

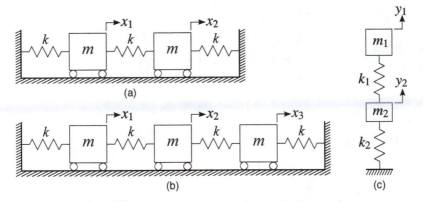

Figure 8.30: *(a) A three-spring, two-degree-of-freedom system, (b) a three-degree-of-freedom system, and (c) a two-spring, two-degree-of-freedom system.*

P8-4 *Modal Analysis of Three-Degree-of-Freedom System.* Consider the mechanical system depicted in Figure 8.30 (b). Determine the modes of vibration.

P8-5 *Bandwidth of a Resonator.* Though the concept of resonator bandwidth was discussed with respect to AC circuits, it can be generally applied to any type of resonator. Recall the simple quarter-car suspension model discussed in Example 8.1 and illustrated in Figure 1.2 (b). Also, remember that the sprung mass, suspension damping, and spring rate are 400 kg, 750 N-s/m, and 34 kN/m. Plot the magnitude of the sinusoidal transfer function relating the vertical vehicle displacement to the road input displacement, and find the resonator bandwidth.

P8-6 *AC Circuit Quality Factor.* A resonator circuit is depicted in Figure 8.31. Given that $R = 5\ \Omega$, $C = 0.01$ F, and $L = 2$ H, plot the magnitude of the sinusoidal impedance, and determine the bandwidth and quality factor.

P8-7 *A Low-Pass Filter.* Figure 8.32 (a) illustrates a simple, first-order, low-pass filter. A low-pass filter is one that passes low frequency dynamics to the output and attenuates the high frequency contributions. The resistance and capacitance are 10 kΩ and 560 pF, respectively. Using MATLAB, generate the Bode plot, and determine the cutoff frequency and bandwidth.

P8-8 *A High-Pass Filter.* The circuit shown in Figure 8.32 (b) is a first-order,

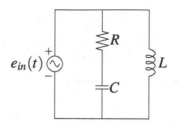

Figure 8.31: *A resonator circuit.*

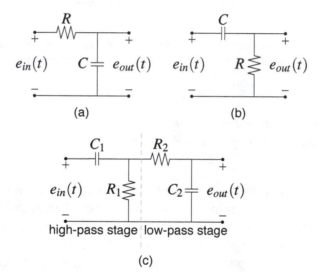

Figure 8.32: *(a) Low-pass filter, (b) high-pass filter, and (c) band-pass filter.*

high-pass filter. It attenuates low frequency inputs. Assume that the resistance and capacitance are 10 kΩ and 15 nF, respectively. Plot the Bode diagram, and determine the corner or break frequency.

P8-9 *A Band-Pass Filter.* A band-pass filter passes signals within a range of frequencies. It attenuates inputs below a lower limit and above an upper limit. A basic band-pass circuit is a tandem combination of a high-pass and low-pass filter as illustrated by Figure 8.32 (c). Assume that the C_1 and R_1 are the same as those used for the high-pass filter in Problem P8-8 and that C_2 and R_2 are the same as those for the low-pass filter in Problem P8-7. Generate the Bode plot, and identify the bandwidth of frequencies that are passed and not attenuated.

P8-10 *Bode Plot of a Mass-Spring-Damper System.* Generate Bode plots for the systems (a) and (b) depicted in Figure 1.15. The plots should be for the responses $v_2(t)$ relative to the input velocity $v(t)$ for system (a) and $x_2(t)$ relative to the input force $F(t)$ for system (b). Assume that $m = 100$ kg, $k = 50$ N/m, $b_1 = 25$ N-s/m, and $b_2 = 75$ N-s/m. Identify any pertinent characteristics like corner frequencies, bandwidths, etc. if they exist.

P8-11 *Bode Plot of a Torsion Plant.* Recall the two-disk, torsion plant from Example 7.3 and Figure 7.17. Plot the Bode diagrams for $\theta_1(t)$ and $\theta_2(t)$ relative to the input torque and identify important characteristics of the two-disk system. What conclusions can you make based on the resultant magnitude and phase plots?

8.11 Challenges

C8-1 *A Toy Electric Car.* Plot the Bode diagram for the toy electric car using parameter values previously provided in Table 4.3. You can make use of the transfer function you previously derived in Chapter 6. What can you discern about the system? Are there any key frequencies or filtering effects?

C8-2 *A Quarter-Car Suspension Model.* You have previously synthesized the impedance bond graph for the quarter-car suspension model in Figure 1.2 (c) and used it to derive the transfer function between the vertical vehicle displacement $y(t)$ and the input road displacement $y_{road}(t)$. Moreover, in Problem P8-5 you plotted the magnitude of the sinusoidal transfer function and determined the resonator bandwidth for the simplified model. Generate the Bode plots for both models for the transfer functions relating the vehicle vertical displacement to the road input displacement. It has been previously mentioned that the suspension absorbs most of the energy at low frequency, and the tire absorbs more energy at high frequency. Explain how the Bode plots demonstrate this. Does the suspension function as a filter? If so, what kind of filter and why? Use the parameters previously provided in Challenge C4-2.

C8-3 *An Electric Drill.* Plot the Bode diagram for the electric drill. You can use the parameters provided Table 4.2 and the transfer function you derived in Chapter 6. Investigate the effect that changing the drill bit has on the resulting Bode plot. Remember that the drill bit can be different lengths and diameters, which will impact its rigidity. Explain your results and what you discovered.

Chapter 9

Classical Control Systems

Thus far in this book we have learned how to model dynamic systems and derive or simulate their transient and steady-state responses. In Chapters 7 and 8 we discussed methods for analyzing the time and frequency domain responses of dynamic systems. The modeling and analysis techniques presented thus far can be expanded upon to design automatic controls, also known as *compensators*. Countless mechanisms and systems are automated. To automate dynamic systems, models are often used to predict the response to control inputs. Analysis is employed to characterize the response. The primary reason for automating a dynamic system is to ensure that the output response has desirable dynamic characteristics. Given the content you have learned in the preceding chapters, ask yourself the following questions:

▷ How do you synthesize a control system?

▷ What criteria are considered when designing a compensator?

▷ How do you characterize the controlled response of a system to assess if it meets design criteria?

▷ What methods exist for designing a control system?

9.1 Introduction

Automatic controls or *automatic control systems* are used to automate a dynamic response. In their simplest form, they utilize a *feedback loop* as depicted in Figure 9.1 to compare a desired dynamic response to the actual or measured response to calculate an error that is used to change the system input in a manner that causes the dynamic system to track the desired response or reach the desired condition at steady-state. The resulting overall system with feedback is called the *closed-loop system*. The error is utilized by the compensator to specify the input necessary to bring the system, also referred to as the *plant*, to the desired output condition. The compensator excites the actuator to generate the input. A sensor is used to measure the output and the signal is fed back to the comparator (or summing block) to determine the error.

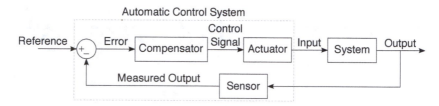

Figure 9.1: *Block diagram of a generic controlled system.*

The most commonly used methods for controlling linear systems can be divided into two major categories – *classical control* and *modern control*. This chapter will focus on the more traditional control methods for linear systems. Entire courses and textbooks are dedicated to the topic of classical control. This chapter is a cursory review of the more important concepts and design methods.

Traditional linear control methods are primarily implemented in the *s*-domain and frequency domain. That is, in classical control, transfer function models are used as opposed to state-space models to design and evaluate control systems. We will explore three primary, classical, compensator design methods – the root locus method, tuning rules, and the Bode plot method. The root locus method is related to placement of poles in the complex plane. Specifically, it visualizes and assesses the impact of compensation on the placement of the dominant, closed-loop poles – the poles that result from implementing the compensator and closing the loop. The largest class of compensators commonly used in Industry are *proportional-integral-derivative* (PID) compensators. They are typically designed and optimized using tuning rules, which we shall discuss later in the chapter. Apart from being used to visualize the frequency response, Bode plots can also be used to design and assess the impact of closed-loop compensation. We will also investigate the use of Bode plots in the design of another form of compensation – lead-lag compensation. The systems that will be examined in this chapter are *single-input-single-output* (or SISO) systems. That is, there is a single output that is controlled using a single input to the system. In the following chapter we will explore methods that are more conducive to *multi-input-multi-output* (or MIMO) problems.

In this chapter, the outcomes and objectives are:

▷ *Objectives:*

1. To understand and assess stability of a controlled system,

2. To design compensators using classical methods, and

3. To understand how to implement some basic controller designs.

▷ *Outcomes:* Upon completing this chapter you will be able to

1. Assess the stability of dynamic systems,

2. Design a compensator using the Root Locus Method,

3. Design a proportional-plus-integral-plus-derivative (PID) controller using tuning rules, and

4. Design lead-lag compensators using Bode plots.

9.2 Block Diagram Algebra

Bond graphs are primarily used to synthesize the models utilized to predict the dynamics of the system, actuator, and sensor in the block diagram. That is, bond graphs are used separately to derive the transfer function or state-space models used in the block diagram for the actuator, system, and sensor. Typically the actuator and sensor respond much more quickly than the system and their steady-state models usually suffice. If the actuator or sensor has no inherent gain, the associated block can be eliminated or a unit gain can be used in its place. This is referred to as *unity feedback* when the sensor has no gain. Unity feedback assumes that the signal is measured directly and instantaneously and sent unaltered to the comparator. For most of the systems we will be analyzing in this chapter, we will assume unity feedback.

Alternatively, a partial block diagram can be incorporated into the bond graph as illustrated by the simple example in Figure 9.2. The figure depicts a compensated first-order torsion plant. The signal associated with the 1-junction (in this case, the angular velocity) is fed back to compare with the desired angular velocity, $\omega_d(t)$, to calculate the error, which is sent to the compensator. The compensator modulates the effort source which varies the input torque to achieve the desired angular velocity.

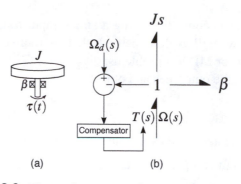

(a)   (b)

Figure 9.2: *First-order torsion plant with feedback control.*

Control-system, block diagrams, however, tend to provide a higher level system overview and are often more convenient for the purpose of compensator design. Take for instance the block diagram in Figure 9.3 which is similar to that in Figure 9.1 except that the transfer functions and signals have been substituted for the labels. The block diagram illustrates that the comparator calculates the difference, $\Delta y(s)$, between the signal of the desired output, $y_d(d)$, and the signal of the measured output, $y_m(s)$. The difference, or error, is fed into the compensator, $G_c(s)$, to calculate the control signal, $c(s)$. The control signal excites the actuator which generates the input, $u(s)$, that will drive the plant, $G(s)$, to adjust the resulting output, $y(s)$. The compensator and plant dynamics are modeled by the transfer functions $G_c(s)$ and $G(s)$, respectively. The actuator and sensor dynamics are represented by the transfer functions $G_a(s)$ and $G_s(s)$. As previously mentioned, impedance bond graphs are utilized to derive the transfer functions for the actuator, plant, and sensor.

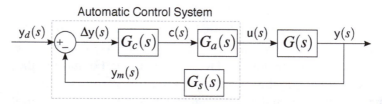

Figure 9.3: *Generic control system block diagram with transfer functions.*

The output signal of each block is simply the input signal times the gain or transfer function inside the block. Take for instance the output $y(s)$ for the generic block diagram in Figure 9.3. The output is related to the input through the following relation:

$$y(s) = G(s)u(s).$$

Similarly, the output can be related to the control signal, $c(s)$, by the following equation:

$$y(s) = G(s)u(s) = G(s)G_a(s)c(s) = G_a(s)G(s)c(s).$$

The *open-loop transfer function* relates the signal error, $\Delta y(s)$, to the feedback or measured signal, $G_s(s)y(s)$. To derive the open-loop transfer function, trace the measured signal, $G_s(s)y(s)$, back around the loop to the error, $y_m(s)$:

$$
\begin{aligned}
y_m(s) &= G_s(s)y(s) \\
&= G_s(s)G(s)\,u(s) \\
&= G_s(s)G(s)G_a(s)\,c(s) \\
&= G_s(s)G(s)G_a(s)G_c(s)\,\Delta y(s)
\end{aligned}
$$

$$\frac{y_m(s)}{\Delta y(s)} = G_s(s)G(s)G_a(s)G_c(s) = G_c(s)G_a(s)G(s)G_s(s). \qquad (9.1)$$

The relationship between the output, $y(s)$, and the error, $\Delta y(s)$, is the *feedforward transfer function*,

$$
\begin{aligned}
y(s) &= G(s)\,u(s) \\
&= G(s)G_a(s)\,c(s) \\
&= G(s)G_a(s)G_c(s)\,\Delta y(s)
\end{aligned}
$$

$$\frac{y(s)}{\Delta y(s)} = G(s)G_a(s)G_c(s) = G_c(s)G_a(s)G(s). \qquad (9.2)$$

If the actuator and sensor transfer functions are unity, the feedforward and open-loop transfer functions are the same,

$$\frac{y_m(s)}{\Delta y(s)} = \frac{y(s)}{\Delta y(s)} = G(s)G_c(s) = G_c(s)G(s).$$

   Closed-loop transfer functions are derived using block diagram algebra. The output, $y(s)$, is related to the reference, $y_d(s)$, by tracking signals through the diagram. Given the transfer functions for the compensator, actuator, plant, and sensor dynamics ($G_c(s)$, $G_a(s)$, $G(s)$, and $G_s(s)$, respectively), the closed-loop transfer function can be determined algebraically:

$$y(s) = G_c G_a G \Delta y(s)$$
$$= G_c G_a G [y_d(s) - G_s y(s)]$$
$$y(s) + G_c G_a G G_s y(s) = G_c G_a G y_d(s)$$
$$[1 + G_c G_a G G_s] y(s) = G_c G_a G y_d(s)$$
$$y(s) = \frac{G_c G_a G}{1 + G_c G_a G G_s} y_d(s)$$
$$\frac{y(s)}{y_d(s)} = \frac{G_c G_a G}{1 + G_c G_a G G_s}. \tag{9.3}$$

Note that "(s)" was dropped from the transfer function variables for the sake of brevity. For many examples, the actuator and sensor models are idealized because, for many systems, the actuator and sensor dynamics are orders of magnitude faster than the plant dynamics. In such cases, the steady-state models of the actuator and sensor are used. A steady-state model simplifies to a simple gain or constant conversion factor. In the case of unity gains for the actuator and sensor, the closed-loop transfer function is

$$\frac{y(s)}{y_d(s)} = \frac{G_c G}{1 + G_c G}. \tag{9.4}$$

9.3 Proportional, Integral, and Derivative Control Actions

Traditional compensators are often composed of proportional, integral, and/or derivative control actions. That is, the control signal is made up of terms that are proportional to the error, its integral, and/or its derivative. PID compensators are of the general form

$$G_c(s) = K_P + \frac{K_I}{s} + K_D s. \tag{9.5}$$

To illustrate the basic PID compensating actions, let us examine the closed-loop response of the speed control system illustrated in Figure 9.4. To begin, let us assume a simple *proportional compensator* (i.e., $G_c = K_P$) with unity gains for the actuator and sensor. Recall from Example 7.1 that the first-order, torsion plant is modeled by the transfer function

$$G(s) = \frac{\Omega(s)}{T(s)} = \frac{1}{Js + \beta}.$$

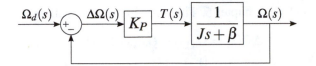

Figure 9.4: *Proportional control of a first-order torsion plant.*

Equation 9.4 is used to find the closed-loop transfer function,

$$G_{cl}(s) = \frac{\Omega(s)}{\Omega_d(s)} = \frac{\dfrac{K_P}{Js+\beta}}{1+\dfrac{K_P}{Js+\beta}} = \frac{K_P}{Js+\beta+K_P}. \tag{9.6}$$

The unit step response is often used to assess the transient and steady-state characteristics for a *set point problem*. A set point control problem is one in which it is desired to transition the system from an initial condition to a final constant steady-state value (or set point). As has been illustrated in Chapter 7, with regards to linear systems, certain characteristics of the step response, like the time constant, do not change regardless of the magnitude of the step input. Since it is convenient, the unit step input is often used instead of a constant input of a magnitude other than unity. It can be shown that the unit step response is

$$\omega(t) = \frac{K_P}{\beta+K_P}\left[1 - e^{-\left(\frac{\beta+K_P}{J}\right)t}\right],$$

which has a steady-state value of

$$\omega_{ss}(t) = \frac{K_P}{\beta+K_P}$$

because the exponential term decays to zero at steady-state. The time constant is

$$T = \frac{J}{\beta+K_P}.$$

Thus, the steady-state error is

$$\Delta\omega_{ss} = \omega_d - \omega_{ss} = 1 - \frac{K_P}{\beta+K_P}.$$

The steady-state error can also be determined using the Final Value Theorem

(Equation 5.7) from Chapter 5. The error in the s-domain is

$$\Delta\Omega(s) = \Omega_d(s) - G_{cl}(s)\Omega_d(s)$$
$$= [1 - G_{cl}(s)]\Omega_d(s)$$
$$= \left[1 - \frac{K_P}{Js + \beta + K_P}\right]\Omega_d(s).$$

For a unit step input (i.e., $\Omega_d(s) = 1/s$), the steady-state error according to the Final Value Theorem is

$$\omega_{ss} = \lim_{s \to 0} s\Delta\Omega(s) = \lim_{s \to 0}\left[1 - \frac{K_P}{Js + \beta + K_P}\right] = 1 - \frac{K_P}{\beta + K_P}.$$

Notice that the control action like the error is constant and nonzero at steady-state. The larger the proportional gain K_P is, the closer the second term is to unity and the smaller the error. Nonetheless, for this particular system, proportion control will always result in some finite, steady-state error as illustrated in Figure 9.5.

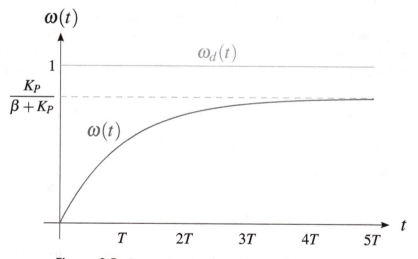

Figure 9.5: *Step response of compensated torsion plant.*

If instead of proportional compensation, *integral compensation* (i.e., $G_c(s) = K_P/s$) is utilized as illustrated in Figure 9.6, the closed-loop transfer function is

$$G_{cl}(s) = \frac{\Omega(s)}{\Omega_d(s)} = \frac{\dfrac{K_I}{s(Js + \beta)}}{1 + \dfrac{K_I}{s(Js + \beta)}} = \frac{K_I}{Js^2 + \beta s + K_I} \tag{9.7}$$

and the error is

$$\Delta\Omega(s) = [1 - G_{cl}(s)]\,\Omega_d(s)$$

$$= \left[1 - \frac{K_I}{Js^2 + \beta s + K_I}\right]\Omega_d(s).$$

Notice that the resultant closed-loop transfer function has a second-order denominator. The roots of this denominator are referred to as the *closed-loop poles* of the compensated system. Use of integral action eliminates the steady-state error:

$$\Delta\omega_{ss} = \lim_{s\to 0} s\,\Delta\Omega(s)$$

$$= \lim_{s\to 0} s\left[1 - \frac{K_I}{Js^2 + \beta s + K_I}\right]\Omega_d(s)$$

$$= \lim_{s\to 0} s\left[1 - \frac{K_I}{Js^2 + \beta s + K_I}\right]\frac{1}{s}$$

$$= \lim_{s\to 0}\left[1 - \frac{K_I}{Js^2 + \beta s + K_I}\right]$$

$$= 1 - \frac{K_I}{K_I} = 0.$$

Thus, for this first-order system, proportional control can reduce steady-state error, but integral control will eliminate it.

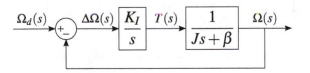

Figure 9.6: *Integral control of first-order torsion plant.*

Derivative compensation responds to time rate of change of actuation error (Ogata 2004). When used in conjunction with proportional control, the derivative control action will respond before the error becomes too large if the error is rapidly changing. This tends to improve the sensitivity of the compensator and increase its stability. The derivative control action is much like the second term in a second-order system which, when positive, dampens the response. Though it can slow the response, added damping tends to improve the stability of the system. Derivative control action is always used in concurrence with proportional or proportional-integral compensation.

Thus far, the discussion has focused on PID compensation of first-order systems. What if instead of controlling the angular velocity, we wish to control the angular position. Because the position is the integral of the angular velocity, the plant transfer function is simply

$$\frac{\Theta(s)}{T(s)} = \left(\frac{1}{Js+\beta}\right)\frac{1}{s} = \frac{1}{s(Js+\beta)}.$$

The corresponding proportional compensation block diagram is given in Figure 9.7. The closed-loop transfer function is

$$G_{cl}(s) = \frac{\Theta(s)}{\Theta_d(s)} = \frac{\dfrac{K_P}{s(Js+\beta)}}{1+\dfrac{K_P}{s(Js+\beta)}} = \frac{K_P}{Js^2+\beta s+K_P}. \qquad (9.8)$$

The resulting closed-loop response is that of a damped second-order system similar to many we have previously examined. Note that the resulting closed-loop transfer function is similar to that of the velocity control with integral compensation (Equation 9.6). Recall the control resulted in zero steady-state error. If, however, the compensated torsion plant is lightly damped and takes many oscillations to settle, it may be necessary to employ a compensator that increases the damping. This can be accomplished by adding derivative compensation.

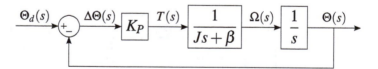

Figure 9.7: *Proportional compensation for position control of torsion plant.*

As was previously explained, derivative action in tandem with proportional action tends to improve stability and hasten attenuation of the response error by increasing damping. The closed-loop transfer function of the block diagram in Figure 9.8 is

$$G_{cl}(s) = \frac{\Theta(s)}{\Theta_d(s)} = \frac{(K_P+K_Ds)\dfrac{1}{s(Js+\beta)}}{1+(K_P+K_Ds)\dfrac{1}{s(Js+\beta)}} = \frac{K_Ds+K_P}{Js^2+(\beta+K_D)s+K_P}.$$

$$(9.9)$$

Note that the compensation adds a zero and enlarges the effective damping of the second-order poles in the denominator (i.e., $\beta + K_D > \beta$). Thus, adding derivative control increases the rate of attenuation as shown by the plot in Figure 9.9.

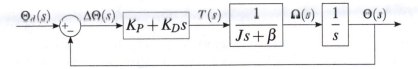

Figure 9.8: *Proportional-plus-derivative compensation for position control of torsion plant.*

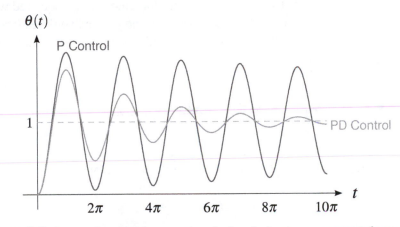

Figure 9.9: *Proportional and proportional-plus-derivative compensated responses.*

9.4 Higher-Order Systems and Dominant Closed-Loop Poles

An automatic control system is implemented to improve the transient and/or steady-state response. The transient step response is primarily evaluated in terms of overshoot and settling time. However, systems are generally higher than second-order. That raises the question, "What about compensating higher-order systems?" In Section 7.5, the idea of dominant poles was introduced. Example 7.4 illustrated how the presence of a zero can negate the impact of a pole on the overall response and how the proximity of a pole to the imaginary axis affects the dominance of that pole. Generally, the closer a pole pair is to the imaginary axis, the longer it takes to attenuate the corresponding dynamics. Recall that the distance of a pair of complex poles,

$$s_{1,2} = -\zeta\omega_n \pm j\omega_n\sqrt{1-\zeta^2},$$

to the imaginary axis is $\zeta\omega_n$. Higher-order systems have multiple pairs of complex poles. If one pair lies farther to the right of the rest (or closer to the imaginary axis), that pair will dominate. The dominance of one complex pair of poles relative to another is thus determined by the ratio of the real parts of those poles. Barring the presence of nearby zeros which would negate the impact of a pole pair, any pair that is five times or more farther from the imaginary axis than the pair closest to the axis will have minimal relative impact. This is due to the fact that settling time (from Equation 7.27) associated with a given pole pair is $t_s = 4T = 4/(\zeta\omega_n)$. Thus, if one pair is five times farther left of the imaginary axis than another, the relative settling time is 20% that of those closest to the axis:

$$\frac{t_{s1}}{t_{s2}} = \frac{4T_1}{4T_2} = \frac{T_1}{T_2} = \frac{\dfrac{1}{5\zeta\omega_n}}{\dfrac{1}{\zeta\omega_n}} = \frac{1}{5} = 20\%.$$

The dynamics of the subordinate pair has a settling time of

$$t_{s1} = 4T_1 = \frac{4}{5\zeta\omega_n},$$

which is less than one time constant of the dominant pair,

$$t_{s1} < T_2 = \frac{1}{\zeta\omega_n}.$$

Hence, the subservient dynamics will have settled to steady-state before the dominant dynamics reaches one time constant (i.e., $T = 1/(\zeta\omega_n)$).

Since a higher-order system can be dominated by a single pair of poles, a compensator can be devised so that the dominant closed-loop poles have the desired dynamics. Specifying the desired dynamics predominantly involves choosing reasonable settling time and overshoot, t_s and M_p. Being that the settling time is inversely proportional to ζ and ω_n, its time can be decreased by increasing either. The maximum overshoot is less obvious. To determine how the overshoot varies, we must examine the second-order step response. The underdamped, second-order step response (Equation 7.39) reaches a maximum at the peak time, t_p. To determine the peak time, we must recognize that the peak occurs at the global maximum where the step response has a slope of zero. Ergo, the peak time can be determined by equating the derivative of the unit step response to zero:

$$\dot{x} = \frac{\omega_n}{\sqrt{1-\zeta^2}} e^{-\zeta\omega_n t} \sin \omega_d t = 0.$$

The derivative is zero when

$$\sin \omega_d t = 0,$$

which occurs when $\omega_d t$ is some multiple of π. The first and tallest peak occurs at the first multiple of π (i.e., $\omega_d t_p = \pi$), and therefore the peak time is

$$t_p = \frac{\pi}{\omega_d} = \frac{\pi}{\omega_n\sqrt{1-\zeta^2}}.$$

The maximum or peak value is attained by substituting the peak time into the step response given in Equation 7.39,

$$x(t_p) = \frac{A}{\omega_n^2}\left[1 - e^{-\zeta\omega_n t_p}\left(\frac{\zeta}{\sqrt{1-\zeta^2}}\sin\omega_d t_p + \cos\omega_d t_p\right)\right]$$

$$= \frac{A}{\omega_n^2}\left[1 - e^{-\frac{\zeta}{\sqrt{1-\zeta^2}}\pi}\left(\frac{\zeta}{\sqrt{1-\zeta^2}}\sin\pi + \cos\pi\right)\right]$$

$$= \frac{A}{\omega_n^2}\left[1 + e^{-\frac{\zeta}{\sqrt{1-\zeta^2}}\pi}\right].$$

Maximum overshoot is the difference between the peak and steady-state values,

$$x(t_p) - x(\infty) = \frac{A}{\omega_n^2}\left[1 + e^{-\frac{\zeta}{\sqrt{1-\zeta^2}}\pi}\right] - \frac{A}{\omega_{n^2}} = \frac{A}{\omega_{n^2}}e^{-\frac{\zeta}{\sqrt{1-\zeta^2}}\pi},$$

and the maximum percent overshoot is

$$M_p = \frac{x(t_p) - x(\infty)}{A/\omega_{n^2}} \times 100 = e^{-\frac{\zeta}{\sqrt{1-\zeta^2}}\pi}. \tag{9.10}$$

The percent overshoot decays with increasing values of damping ratio. For damping ratios greater than unity, there is no overshoot. This is illustrated in Figure 9.10.

Given the relations for the settling time and overshoot, the damping ratio can be increased between 0 and 1 to decrease both the settling time and

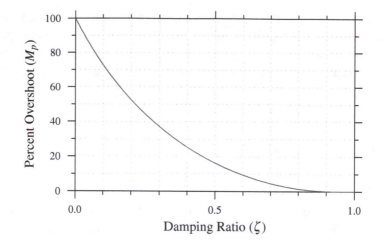

Figure 9.10: *Percent overshoot as a function of damping ratio, ζ.*

overshoot. The damping ratio should be limited to the range of 0.4 to 0.8 (Ogata 2004). Too low a damping ratio results in excessive overshoot and too high results in an increased settling time or slowed response. As damping ratio approaches and exceeds unity, the settling time begins to increase and the response takes longer to reach steady-state. As Figure 9.10 illustrates, a damping ratio between 0.4 and 0.8 will limit the overshoot to between 25% and 2.5%.

9.5   Static Error Constants

In the previous section, the compensators were assessed utilizing the unit step response of the closed-loop system. This is a valid approach for evaluating the transient characteristics of set point problems. However, compensators are also employed to improve the steady-state response. The steady-state is gauged using principally the steady-state error. Furthermore, not all control problems are set point problems. Some controls are designed for *tracking problems* or problems where the set point is varied and it is important how well the response tracks the input as the set point varies. Depending on how the input is varied, a time-varying (instead of constant) input like a ramp or parabola may be used to evaluate the transient characteristics of the response. Figure 9.11 depicts the steady-state error for the ramp and parabolic responses for a second-order system. The ramp response reaches steady-state when it continues to rise at a constant rate. The parabolic response reaches steady-state when it stays parallel to the parabolic input.

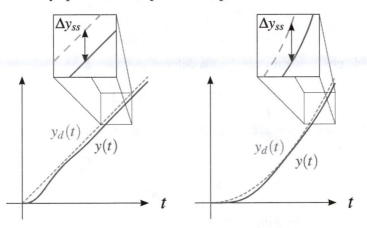

Figure 9.11: *Steady-state errors for ramp and parabolic inputs.*

Error constants are used to quantify and compare the relative accuracy of controlled systems. To derive these constants, we examine the closed-loop response. Take for example the block diagram in Figure 9.12. For simplicity, assume that $G_f(s)$ is the overall feedforward transfer function, which includes compensation, actuation, and plant dynamics and which the system has unity feedback. The error is

$$\Delta y(s) = y_d(s) - y(s)$$
$$= y_d(s) - G_f(s)\Delta y(s)$$
$$\Delta y(s) + G_f(s)\Delta y(s) = y_d(s)$$
$$[1 + G_f(s)]\Delta y(s) = y_d(s)$$
$$\Delta y(s) = \frac{1}{1 + G_f(s)} y_d(s).$$

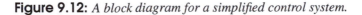

Figure 9.12: *A block diagram for a simplified control system.*

The *static position error constant*, K_p, (not to be confused with K_P the proportional gain), is used as a metric for the closed-loop, unit-step response.

The use of the word "position" is somewhat of a misnomer. Using the Final Value Theorem (Equation 5.7), the steady-state error of the unit step response is

$$\Delta y_{ss} = \lim_{s \to 0} s \frac{1}{1 + G_f(s)} \frac{1}{s} = \lim_{s \to 0} \frac{1}{1 + G_f(s)} = \frac{1}{1 + G_f(0)}.$$

The static position error constant is defined as

$$K_p = \lim_{s \to 0} G_f(s) = G_f(0). \tag{9.11}$$

Thus, the steady-state step response error can be specified in terms of the static position error constant

$$\Delta y_{ss} = \frac{1}{1 + K_p}. \tag{9.12}$$

Given a unit-ramp input, the error is

$$\Delta y(s) = \frac{1}{1 + G_f(s)} \frac{1}{s^2},$$

making the steady-state error

$$\begin{aligned}
\Delta y_{ss} &= \lim_{s \to 0} s \frac{1}{1 + G_f(s)} \frac{1}{s^2} \\
&= \lim_{s \to 0} \frac{1}{1 + G_f(s)} \frac{1}{s} \\
&= \lim_{s \to 0} \frac{1}{s + s G_f(s)} \\
&= \lim_{s \to 0} \frac{1}{s G_f(s)}.
\end{aligned}$$

The *static velocity error constant* is

$$K_v = \lim_{s \to 0} s G_f(s) \tag{9.13}$$

so that the steady-state error in terms of K_v is

$$\Delta y_{ss} = \frac{1}{K_v}. \tag{9.14}$$

In a similar fashion, it can be shown that the *static acceleration error constant*, K_a, used in assessing the unit-parabola, response error, is

$$K_a = \lim_{s \to 0} s^2 G_f(s) \tag{9.15}$$

and the corresponding steady-state error

$$\Delta y_{ss} = \frac{1}{K_a}.$$ (9.16)

Note that in each case – step, ramp, or parabolic input – the greater the error constant the smaller the error.

Example 9.1

Determine the static error constants and respective steady-state errors for the torsion plant, position control system depicted in Figure 9.7. Assume that the rotational inertia is 0.0104 kg-m^2 and the damping constant is 0.01 N-m-s/rad. Evaluate the steady-state responses for two distinct proportional gains of $K_P = 0.01$ and 0.1.

Solution. The open-loop transfer function for the position control system is

$$G_f(s) = K_P \frac{1}{Js+\beta} \frac{1}{s} = \frac{K_P}{s(Js+\beta)}.$$

Do not confuse the proportional gain K_P with the static position error constant K_p. The static position error constant is

$$K_p = \lim_{s \to 0} \frac{K_P}{s(Js+\beta)} = \frac{K_P}{0} = \infty,$$

and therefore the steady-state error for a unit-step input is

$$\Delta\theta_{ss} = \frac{1}{1+K_p} = \frac{1}{\infty} = 0.$$

The system can compensate with no error when the input is a step function.

The static velocity error constant is

$$K_v = \lim_{s \to 0} s \frac{K_P}{s(Js+\beta)} = \lim_{s \to 0} \frac{K_P}{Js+\beta} = \frac{K_P}{\beta}.$$

The corresponding steady-state error is

$$\Delta\theta_{ss} = \frac{1}{K_v} = \frac{\beta}{K_P}.$$

Therefore, given a unit-ramp input, steady-state error remains constant.

For a parabolic input, the static acceleration constant and correlating steady-state error are

$$K_a = \lim_{s \to 0} s^2 \frac{K_P}{s(Js+\beta)} = \lim_{s \to 0} \frac{K_P s}{Js+\beta} = 0$$

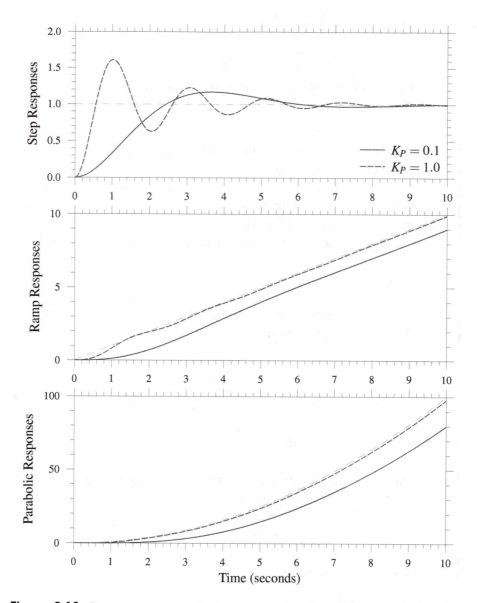

Figure 9.13: *Step, ramp, and parabolic responses for PD position control of torsion plant.*

and

$$\Delta\theta_{ss} = \frac{1}{0} = \infty,$$

respectively. Hence, this system cannot track a parabolic input indefinitely.

At first glance, it may seem that this system is poorly designed for a tracking problem. However, the static error constant and stead-state errors only quantify performance at steady-state. They do not specify how rapidly the error response reaches its steady-state performance. To illustrate this phenomenon, we examine the second part of the problem statement which asks for the performance given two different proportional compensator gains. To assess the set point and tracking performance, we must plot the dynamic responses for the step, ramp, and parabolic inputs. Figure 9.13 shows the responses and illustrates that for both cases, the steady-state, unit-step, response error approaches zero. The ramp responses reach constant errors. The steady-state error for the first compensator ($K_P = 0.01$ N-m/rad) is

$$\Delta \theta_{ss} = \frac{1}{K_v} = \frac{\beta}{K_P} = \frac{0.01}{0.01} = 1 \text{ rad}$$

and

$$\Delta \theta_{ss} = \frac{\beta}{K_P} = \frac{0.01}{0.1} = 0.1 \text{ rad}$$

for the second compensator ($K_P = 0.1$ N-m/rad). Though the parabolic responses approach infinite error at steady-state, the steady-state error increases more rapidly for $K_P = 0.01$ than $K_P = 0.1$. Moreover, at $K_P = 0.1$ the error increases slowly and is relatively negligible even after 10 seconds. If the input follows a parabolic trajectory for short periods and then returns to a constant value, the position control will likely perform satisfactorily for gains greater than 0.1 (i.e., $K_P > 0.1$).

9.6 Stability Analysis

Though the transient and steady-state characteristics are important in assessing the performance of a compensated system, perhaps the more important concern is whether the system is stable or not. Recall from Equation 7.49 that the partial fraction expansion of a higher-order system is generally of the form

$$x(s) = \frac{a}{s} + \sum_{j=1}^{q} \frac{a_j}{s + p_j} + \sum_{k=1}^{r} \frac{b_k(s + \zeta_k \omega_k) + c_k \omega_k \sqrt{1 - \zeta_k^2}}{s^2 + 2\zeta_k \omega_k s + \omega_k^2},$$

and from Equation 7.50 that the corresponding response is

$$x(t) = a + \sum_{j=1}^{q} a_j e^{-p_j t} + \sum_{k=1}^{r} b_k e^{-\zeta_k \omega_k t} \cos \omega_k \sqrt{1 - \zeta_k^2}\, t$$

$$+ \sum_{k=1}^{r} c_k e^{-\zeta_k \omega_k t} \sin \omega_k \sqrt{1 - \zeta_k^2}\, t.$$

Though physically, the exponential terms usually decay, for a compensated system, the closed-loop transfer function can result in exponential terms with positive exponents that increase with time and potentially cause an instability or unbounded response.

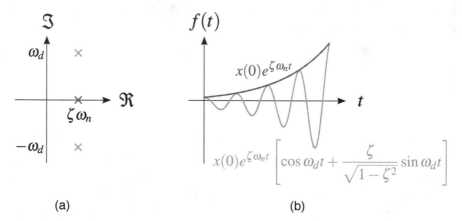

(a)   (b)

Figure 9.14: *(a) Unstable first- and second-order poles and (b) corresponding responses.*

As has been illustrated, the numerator and denominator of a higher-order transfer function can be factored into first- and second-order terms. If any of the poles of the closed-loop transfer function have a positive real part, the matching factor will result in a term in the partial fraction expansion that increases exponentially. Take for example the first- and second-order factors

$$\frac{1}{s - \zeta \omega_n} \quad \text{and} \quad \frac{1}{s^2 - 2\zeta \omega_n + \omega_n^2}$$

which have unstable poles at $s = \zeta \omega_n$ and $s_{1,2} = \zeta \omega_n \pm j \omega_d$, respectively. Figure 9.14 depicts the poles and their corresponding natural responses. If these poles are factors of a closed-loop transfer function, the pole at $s = \zeta \omega_n$ will contribute

$$x(t) = x(0) e^{\zeta \omega_n t}$$

to the closed-loop, natural response. The unstable, complex-conjugate pole pair adds

$$x(t) = x(0)e^{\zeta \omega_n t}\left[\cos \omega_d t + \frac{\zeta}{\sqrt{1-\zeta^2}}\sin \omega_d t\right]$$

to the natural response. If either contribution increases more rapidly than the remaining terms in the closed-loop response, the system will become unbounded or unstable. Figures 9.15 and 9.16 illustrate how the time response contribution of a pair of unstable poles varies as the poles move in the right half of the complex plane. As both figures demonstrate, the response becomes increasingly unstable as the poles move away from the imaginary axis or farther to the right. Left of the imaginary axis, proximity to the axis resulted in longer settling times. Right of the imaginary axis, the response becomes more rapidly unstable farther from the axis.

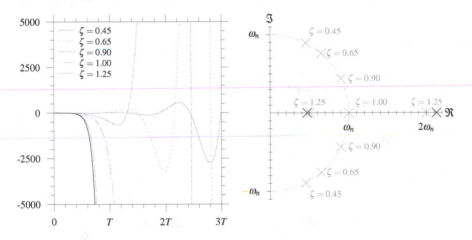

Figure 9.15: *Variation of unstable second-order response with damping constant.*

Thus, to determine if a system is stable, one must discern if any closed-loop poles are in the right half of the complex plane (i.e., the poles have positive real parts). Moreover, if a pole (or poles) are found to be in the right-half plane, we must then ascertain their relative instability. If the transfer function can be readily factored, the unstable poles can be easily determined. However, this is not always the case. One means of determining if any of the poles lie in the right half of the complex plane without factoring the transfer function is *Routh-Hurwitz stability criterion*. Routh-Hurwitz stability criterion is used to determine the absolute stability directly from the coefficients of the characteristic equation (Ogata 2002). It allows one to determine if a system is

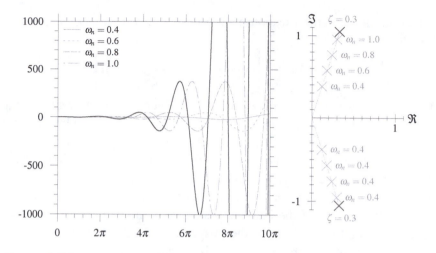

Figure 9.16: *Variation of unstable second-order response with natural frequency.*

unstable and how many, if any, poles are in the right-half plane. The following is the process for discerning the absolute stability using Routh-Hurwitz
stability criterion (Ogata 2002):

1. *Identify the characteristic equation.*   Given the closed-loop transfer
   function

$$G_{cl}(s) = \frac{b_0 s^m + b_1 s^{m-1} + \cdots + b_{m-1} s + b_m}{a_0 s^n + a_1 s^{n-1} + \cdots + a_{n-1} s + a_n}$$

   the characteristic equation is

$$a_0 s^n + a_1 s^{n-1} + \cdots + a_{n-1} s + a_n = 0$$

   where it is assumed that $a_n \neq 0$ (or that any zero roots have been removed).

2. *Determine if any coefficients are negative.* Should any coefficient be
   negative while at least one coefficient is positive, there exists a root or
   roots that are either on the imaginary axis or in the right-half plane.
   Under such circumstances, the system is unstable. For the system to be
   stable, all the coefficients of the characteristic equation must be positive. If the characteristic equation is factored into its base first- and
   second-order factors (e.g., $s + a$ and $s^2 + bs + c$), each factor will have
   roots with negative real part if and only if the coefficients are positive
   (i.e., $a > 0$ or $b > 0$ and $c > 0$). Positive coefficients are a necessary
   but not sufficient condition for absolute stability.

3. *Generate coefficient array.* Even if all the coefficients are positive, the system may still be unstable or have a root in the right-half plane. To determine how many poles have a positive real part, one must prepare the following array:

$$
\begin{array}{c|cccccc}
s^n & a_0 & a_2 & a_4 & a_6 & \cdots \\
s^{n-1} & a_1 & a_3 & a_5 & a_7 & \cdots \\
s^{n-2} & b_1 & b_2 & b_3 & b_4 & \cdots \\
s^{n-3} & c_1 & c_2 & c_3 & c_4 & \cdots \\
\vdots & \vdots & \vdots & \vdots \\
s^2 & e_1 & e_2 \\
s^1 & f_1 \\
s^0 & g_1
\end{array}
$$

where

$$b_1 = \frac{a_1 a_2 - a_0 a_3}{a_1}$$

$$b_2 = \frac{a_1 a_4 - a_0 a_5}{a_1}$$

$$b_3 = \frac{a_1 a_6 - a_0 a_7}{a_1}$$

$$\vdots$$

$$b_1 = \frac{b_1 a_3 - a_1 b_2}{b_1}$$

$$b_2 = \frac{b_1 a_5 - a_1 b_3}{b_1}$$

$$b_3 = \frac{b_1 a_7 - a_1 b_4}{b_1}$$

$$\vdots$$

$$d_1 = \frac{c_1 b_2 - b_1 a_2}{c_1}$$

$$d_2 = \frac{c_1 b_3 - b_1 a_3}{c_1}$$

$$\vdots$$

The array is generated such that it includes rows for the *n*th power of *s* down to the constant. It should be noted that a row can be scaled to make the arithmetic more convenient without modifying the conclusions made about the absolute stability.

4. *Count the sign changes.* Routh-Hurwitz stability criterion states the number of poles in the right-half plane is equal to the number of sign changes in the first column of the coefficient array,

$$a_0 \quad a_1 \quad b_1 \quad c_1 \quad \ldots \quad e_1 \quad f_1 \quad g_1.$$

The actual values are unimportant in determining the number of unstable poles. Only the sign changes are necessary. Thus, the necessary and sufficient conditions for stability are (1) the coefficients of the characteristic equation must be positive, and (2) the first column of the coefficient array should have no sign changes.

Example 9.2

Show that the proportional position control system from Example 9.1 is a stable system.

Solution. Remember from Equation 9.8 that the closed-loop transfer function is

$$G_{cl}(s) = \frac{K_P}{Js^2 + \beta s + K_P},$$

which has a characteristic equation

$$a_0 s^2 + a_1 s + a_3 = Js^2 + \beta s + K_P = 0.$$

Because the denominator is second order, our coefficient array will contain rows for s^2, s^1, and s^0:

$$
\begin{array}{ccc}
s^2 & J & K_P \\
s^1 & \beta & 0 \\
s^0 & b_1 &
\end{array}
$$

where

$$b_1 = \frac{\beta K_P - J \cdot 0}{\beta} = K_P.$$

Elements of the first column of the array are thus J, β, and K_P. Since each is positive, there is no sign change, and the closed-loop transfer function has no poles in the right-half plane.

Example 9.3

Determine whether the following system is a stable system, and if the system is unstable, specify how many of the system poles are in the right-half plane.

$$G(s) = \frac{1}{5s^4 + 4s^3 + 3s^2 + 2s + 1}.$$

Solution: Following the procedure previously outlined, we first identify the characteristic equation

$$5s^4 + 4s^3 + 3s^2 + 2s + 1 = 0.$$

There are no missing coefficients so we can continue. Then, we check to see if any of the coefficients are negative. All the coefficients are positive; hence, we continue to generate the coefficient array:

$$
\begin{array}{c c c c}
s^4 & 5 & 3 & 1 \\
s^3 & 4 & 2 & 0 \\
s^2 & b_1 & b_2 & \\
s^1 & c_1 & 0 & \\
s^0 & d_1 & &
\end{array}
$$

where

$$b_1 = \frac{4(3) - 5(2)}{4} = \frac{1}{2}$$

$$b_2 = \frac{4(1) - 5(0)}{4} = 1.$$

Remember that we can scale any row without changing the outcome of the analysis. It is convenient to scale b_1 and b_2 by a factor of two so that the resulting coefficient array becomes

$$
\begin{array}{c c c c}
s^4 & 5 & 3 & 1 \\
s^3 & 4 & 2 & 0 \\
s^2 & 1 & 2 & \\
s^1 & c_1 & 0 & \\
s^0 & d_1 & &
\end{array}
$$

The two remaining coefficients are

$$c_1 = \frac{1(2) - 4(2)}{1} = -2$$

and

$$d_1 = \frac{-2(2) - 1(0)}{-2} = 2.$$

The coefficients of the first column of the array are 5, 4, 1, -2, 1. The two sign changes that result indicate two unstable poles.

Occasionally, when using the Routh-Hurwitz stability criterion one may find that a zero can result in the first column of the coefficient array with no apparent sign changes. Since we cannot divide by zero, additional analysis is required. When a zero results in the first column, it must be replaced by a small positive number, ε, and the remaining array evaluated. Take for example the following characteristic equation,

$$2s^3 + 3s^2 + 2s + 3 = 0.$$

The corresponding coefficient array is

$$
\begin{array}{ccc}
s^3 & 2 & 2 \\
s^2 & 3 & 3 \\
s^1 & 0 & 0 \\
s^0 & \infty &
\end{array}.
$$

If, instead, the zero is replaced by a small positive number ε, the coefficient array becomes

$$
\begin{array}{ccc}
s^3 & 2 & 2 \\
s^2 & 3 & 3 \\
s^1 & \varepsilon & 0 \\
s^0 & 3 &
\end{array}.
$$

If there is no sign change, as in this case, the result indicates that there exists a marginally stable pair of poles on the imaginary axis. The factors of the characteristic equation are

$$2s^3 + 3s^2 + 2s + 3 = 2(s + 1.5)(s^2 + 1) = 0,$$

which indeed has roots on the imaginary axis,

$$s_{1,2} = \pm j.$$

If instead there does result a sign change, then an unstable pole exists in the right-half plane.

Should all the coefficients in a given row be zero, then there exist roots of equal magnitude that are mirror images about the real or imaginary axis as depicted in Figure 9.17. In such cases, an auxiliary polynomial must be formed from the coefficients in the prior row and derivative of the resultant polynomial. To illustrate, take the following example discussed in detail in (Ogata 2004):

$$s^5 + 2^4 + 24s^3 + 48s^2 - 25s - 50 = 0.$$

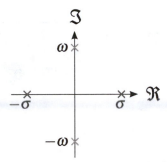

Figure 9.17: *Radially opposed pole pairs.*

For this characteristic equation, the coefficient array is

$$
\begin{array}{llll}
s^5 & 1 & 24 & -25 \\
s^4 & 2 & 48 & -50 \quad \leftarrow \text{Coefficients for auxilary polynomial } P(s). \\
s^3 & 0 & 0 &
\end{array}
$$

The second row is used to form the auxiliary polynomial

$$P(s) = 2s^4 + 48s^2 - 50,$$

which has a derivative

$$\frac{dP(s)}{ds} = 8s^3 + 96.$$

The auxiliary polynomial and its derivative form the first two rows of the coefficient array:

$$
\begin{array}{lll}
s^5 & 1 & 24 \quad -25 \\
s^4 & 2 & 48 \quad -50 \\
s^3 & 8 & 96 \qquad\quad \leftarrow \text{Coefficients of } dP(s)/ds. \\
s^2 & 24 & -50 \\
s^1 & 112.7 & 0 \\
s^0 & -50 &
\end{array}
$$

The first column has one sign change. Given the rows of zeros and the single sign change that resulted using the auxiliary array, this characteristic equation has one pair of poles that are radially opposed and an unpaired single pole in the right-half plane. MATLAB can be used to find the roots and confirm the results of the stability analysis.

```
>> roots([1 2 24 48 -25 -50])

ans =

   0.0000 + 5.0000i
   0.0000 - 5.0000i
   1.0000
  -2.0000
  -1.0000
```

This particular characteristic equation has a pair of imaginary poles at $s = \pm j5$ and an unstable pole at $s = 1$. It is important to note that Routh-Hurwitz stability criterion is useful to determine whether a system is stable and how many poles are in the right-half plane when a characteristic equation is not readily factored or numerical tools are not available. The use of the roots command brings up a point. MATLAB can be readily used to determine stability of a system. A number of commands can be employed to determine stability. In addition to the roots command, given a transfer function, the pole and pzmap can be used to determine if a system has unstable poles. Take for instance the plant model

$$G(s) = \frac{s^2 + 2s + 5}{s^5 + 2^4 + 24s^3 + 48s^2 - 25s - 50}.$$

Using MATLAB we can determine the poles and plot the pole-zero map to see if any poles exist in the right-half plane.

```
>> pole(G)

ans =

   0.0000 + 5.0000i
   0.0000 - 5.0000i
   1.0000
  -2.0000
  -1.0000

>> pzmap(G)
```

A detailed depiction of the pole-zero map is provided in Figure 9.18. As the figure shows, two marginally stable poles exist on the imaginary axis and one pole is in the right-half plane.

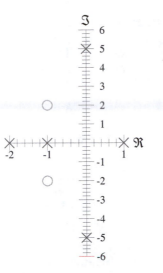

Figure 9.18: *Pole-zero map of* $G(s) = (s^2 + 2s + 5)/(s^5 + 2^4 + 24s^3 + 48s^2 - 25s - 50)$.

Example 9.4

Given a control system with feedforward transfer function

$$G_f(s) = \frac{K}{s(s+1)(s+3)}$$

and unity feedback (i.e., $G_s = 1$), determine the range of gain K that will result in stable response.

Solution. We can use the Routh-Hurwitz stability criterion to determine the range. First we must find the closed-loop transfer function using Equation 9.3:

$$G_{cl} = \frac{\dfrac{K}{s(s+1)(s+3)}}{1 + \dfrac{K}{s(s+1)(s+3)}} = \frac{K}{s(s+1)(s+3) + K}.$$

The characteristic equation is

$$s(s+1)(s+3) + K = 0$$
$$s(s^2 + 4s + 3) + K = 0$$
$$s^3 + 4s^2 + 3s + K = 0.$$

Given the characteristic equation, we can generate the coefficient array,

$$
\begin{array}{ccc}
s^3 & 1 & 3 \\
s^2 & 4 & K \\
s^1 & (12-K)/4 & 0 \\
s^0 & K &
\end{array}
$$

which is a function of the gain. If we assume that $K > 0$, we require that

$$\frac{12-K}{4} > 0$$
$$12 - K > 0$$
$$K < 12.$$

Ergo, for the system to be stable, the gain must be between 0 and 12 (i.e., $0 < K < 12$).

9.7   Root-Locus Analysis

It is already known that the placement of the dominant poles dictates the overall response of the system. Compensators are used to vary the placement of these poles and adjust the characteristics of the system response. We know that the placement of the dominant poles primarily determines properties such as the settling time and overshoot of the step response. By varying the gains of the compensator, we can move the dominant closed-loop poles to the desired damping ratio and natural frequency to achieve specified overshoot and settling time. W. R. Evans devised a graphical approached called the *root-locus method* to illustrate how the placement of the dominant closed-loop poles varies with changes in the compensator gains (Evans 1948, 1950). This is one of the most commonly used classical methods for designing compensators. The method can be used to estimate the effect dominant pole location will have on the overall response. Though MATLAB is well equipped to generate root-locus plots, it is useful to understand the steps involved in sketching a plot. This helps one garner intuition about the impact of placing additional poles and zeros through closed-loop compensation. PID compensators, for example, can be used to place an additional two zeros and a pole:

$$G_c(s) = K_P + \frac{K_I}{s} + K_D s$$

$$= \frac{K_D s^2 + K_P s + K_I}{s}$$

$$= K_D \frac{s^2 + \left(\dfrac{K_P}{K_D}\right) s + \dfrac{K_I}{K_D}}{s}.$$

In this section we will explore the basis of the root-locus method, how it is sketched, and its use in selecting compensator gains.

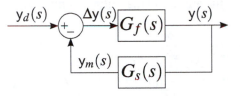

Figure 9.19: *A block diagram for a compensated system with feedforward and feedback dynamics.*

The method is based on the basic premise that there exist loci that satisfy the characteristic equation. Because the closed-loop compensation introduces an overall gain, like K_D in the PID controller above, the locus is the set of all values of s that satisfy the characteristic equation as the overall gain is varied from 0 to ∞. Take the compensated system in Figure 9.19. The closed-loop transfer function is

$$\frac{y(s)}{y_d(s)} = \frac{G_f(s)}{1 + G_f(s)G_s(s)}. \tag{9.17}$$

Its corresponding characteristic equation is

$$
\begin{aligned}
0 &= 1 + G_f(s)G_s(s) \\
&= 1 + \frac{K(s+z_1)(s+z_2)\dots(s+z_m)}{(s+p_1)(s+p_2)\dots(s+p_n)} \\
&= 1 + K\frac{A(s)}{B(s)}
\end{aligned}
\tag{9.18}
$$

where in general $G_f(s)G_s(s)$ has m zeros and n poles and where $A(s)$ and $B(s)$ are the numerator and denominator of the open-loop, transfer function

$G_f(s)G_s(s)$. Note that the open-loop, transfer function can be generally factored into the form

$$G_f(s)G_s(s) = K\frac{(s+z_1)(s+z_2)\ldots(s+z_m)}{(s+p_1)(s+p_2)\ldots(s+p_n)}$$

where K is the overall gain after the compensation is applied and the sensor dynamics accounted for. The characteristic equation can be alternatively written as

$$G_f(s)G_s(s) = -1. \tag{9.19}$$

In the s plane, $G_f(s)G_s(s) = -1$ is located on the real axis at -1 and makes an angle of $180°$ with the positive real axis. These identities form the basis for two conditions – the angle and magnitude conditions. The *angle condition* requires that

$$\angle G_f(s)G_s(s) = \sum \phi_i - \sum \theta_j$$
$$= 180°(2k+1) \quad (k=0,1,2,\ldots), \tag{9.20}$$

where ϕ_i are angles associated with the zeros and θ_i are those associated with the poles of the open-loop transfer function. Lines are drawn from each pole and zero to a test point with value s. The angles are measured counterclockwise from the right facing horizontal. The angles for the lines drawn from each zero to the test point are designated ϕ_i, and those from each pole to the test point are labeled θ_j where $i = 1,\ldots,m$ and $j = 1,\ldots,n$. The *magnitude condition* requires

$$|G_f(s)G_s(s)| = 1. \tag{9.21}$$

The values of s that fulfill both the angle and magnitude conditions are the roots of the characteristic equation, or the closed-loop poles, and a plot of those points that satisfy only the angle condition compose the root locus. To sketch the root locus, follow these procedures (Ogata 2002):

1. *Locate the poles and zeros.* The root-locus branches start from the poles and terminate at zeros. Note that these are the poles and zeros of $G_f(s)G_s(s)$ and not the closed-loop transfer function $y(s)/y_d(s)$. Also note that the root locus is symmetric about the real axis.

2. *Determine the loci on the real axis.* Test points between all roots on the real axis and determine if the sum of angles is an odd integer multiple of $180°$. Alternatively, if the total number of *real* poles and *real* zeros

to the right of the test point is an odd number, the test point lies on the
root locus. This is because the poles and zeros that lie along the real
axis will contribute either $0°$ or $180°$. The remaining poles and zeros
are generally complex and will exist in pairs with angles that sum to
$360°$ which is the equivalent of $0°$. An example is depicted in Figure
9.20. The example has four poles (including a complex conjugate pair)
and two complex zeros. If the angles for the complex pole pair and
complex zero pair are summed as demonstrated in Figure 9.20 (b) and
(c), they form a complete circle or $360°$. The shortcut is to simply
count the total number of real poles and zeros to the right of the test
point along the real axis. If the count is an odd number, the test point
will lie in the root locus. Each pole or zero to the right will contribute
$\pm180°$ and each to the left will contribute $0°$. An even number to the
right will sum to an integer multiple of $360°$ and will not satisfy the
angle condition.

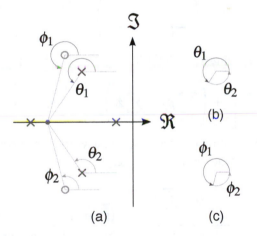

Figure 9.20: *An illustration of a test point along the real axis.*

3. *Determine the asymptotes.* Assuming that the number of poles of $G_f(s)G_s(s)$
   is greater than the number of zeros (i.e., $n > m$) the number of asymptotes will equal $n - m$ and the angles of those asymptotes will be

$$\frac{180°(2k+1)}{n-m} \quad (k=0,1,2,\ldots,n-m-1). \qquad (9.22)$$

This can be shown by examining the behavior of $G_f(s)G_s(s)$ for large
values of s (i.e., $s \to \infty$). At large values of s, the transfer function will

approach asymptotes. Given that

$$G_f(s)G_s(s) = \frac{K(s+z_1)(s+z_2)\ldots(s+z_m)}{(s+p_1)(s+p_2)\ldots(s+p_n)}$$

$$= \frac{K\left(s^m + \sum_{i=1}^{m} z_i s^{m-1} + \ldots + \prod_{i=1}^{m} z_i\right)}{s^n + \sum_{j=1}^{n} p_j s^{n-1} + \ldots + \prod_{j=1}^{n} p_j}$$

$$= \frac{K}{s^{n-m} + \left(\sum_{j=1}^{n} p_j - \sum_{i=1}^{m} z_i\right) s^{n-m-1} + \ldots},$$

the characteristic equation can be re-written as

$$s^{n-m} + \left(\sum_{j=1}^{n} p_j - \sum_{i=1}^{m} z_i\right) s^{n-m-1} + \ldots = -K,$$

which can be approximated for large values of s by

$$\left[s + \left(\sum_{j=1}^{n} p_j - \sum_{i=1}^{m} z_i\right) \div (n-m)\right]^{n-m} = 0.$$

The above equation represents $n-m$ asymptotes at equal angular intervals in the s plane with an abscissa,

$$s = \sigma_a = -\frac{\sum_{j=1}^{n} p_j - \sum_{i=1}^{m} z_i}{n-m} = +\frac{(\text{sum of poles}) - (\text{sum of zeros})}{n-m}.$$

$$(9.23)$$

To summarize, there are $n-m$ asymptotes at angles $180°(2k+1)/(n-m)$ centered at the abscissa σ_a.

4. *Determine the breakaway and break-in points.* From the characteristic equation (Equation 9.19)

$$-1 = G_f(s)G_s(s)$$

$$= K\frac{A(s)}{B(s)}$$

$$K = -\frac{B(s)}{A(s)}.$$

Breakaway and break-in points correspond to repeated roots of the characteristic equation. That is where two poles or zeros come together and split. The breakaway and break-in points can be determined from

$$\frac{dK}{ds} = -\frac{B'(s)A(s) - B(s)A'(s)}{A^2(s)} = 0,\qquad (9.24)$$

where $A'(s) = dA(s)/ds$ and $B'(s) = dB(s)/ds$.

5. *Find the departure angle (arrival angle) of the root locus from a complex pole (to a complex zero).* Recall that a test point s that is part of the root locus must satisfy the angle condition (Equation 9.20). Imagine that you are testing a point near the complex pole (or complex zero) to determine if it is part of the root loci so that you know at which angle the loci departs from the pole (arrives at the zero). If you test a point in the vicinity of the complex pole (complex zero) the angles of the vectors drawn from each other pole and zero to this test point will be approximately equal to those for vectors drawn to the actual pole (zero) around which you are testing. You can sum the angles for the remaining poles and zeros and subtract it from 180° to determine the angle of departure (arrival). Therefore,

$$\angle \text{ of departure from a complex pole} = 180°$$
$$-(\text{sum of angles to the complex pole in question from other poles})$$
$$+(\text{sum of angles to the complex pole in question from zeros})$$
$$(9.25)$$

$$\angle \text{ of arrival to a complex zero} = 180°$$
$$-(\text{sum of angles to the complex zero in question from other zeros})$$
$$+(\text{sum of angles to the complex zero in question from poles}).$$

$$(9.26)$$

6. *Find the points where the root locus intersects the imaginary axis.* This step helps refine the sketch of the root locus. These points can be determined using Routh-Hurwitz stability criterion or by substituting $s = j\omega$ into the characteristic equation (Equation 9.18) and solving for K and ω where ω will be the points on the imaginary axis where the root locus crosses. To solve for K and ω, the real and imaginary parts of the characteristic equation are equated to zero. The values of ω that satisfy

the solution are the points where the root loci cross the imaginary axis, and the solution K is the corresponding gain that makes the system marginally stable. Beyond this gain, the system becomes unstable.

7. *Determine the behavior of the root locus in the neighborhood of the origin.* Test points in and around the origin of the complex plane. Greater detail is required in this region than is required near the poles and zeros or out at the asymptotes.

8. *Determine the closed-loop poles.* The overall gain K is associated with a specific pair of dominant closed-loop poles. If, as indicated in step 4, the characteristic equation can be written as $K = -B(s)/A(s)$ then the magnitude condition requires

$$K = \frac{|B(s)|}{|A(s)|} = \frac{|s+p_1||s+p_2|\ldots|s+p_n|}{|s+z_1||s+z_2|\ldots|s+z_m|}.$$

Hence, the magnitude condition can be applied to find the closed-loop poles associated with a specific value for the gain K.

The above process is best illustrated through an example. The following examples will you help understand the individual steps used to sketch the root locus.

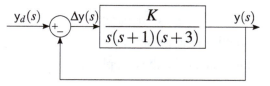

Figure 9.21: *Block diagram for a control system with simple gain compensation.*

Example 9.5
Generate the root locus for the system depicted in Figure 9.21, which has a feedforward transfer function of

$$G_f(s) = \frac{K}{s(s+1)(s+3)}$$

and unity feedback (i.e., $G_s(s) = 1$). Also, find the gain necessary to get a set of closed-loop poles with an associated maximum overshoot of no more than 20%. Confirm that the gain you derive garners the desired overshoot.

Solution. Following the process previously described, the steps are:

1. *Locate the poles and zeros.* We begin by identifying the poles and zeros. Being that the transfer function is already factored, it is easy to determine the poles and zeros. There are no zeros and the poles are located at $s = 0, -1$, and -3.

2. *Determine the loci on the real axis.* Next we determine what parts of the real axis are in the root locus. Using the shortcut, we can test a point right of $s = 0$, between $s = 0$ and $s = -1$ ($-1 < s < 0$), between $s = -1$ and $s = -3$ ($-3 < s < -1$), and left of $s = -3$ ($s < 3$). Nothing right of $s = 0$ is in the root locus because there are 0 poles and zeros to the right. Between $s = 0$ and $s = -1$ there is one pole to the right which is an odd number. Therefore, any part of the axis between $s = 0$ and $s = -1$ will be part of the root locus. For $-3 < s < -1$ the number of poles to the right is two, which is an even number. Left of $s = -3$, however, all points will have three poles to the right. The parts of the real axis that are part of the root locus are $s < -3$ and $-1 < s < 0$.

3. *Determine the asymptotes.* Because there are no zeros and three poles, the root locus will have three asymptotes,

$$n - m = 3 - 0 = 0,$$

with angles
$$\frac{180°(2k+1)}{3} \quad (k = 0, 1, 2)$$
or $60°$, $180°$, and $300°$. The abscissa is located at

$$s = -\frac{\sum\limits_{j=1}^{n} p_j - \sum\limits_{i=1}^{0} z_i}{n - m} = -\frac{(0+1+3) - 0}{3} = -\frac{4}{3}.$$

4. *Determine the breakaway and break-in points.* Being that there are three asymptotes there must exist a breakaway point between $s = 0$ and $s = -1$. To determine this breakaway point we find where

$$\frac{dK}{ds} = 0.$$

First we must solve for K(s):

$$K = -\frac{B(s)}{A(s)} = -s(s+1)(s+3) = s(s^2 + 4s + 3) = -(s^3 + 4s^2 + 3s).$$

The derivative is

$$\frac{dK}{ds} = -(3s^2 + 8s + 3)$$

which has roots

$$s_{1,2} = \frac{-8 \pm \sqrt{8^2 - 4(3)(3)}}{2(3)} = \frac{8 \pm \sqrt{64 - 36}}{6} = -2.2153 \text{ and } -0.4514 \,.$$

Only the root $s_2 = -0.4514$ is part of the root locus so this must be the breakaway point.

5. *Find the departure angle (arrival angle) of the root locus from a complex pole (to a complex zero).* There are no complex poles or zeros. Thus, there are not departure or arrival angles to calculate.

6. *Find the points where the root locus intersects the imaginary axis.* If we substitute $s = j\omega$ into the characteristic equation, we can determine where the root locus crosses the imaginary axis:

$$\begin{aligned}
0 &= 1 + \frac{K}{j\omega(j\omega + 1)(j\omega + 3)} = 1 + \frac{K}{\omega[-4\omega + j(3 - \omega^2)]} \\
&= 1 + \frac{K}{\omega}\left[\frac{-4\omega - j(3 - \omega^2)}{16\omega^2 + (3 - \omega^2)^2}\right] = 1 + \frac{K}{\omega}\left[\frac{-4\omega - j(3 - \omega^2)}{\omega^4 + 10\omega^2 + 9}\right] \\
&= \left[1 - \frac{4K}{\omega^4 + 10\omega^2 + 9}\right] - j\frac{K}{\omega}\left[\frac{3 - \omega^2}{\omega[\omega^4 + 10\omega^2 + 9]}\right].
\end{aligned}$$

Equating the real and imaginary part to zero yields

$$0 = \left[1 - \frac{4K}{\omega^4 + 14\omega^2 + 9}\right]$$
$$4K = \omega^4 + 10\omega^2 + 9$$
$$K = \frac{\omega^4 + 10\omega^2 + 9}{4}$$

and

$$0 = \frac{K}{\omega}\left[\frac{3 - \omega^2}{\omega[\omega^4 + 10\omega^2 + 9]}\right]$$
$$= 3 - \omega^2$$
$$\omega^2 = 3$$

which has solutions

$$\omega = \sqrt{3}$$

and

$$K = \frac{3^2 + 10(3) + 9}{4} = \frac{9 + 30 + 9}{4} = 12.$$

Therefore, the root loci cross the imaginary axis at

$$s_{1,2} = \pm j\sqrt{3}.$$

Recall from Example 9.4 that the range of gains that ensure a stable closed-loop response is $0 < K < 12$.

7. *Determine the behavior of the root locus in the neighborhood of the origin.* We have already determined that all the points between -1 and 0 on the real axis are part of the root locus and that there exists along this segment a breakaway point. The points right of 0 on the real axis are not part of the root locus.

8. *Determine the closed-loop poles.* The design criteria state that the maximum overshoot should not exceed 20%. In Section 9.4, the maximum overshoot was derived as a function of damping ratio. According to Figure 9.10 an overshoot of 20% corresponds to a damping ratio between 0.4 and 0.5. If we select a damping ratio of 0.5, the line of constant damping makes an angle of 60° with the negative real axis as shown in Figure 7.10:

$$\zeta = \cos\theta$$
$$\zeta = \cos^{-1}\zeta$$
$$\zeta = \cos^{-1}0.5$$
$$\zeta = 60° \text{ or } (\pi/3) \text{ rad}.$$

According to the figure, the root locus intersects the $\zeta = 0.5$ line of constant damping at approximately

$$s_{1,2} = -0.375 \pm j0.65.$$

In the next section, it will be demonstrated how the near exact value used can be attained using the data cursor and `rlocus` command in MATLAB. Thus, the controller gain is

$$K = |s||s+1||s+3|_{s=-0.375+j0.65} \approx 1.83.$$

Notice that there is only one point along the top and bottom segments that satisfies the damping ratio specified to meet the overshoot criteria. The closed-loop poles have an associated natural frequency of

$$\omega_n = \sqrt{(-0.375)^2 + (0.65)^2} = 0.75 \text{ rad/s}.$$

The settling time of the dominant closed-loop poles will be

$$t_s = \frac{4}{\zeta \omega_n} = \frac{4}{(0.5)(0.75 \text{ rad/s})} \approx 10.7 \text{ s}.$$

The root locus is plotted in Figure 9.22. As the figure demonstrates, the poles are located at $s = 0$, -1, and -3. The parts of the real axis included in the locus include $-3 < s$ and $-1 < s < 0$. The loci that start at the open loop poles at $s = 0$ and $s = -1$ travel toward each other and then break away just right of $s = -0.5$ (at $s = -0.4514$ to be more exact). The loci beginning at $s = -3$ travel out to the left infinitely. The locus crosses the imaginary axis at $s = \pm\sqrt{3} \approx \pm 1.73$. The locus approaches three asymptotes at angles of $\pm 60°$ and $180°$ centered at an abscissa of approximately $s = -1.33$. The root locus plot also indicates the gain value at various points.

The closed-loop step response is plotted in Figure 9.23. The response was generated using $K = 1.83$ which correlates with the damping ratio of $\zeta = 0.5$. Notice that the overshoot does not exceed 20% and that the settling time is approximately 10.7 s.

Note that in the previous example there is a limited range of gains K that produces a stable response. Furthermore, as the gain is increased between 0 and 12 the damping ratio decreases, but the natural frequency increases. Recall from Figure 7.11 how the damping decreases as the poles rotate around from the imaginary axis to the real axis. Also remember from Figure 7.12 how the natural frequency increases as the pole moves farther away from the origin. Adjusting the gain will garner limited improvement to the settling time because the damping ratio and natural frequency change in opposing directions as the gain is increased. If we could pull the asymptotes at $\pm 60°$ to a vertical position, we may be able to drag the root locus such that the loci leaving the poles at $s = -1$ and $s = 0$ will travel vertically or diagonally in the left-half plane so that a greater range of gains will produce a stable response. This can be accomplished by adding poles and zeros through compensation.

Take as a case study the problem discussed in Example 9.5. If, as illustrated in the block diagram in Figure 9.24, we add a zero, $s + z$, to the system somewhere left of the imaginary axis, it will negate or minimize the effect of

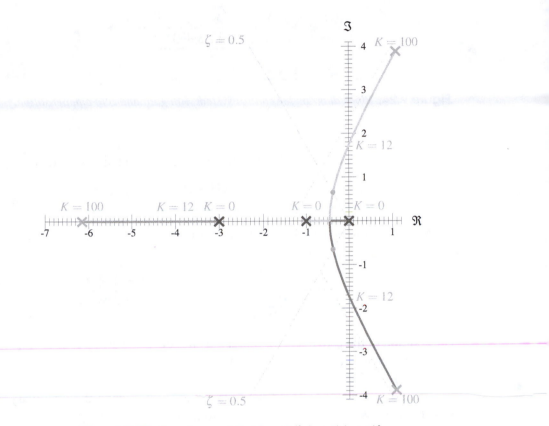

Figure 9.22: *Root locus of $G_f(s) = K/[s(s+1)(s+3)]$.*

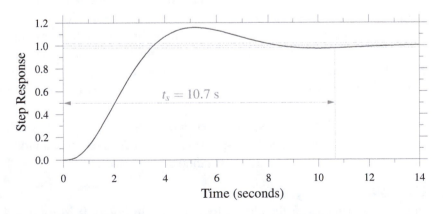

Figure 9.23: *Closed-loop step response for $G_f(s) = 1.83/[s(s+1)(s+3)]$.*

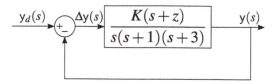

Figure 9.24: *Block diagram for control system with gain and zero compensation.*

one of the real-valued poles. The feedforward transfer function would be

$$G_f(s) = \frac{K(s+z)}{s(s+1)(s+3)}$$

and the corresponding closed-loop transfer function is

$$G_{cl}(s) = \frac{K\dfrac{s+z}{s(s+1)(s+3)}}{1+K\dfrac{s+z}{s(s+1)(s+3)}} = \frac{K(s+z)}{s^3+4s^2+(3+K)s+Kz}.$$

The addition of the zero does two things: (1) it reduces the number of asymptotes from three to two, and (2) it changes the abscissa of those asymptotes. Refer to Figure 9.25.

Based on Equation 9.23, as the zero is moved left along the imaginary axis, the abscissa moves right. If the zero moves out far enough along the negative real axis, it will reach a point at which it has negligible impact on the root locus, and the locus will look like that in Figure 9.22. Notice the lines of constant damping $\zeta = 0.5$ labeled in both the top and bottom halves. As the zero moves closer to the imaginary axis the root locus plots cross the lines farther and farther from the origin resulting in higher and higher natural frequencies. Poles and zeros can be placed strategically to change the shape of the root locus and force it through the desired dominant closed loop poles with the preferred associated overshoot and settling time (or damping ratio and natural frequency).

As Figure 9.26 demonstrates, the closed-loop step response tends to settle more quickly as the zero nears the origin (i.e., $z \to 0$). The responses plotted in Figure 9.26 all remain under 20% overshoot.

Example 9.6
Draw the root locus for the closed-loop system in Figure 9.24 for $z = 4$, and determine the overall gain K associated with dominant closed-loop poles that have a damping ratio of $\zeta = 0.5$.

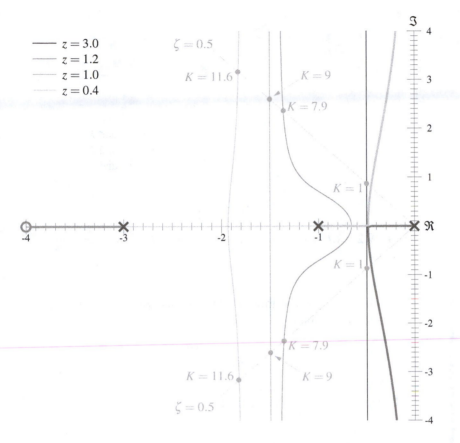

Figure 9.25: *Root locus of $G_f(s) = K(s+z)/[s(s+1)(s+3)]$.*

Solution. The root locus for $z = 4$ is plotted in Figure 9.25. However, we will follow the steps for sketching the root locus to confirm that plot illustrated in Figure 9.25 matches the results garnered using the process. The steps for drawing the root locus are:

1. First, we identify the poles and zeros. There are poles located at $s = 0$, -1, and -3. There exists one zero at $s = -4$.

2. The parts of the real axis that are part of the root locus are $-1 < s < 0$ and $-4 < s < -3$. Between $s = -1$ and $s = 0$ the number of poles and zeros to the right is one which is an odd number (one pole and no zeros). For points between $s = -4$ and $s = -3$ there are three poles and no zeros to the right (again, an odd number).

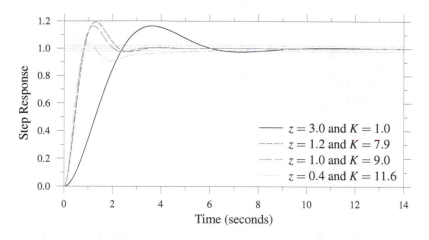

Figure 9.26: *Closed-loop step responses for various gains and zeros.*

3. Given that there are three poles and one zero, the root locus will have two asymptotes,

$$n - m = 3 - 1 = 2.$$

The asymptotes are centered at

$$s = \sigma_a = -\frac{\displaystyle\sum_{j=1}^{n} p_j - \sum_{i=1}^{m} z_i}{n-m} = -\frac{(0+1+3)-4}{3-1} = 0.$$

Thus, the root locus will asymptotically approach the imaginary axis top and bottom.

4. The breakaway point will occur along the real axis at a value of s that satisfies

$$\frac{dK}{ds} = 0$$

where

$$K = -\frac{B(s)}{A(s)} = -\frac{s(s+1)(s+3)}{s+4}.$$

Setting the derivative to zero results in

$$\frac{dK}{ds} = -\frac{(3s^2 + 8s + 3)(s+4) - (s^3 + 4s^2 + 3s)}{(s+4)^2} = 0$$

$$\frac{(3s^3 + 20s^2 + 25s + 12) - (s^3 + 4s^2 + 3s)}{(s+4)^2} =$$

$$\frac{2s^3 + 16s^2 + 32s + 12}{(s+4)^2} =$$

$$2s^3 + 16s^2 + 32s + 12 = 0.$$

The roots of the numerator are $s = -5.0861$, -2.4280, -0.4859. Only one of those points, $s = -0.4859$, exists on the root locus. Hence, the breakaway point will be between the two rightmost poles at $s = -0.4859$.

5. There are no complex poles or zeros.

6. It has already been determined that the root locus will asymptotically approach and not cross the imaginary axis.

7. It has also already been determined that in the vicinity of the origin, the part of the real axis between $s = -1$ and $s = 0$ is included in the locus.

8. As Figure 9.25 indicates, the root locus crosses $\zeta = 0.5$ at $s = -0.455 \pm j0.788$. The gain at that point along the root locus is

$$K = \left.\frac{|s||s+1||s+3|}{s+4}\right|_{s=-0.455+j0.788} \approx 0.640.$$

The associated natural frequency is thus

$$\omega_n = \sqrt{(-0.455)^2 + (0.788)^2} \approx 0.910.$$

The corresponding settling time is

$$t_s = \frac{4}{\zeta \omega_n} = \frac{4}{0.5(0.910 \text{ rad/s})} \approx 8.8 \text{ s}.$$

Figure 9.27 shows the closed-loop step response use a gain of $K = 0.640$. The overshoot does not exceed 20% and the settling time is approximately 8.8 s.

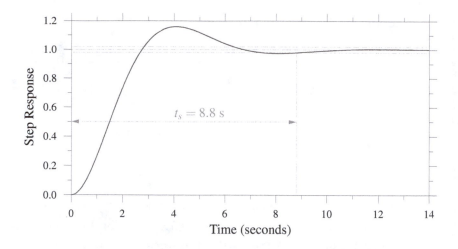

Figure 9.27: *Closed-loop step response for* $G_f(s) = 0.640(s+4)/[s(s+1)(s+3)]$.

Thus far, none of the examples have dealt with complex poles or zeros. The following includes a pair of complex poles to illustrate how to calculate the departure angles from the poles.

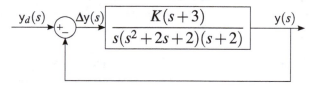

Figure 9.28: *A control system with a plant that has complex poles.*

Example 9.7

Draw the root locus for control system in Figure 9.28. The feedforward transfer function is

$$\frac{K(s+3)}{s(s^2+2s+2)(s+2)}.$$

Also, find the gain K necessary to limit the overshoot to less than 10%.

Solution. To root locus is drawn by completing the following steps:

1. The poles and zeros are readily identified because the feedforward transfer function is already factored. The only complexity is the pair of complex poles. Those can be determined by applying the

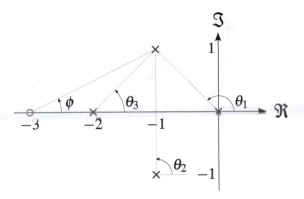

Figure 9.29: *Calculation of departure angle from complex pole.*

quadratic formula. There is one zero a $s = -3$. There are two real-valued poles at $s = 0$ and -2. The complex poles are at

$$s_{1,2} = \frac{-2 \pm \sqrt{2^2 - 4(1)(2)}}{2(1)} = -1 \pm j.$$

2. By testing points along the real axis we will find that the parts of the real axis that satisfy the angle condition are $s < -2$ and $-1 < s < 0$. For all points left of $s = -2$ there is a total of 3 zeros and poles to the right. For points contained within $-1 < s < 0$ there is just one pole to the right.

3. There are one zero and four poles; therefore, the number of asymptotes will be three. The abscissa will be at

$$s = \sigma_a = -\frac{\displaystyle\sum_{j=1}^{n} p_j - \sum_{i=1}^{m} z_i}{n - m} = -\frac{[0 + (1+j) + (1-j) + 2] - 3}{4 - 1} = \frac{1}{3}.$$

4. The break away points are found by determining the values of s that satisfy

$$\frac{dK}{ds} = 0$$

where

$$K = -\frac{B(s)}{A(s)} = -\frac{s(s^2 + 2s + 2)(s + 2)}{s + 3} = -\frac{s^4 + 4s^3 + 6s^2 + 4s}{s + 3}.$$

The derivative is

$$\frac{dK}{ds} = -\frac{(4s^3 + 12s^2 + 12s + 4)(s+3) - (s^2 + 4^3 + 6s^2 + 4s)}{(s+3)^2} = 0$$

$$-\frac{(4s^4 + 24s^3 + 48s^2 + 40s + 12) - (s^2 + 4^3 + 6s^2 + 4s)}{s^2 + 6s + 9} =$$

$$-\frac{3s^4 + 20s^3 + 42s^2 + 36s + 12}{s^2 + 6s + 9}.$$

The roots of

$$3s^4 + 20s^3 + 42s^2 + 36s + 12$$

are $s = -3.6487$, -1.5391, and $-0.7394 \pm j0.4069$. Both real roots are points along the real axis that are part of the root locus. The loci emanate from poles at $s = 0$ and $s = -1$, meet at $s = -1.5391$, and break away. As they depart from the real axis, they feel the presence of the zero at $s = -3$ and bend back into the real axis until they break back in at $s = -3.6487$.

5. There are two complex poles at $s = -1 \pm j$. Therefore, we must determine the departure angle of the loci from each of the complex poles. We employ Equation 9.25 to calculate the departure angles from each of the poles. Figure 9.29 illustrates the relative positions and angles of the poles and zero to the pole at $s = -1 + j$. The angles are

$$\theta_1 = 135°,$$
$$\theta_2 = 90°,$$
$$\theta_3 = 45°, \text{ and}$$
$$\phi = 26.6°.$$

Hence, the departure angle from the topmost pole is

$$180 - (135° + 90° + 45°) + 26.6° = -63.4°.$$

Because the root locus is symmetric about the real axis, the departure angle from the bottommost pole will be $63.4°$.

6. By substituting $s = j\omega$ into the characteristic equation, we can calculate the frequency and gain when the root locus crosses the imag-

inary axis,

$$0 = 1 + \frac{K(s+3)}{s(s^2+2s+2)(s+2)}$$

$$-1 = \frac{K(s+3)}{s^4+4s^3+6s^2+4s}$$

$$-(s^4+4s^3+6s^2+4s) = K(s+3)$$

$$(\text{substitute } s = j\omega)$$

$$-[(j\omega)^4+4(j\omega)^3+6(j\omega)^2+4(j\omega)] = K(j\omega+3)$$

$$-[\omega^4 - j4\omega^3 - 6\omega^2 + j4\omega] = 3K + jK\omega$$

$$(6\omega^2 - \omega^4) + j(4\omega^3 - 4\omega) = 3K + jK\omega.$$

Equating the real and imaginary parts on both sides of the equation yields

$$6\omega^2 - \omega^4 = 3K$$

$$0 = \omega^4 - 6\omega^2 + 3K$$

and

$$4\omega^3 - 4\omega = K\omega$$

$$4\omega^2 - 4 = K. \tag{9.27}$$

Substituting the second result into the first garners

$$0 = \omega^4 - 6\omega^2 + 3(4\omega^2 - 4)$$

$$0 = \omega^4 + 6\omega^2 - 12,$$

which has roots $\pm j2.7536$ and ± 1.2580. We seek the real roots of this equation because, though s is generally complex, the frequencies themselves are real-valued. Thus, the root locus will cross at $s = \pm j1.2580$.

7. We already know that in the vicinity of the origin the root loci emanate from the pole at $s = 0$ out to the left.

8. Figure 9.30 depicts the root locus for this system. According to Figure 9.10, if the desired overshoot is less than 10%, the necessary

damping associated with the closed-loop pole should be approximately 0.6. As the figure illustrates, the root locus passes through $\zeta = 0.6$ at $s = -0.536 \pm j0.715$. The gain at those points is

$$K = \left.\frac{|s||s^2 + 2s + 2||s + 2|}{|s + 3|}\right|_{-0.536 + j0.715} = 0.549.$$

The natural frequency and settling time of the closed-loop poles are

$$\omega_n = \sqrt{(-0.536)^2 + (0.715)^2} \approx 0.8936$$

and

$$t_s = \frac{4}{\zeta \omega_n} = \frac{4}{(0.6)(0.8936 \text{ rad/s})} \approx 9.3 \text{ s}.$$

The step response, which is plotted in Figure 9.31, exhibits less than 10% overshoot and a settling time under 9.3 s.

9.8 Plotting the Root Locus Using MATLAB

The practice of sketching root locus plots helps develop engineering intuition and understanding of how pole and zero placement impacts the path the loci trace and ultimately how the closed-loop response can be modified to achieve a response with the desired characteristics. However, numerical methods are often utilized to attain an accurate depiction of the root locus. As such, MATLAB can readily generate accurate root locus plots that can be manipulated using plot tools such as the data cursor to analyze characteristics of the root locus.

The MATLAB command rlocus(sys) generates the root locus plot of a LTI object sys. The object can be defined using either the tf, ss, or zpk. The root locus command can be invoked with an additional input and several outputs. In its simplest form without outputs it generates a plot that is automatically scaled by MATLAB and where the chosen gain values for the individual loci are also automatically specified. Alternatively, the gain values for the root locus plot can be manually specified as an array of monotonically increasing values and passed to MATLAB.

```
>> rlocus(sys,K)
```

When invoked with outputs,

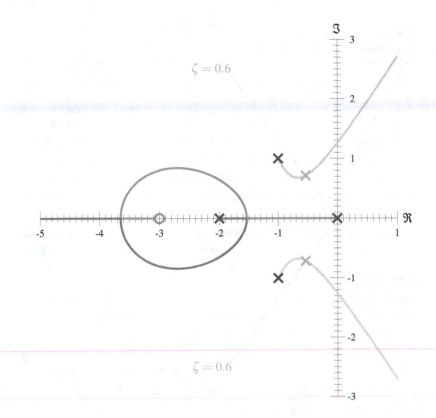

Figure 9.30: *Root locus of system with complex poles.*

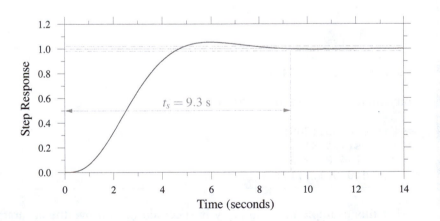

Figure 9.31: *Closed-loop step response of $G_f = 0.549(s+3)/[s(s^2+2s+2)(s+2)]$.*

```
>> [R,K] = rlocus(sys)
>> plot(real(R'),imag(R'))
```

the command returns the complex loci and corresponding gains. The individual loci are stored to the array R where each row represents the loci emanating from a specific pole as the gain is varied from 0 to ∞. The gains are stored to the row array K where each element corresponds to a column in the array of loci. When the outputs are used, the root locus plot is suppressed. To plot the locus manually, the array R must be transposed and the real and imaginary parts of the loci separated. However, this will only plot the loci and not the poles and zeros.

The following example will demonstrate how the root locus is generated in MATLAB and how the data cursor can be used to analyze the root locus.

Example 9.8

Using MATLAB, plot the root locus for the system in Example 9.7.

Solution. To plot the root locus, we must define the system as a transfer function. Because the transfer function is specified in terms of zeros and poles, it is most convenient to use the zpk command. Though a gain can be specified, the value for K that is used is unity because we want to generate the root locus and specifying a value for the overall gain will scale the gains on the root locus plot.

```
>> sys = zpk(-3,[0 -1+j -1-j -2],1)

sys =

            (s+3)
   ---------------------
   s (s+2) (s^2 + 2s + 2)

Continuous-time zero/pole/gain model.

>> K = 0:0.001:200;
>> rlocus(sys,K)
>> axis([-4 0 -2 2])
>> sgrid(0:0.1:1,1:10)
```

For this example the gain array is specified to improve the accuracy within the plotted region. The plot is zoomed to display the range -4 to 1 along the real axis and -2 to 2 along the imaginary axis. An s-plane grid is generated at increments of 0.1 for ζ and 1 rad/s for ω_n.

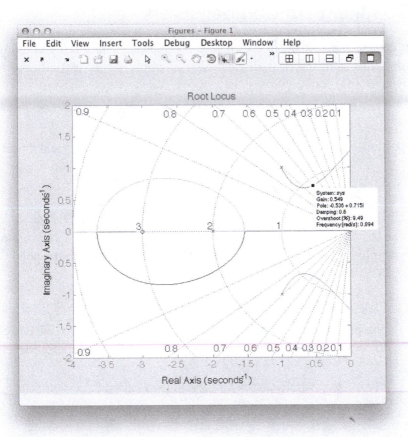

Figure 9.32: *Screenshot of MATLAB with data cursor enabled.*

Figure 9.32 shows the root locus plot generated by MATLAB using the commands detailed above. Notice that the data cursor tool is highlighted in the tool bar and that a cursor has been placed at $s = -0.536 + j0.715$. As the cursor is moved along the root locus, a data window appears that provides pertinent information about the currently selected point along the locus. The information provided includes the gain (K), pole location, damping (ζ), overshoot (M_p), and natural frequency (ω_n). Remember that each gain is associated with a dominant closed-loop pole and that the pole has inherent characteristics including damping ratio and natural frequency. The data tip illustrated in the figure corresponds to $\zeta = 0.6$ which was used in the previous example.

9.9 Tuning PID Compensators

The control mechanism in Example 9.5 is basically proportional compensation. The compensation discussed in Example 9.6 is proportional-plus-derivative (PD) control. Notice the zero we added was of the form

$$K(s+z) = Ks + Kz = K_Ds + K_P.$$

Proportional-plus-integral (PI) compensation adds both a zero and a pole:

$$G_c(s) = K_P + \frac{K_I}{s} = \frac{K_Ps + K_I}{s} = K_P \frac{s + \frac{K_I}{K_P}}{s} = K \frac{s+z}{s}.$$

A pole is added at $s = 0$ and a zero at $s = z$. Complete proportional-plus-integral-plus-derivative (PID) compensation adds two zeros and a pole:

$$G_c(s) = K_P + \frac{K_I}{s} + K_Ds = \frac{K_Ds^2 + K_Ps + K_I}{s}.$$

For tuning purposes, as Figure 9.33 shows, PID compensation is specified in terms of the integral and derivative time constants (T_I and T_D, respectively),

$$G_c(s) = K_P \left[1 + \frac{1}{T_Is} + T_Ds\right]$$

where

$$K_I = \frac{K_P}{T_I} \quad \text{and} \quad K_D = K_PT_D.$$

Therefore, a PID compensator can add a total of two zeros, which may be complex, and a pole at $s = 0$. The compensator generates an output composed of terms proportional to the error, its integral, and its rate of change.

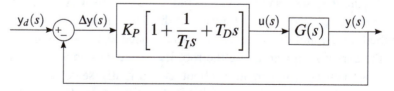

Figure 9.33: *Closed-loop system with PID compensation*

The majority of control systems employed in Industry use one form or another of PID compensation (Ogata 2004). These systems are typically tuned when in operation to optimize the closed-loop performance. Controller design is employed to establish a functional baseline, but since models are susceptible to inaccuracies and systems are vulnerable to un-modeled disturbances, tuning rules are used to set the compensator gains. A number of manual and automatic methods exist for PID compensator tuning. Moreover, gain scheduling extends the usefulness of PID compensation by optimizing the controller gains for various operating ranges (Rugh and Shamma 2000). This also extends the use of PID compensation to nonlinear systems. The nonlinear system model is linearized at a number of operating points and a PID compensator is tuned for each of those operating points. The controller enables the appropriate compensator depending on the operating range.

If a reasonable model can be derived for the system, various tuning methods can be employed to adjust the gains to meet transient and steady-state design criteria (Ogata 2002). If, on the other hand, a model cannot be readily devised, the tuning must be done *in situ*. *Controller tuning* is the process of optimizing gains to meet design criteria. In the 1940s, Ziegler and Nichols (Ziegler and Nichols 1942) proposed a set of tuning rules for PID controllers. These rules were developed for experimental optimization of the compensator gains. Regardless, the rules are still applicable to model-based tuning. The basic procedure is based on the gain necessary to achieve an oscillatory, closed-loop, step response. Numerous tuning rules have been proposed and practiced since the introduction of the Ziegler-Nichols rules. In Ziegler and Nichols (1942), the authors proposed two methods for tuning PID compensators – the *closed-loop method* and the *open-loop method*.

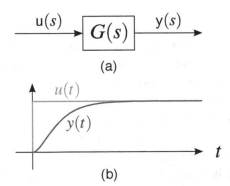

(a)

(b)

Figure 9.34: *(a) Open-loop system and (b) its unit step response.*

The open-loop method.   The open-loop method is applicable when the plant has an "S"-shaped unit step response like that of an overdamped second-order system. In the absence of an integrator term (i.e., a pole at $s = 0$) or complex conjugate poles (i.e., underdamped factors) the open-loop step response may look like that of an overdamped system. In such cases the following procedure can be used to tune the PID compensator gains:

1. *Excite the open-loop system with a step input.* Figure 9.34 illustrates an open-loop step response. If tuning online with the actual system, the loop must be opened, and the plant should be excited with a constant input to measure the step response. If the approach is model-based, one must simulate the step response of the plant.

2. *Determine the time constants.* When the open-loop step response is S-shaped, the open-loop transfer function can be approximated as

$$\frac{y(s)}{u(s)} = \frac{Ke^{-Ls}}{Ts+1}.$$

Given a unit-step input, the approximate response is

$$y(s) = \left[\frac{Ke^{-Ds}}{Ts+1}\right]\frac{1}{s} = \frac{Ke^{-Ds}}{T}\left[\frac{1}{s+1/T}\right]\frac{1}{s}$$

$$= Ke^{-Ds}\left[\frac{1}{s} - \frac{1}{s+1/T}\right]$$

$$y(t) = K1(t-D)\left[1 - e^{-t/T}\right],$$

which resembles the delayed step response of a first-order system where the response is delayed an amount $t = D$. As illustrated in Figure 9.35, D is the delay time and T is the time constant. Here K is the steady-state value. These parameters are taken from the measured or simulated step response. The delay time and time constant are determined by drawing a tangent line through the inflection point of the S-shaped response. The delay time is measured from initiation to the instant where the tangent line crosses the value of the initial output (often this is zero). The time constant is measured from the delay time to the instant that the tangent line crosses the steady-state output value.

3. *Calculate the compensator parameter values.* Ziegler and Nichols suggested setting the compensator parameters according to the rules in Table 9.1 using the delay time, time constant, and steady-state value.

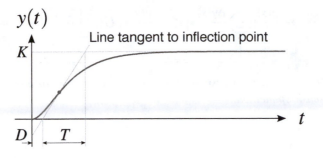

Figure 9.35: *S-shaped response curve.*

Table 9.1: *Ziegler-Nichols open-loop method tuning rules.*

Compensation	K_P	T_I	T_D
P	T/D	∞	0
PI	$0.9T/D$	$3.3D$	0
PID	$1.2T/D$	$2D$	$0.5D$

When using the open-loop method, the PID compensator has repeated, real-valued zeros and a pole at $s = 0$:

$$
\begin{aligned}
G_c &= K_P \left[1 + \frac{1}{T_I s} + T_D s \right] \\
&= 1.2 \frac{T}{D} \left[1 + \frac{1}{2Ds} + 0.5Ds \right] \\
&= 1.2 \frac{T}{D} \left[\frac{D^2 s^2 + 2Ds + 1}{2Ds} \right] \\
&= 0.6T \; \frac{s^2 + \dfrac{2s}{D} + \dfrac{1}{D^2}}{s} \\
&= 0.6T \; \frac{\left(s + \dfrac{1}{D} \right)^2}{s}.
\end{aligned}
$$

The repeated zeros are located at $s = -1/D$.

The closed-loop method.   The closed-loop method can be employed if a sustained oscillatory response can be attained through proportional com-

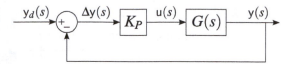

Figure 9.36: *Closed-loop system with proportional gain for use with Ziegler-Nichols closed-loop method.*

pensation. The procedure is applied to the closed-loop system initially setting the integral and derivative time constants to ∞ and 0, respectively, so that the PID compensator simplifies to proportional control as depicted in Figure 9.36. The process for tuning using the closed-loop method is as follows:

1. *Increase the proportional gain until sustained oscillations are attained.* Excite the system with a step change in the set point, for example, and adjust the gain until the system oscillates like the response in Figure 9.37. The value of the proportional gain at the point of sustained oscillations is the critical gain, K_{cr}. For a model-based approach, step 6 of the process for drawing the root locus can be used to determine the frequencies at which the root locus crosses the imaginary axis and the corresponding gain. The critical gain can also be determined by plotting the root locus and using the data cursor in MATLAB to find the gain when the root locus crosses the imaginary axis. Alternatively, the Routh-Hurwitz stability criterion can be used to analytically determine the critical gain.

2. *Measure the critical period of oscillation.* Once an oscillatory response is achieved, the period of oscillation must be measured. This is the critical period of oscillation T_{cr}. Numerically, one can measure the period of oscillation by plotting the closed-loop step response and using two data tips – one placed at the beginning of an oscillation and the other at the end. The period of oscillation would then be the difference between the x-position values for each of the data tips.

3. *Calculate the compensator parameter values.* Based on the measured or numerically attained values of the critical gain and period of oscillation, the compensator parameters are calculated using the equations provided in Table 9.2.

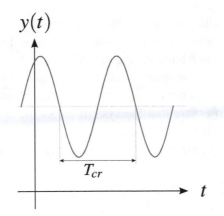

Figure 9.37: *Period of oscillation attained using critical gain.*

Table 9.2: *Ziegler-Nichols closed-loop method tuning rules.*

Compensation	K_P	T_I	T_D
P	$0.5K_{cr}$	∞	0
PI	$0.45K_{cr}$	$0.83T_{cr}$	0
PID	$0.6K_{cr}$	$0.5T_{cr}$	$0.125T_{cr}$

This method also results in a compensator with two real-valued zeros and a pole at the origin:

$$G_c = K_P \left[1 + \frac{1}{T_I s} + T_D s \right]$$

$$= 0.6K_{cr} \left[1 + \frac{1}{0.5T_{cr} s} + 0.125T_{cr} s \right]$$

$$= 0.6K_{cr} \left[\frac{\dfrac{T_{cr}^2 s^2}{16} + \dfrac{T_{cr} s}{2} + 1}{\dfrac{T_{cr} s}{2}} \right]$$

$$= 0.6K_{cr}T_{cr} \left(\frac{2}{16} \right) \frac{s^2 + \dfrac{8s}{T_{cr}} + \dfrac{16}{T_{cr}^2}}{s}$$

$$= 0.075K_{cr}T_{cr} \frac{\left(s + \dfrac{4}{T_{cr}} \right)^2}{s}.$$

The repeated, real-valued zeros are placed at $s = -4/T_{cr}$.

The Ziegler-Nichols tuning rules should be considered a starting point for fine tuning. If applicable, they generally result in closed-loop systems that exhibit 10%-60% overshoot. Note that the tuning rules are devised to attain an acceptable response that may be near optimal but further fine tuning may be necessary to ensure that the system meets the design criteria. For a model-based approach, the root locus can be used first to adjust and achieve dominant closed-loop poles with the desired characteristics.

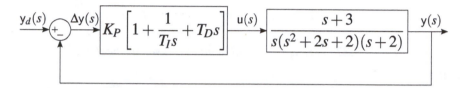

Figure 9.38: *Block diagram for system from Example 9.7 with PID compensation.*

Example 9.9

Figure 9.38 shows the system from Example 9.7 with PID compensation. Using Ziegler-Nichols turning rules, design a PID compensator to control the output.

Solution. Because the system has an integrator term, we must use the closed-loop method. The first step is to determine the critical gain. In Example 9.7, we determined that the root locus crossed the imaginary axis at $s = \pm j\,1.2580$ or at a frequency of $\omega = 1.2580$ rad/s. This is the natural frequency of the response at the point of marginal stability. Using Equation 9.27, the critical gain is

$$K_{cr} = 4\omega_{cr}^2 - 4 = 4(\omega_{cr}^2 - 1) = 4[(1.2580 \text{ rad/s})^2 - 1] \approx 2.33.$$

To determine the critical period of oscillation, we can plot the step response and measure the period of oscillation at steady-state. We can also determine the critical period analytically by using the natural frequency at the point of marginal stability. The critical period of oscillation as shown in Figure 9.39 is

$$T_{cr} = \frac{2\pi}{\omega_{cr}} = \frac{2\pi}{(1.2580 \text{ rad/s})} \approx 5.0 \text{ s}.$$

Knowing the critical gain and period of oscillation the compensator constants can be calculated:

$$K_P = 0.5K_{cr} = 0.6(2.33) = 1.398$$
$$T_I = 0.5T_{cr} = 0.5(5.0) = 2.5$$
$$T_D = 0.125T_{cr} = 0.125(5.0) = 0.625$$

making the PID compensator

$$G_c(s) = K_P \left[1 + \frac{1}{T_I s} + T_D s \right] = 1.398 \left[1 + \frac{1}{2.5s} + 0.625s \right]$$

or

$$G_c(s) = 0.075 K_{cr} T_{cr} \frac{\left(s + \frac{4}{T_{cr}} \right)^2}{s}$$

$$= 0.075(2.33)(5.0) \frac{\left(s + \frac{4}{5.0} \right)^2}{s}$$

$$= 0.8737 \frac{(s+0.8)^2}{s}.$$

Using the PID compensator above, the resultant closed-loop step response plotted in Figure 9.40 has an overshoot of approximately 50%, which is in the expected range. It is interesting to note, however, that the resultant overshoot is greater than that attained using only a proportional gain of $K = 0.549$ derived in Example 9.7 to achieve an overshoot of approximately 5% and settling time of approximately 9.3 s. As Figure 9.40 shows, the step response garnered using PID compensation derived using the tuning rule has significantly more overshoot but similar settling time to the response derived in Example 9.7 labeled "Root-locus tuned P control." The closed-loop Ziegler-Nichols rules specify a proportional gain of $0.5K_{cr} = 2.33/2$ for proportional control which is higher than the gain used in Example 9.7. The P-compensated system with the gain suggested using the Ziegler-Nichols tuning rule is also plotted in Figure 9.40. The proportional gain specified by the tuning rule results in more overshoot and longer settling time than the gain derived in Example 9.7. As was mentioned, the tuning rules are meant as a starting point to achieve an acceptable performance. Optimal performance will likely require fine tuning. If a reasonable model is readily available, much of the online fine tuning can be circumvented or done offline and a suitable design can be achieved more quickly.

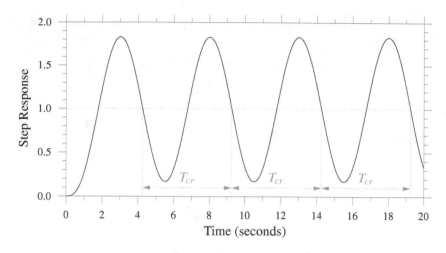

Figure 9.39: *Step response using critical gain.*

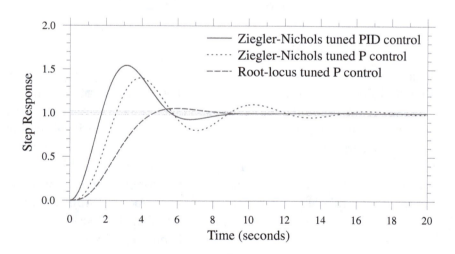

Figure 9.40: *Closed-loop step response of PID compensated system.*

9.10   Relating the Transient and Frequency Responses

In this section we explore how the characteristics of the transient response can be related to those of the frequency response and how system stability is ascertained from the Bode plot.

9.10.1 Stability Analysis in Frequency Domain

System stability can be determined in terms of gain and phase margins. Recall the magnitude and angle conditions used to derive the root locus. Remember that the root locus satisfies the angle condition, and that the magnitude condition correlates to a specific overall gain and set of closed-loop poles. Also remember that the part of the root locus left of the imaginary axis encompasses all loci that are stable. Those loci on the imaginary axis are marginally stable, and those right of the axis are unstable.

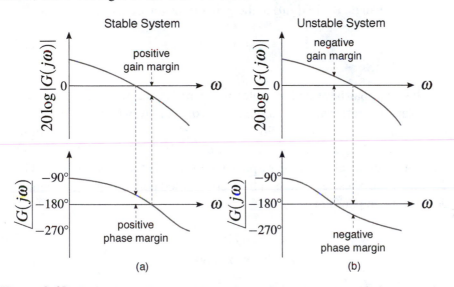

Figure 9.41: *Bode plots of (a) a stable system with positive gain and phase margins and (b) an unstable system with negative gain and phase margins.*

The *gain margin* is defined as the additional gain required at the *phase crossover frequency*, ω_p, to make the system marginally stable. It is the reciprocal of the magnitude at the phase crossover frequency where the phase plot crosses $-180°$,

$$K_g = \frac{1}{|G(j\omega_p)|}.$$ (9.28)

The gain margin can be readily obtained from the root locus plot. The magnitude plot in Figure 9.41 (a) illustrates a positive gain margin and (b) depicts a negative gain margin. Note that the gain margin is measured at the frequency that the phase angle passes through $-180°$ which correlates to the angle condition. Positive gain margin is measured down from 0 dB and negative gain margin up. Recall that 0 dB is equivalent to a magnitude of unity.

Phase margin is the additional phase angle required at the *gain crossover frequency*, ω_g, to make the system marginally stable. The phase margin is measured at the gain crossover frequency, which is where the magnitude passes through 0 dB or where $|G(j\omega)| = 1$. (Recall the magnitude condition.) Positive phase margin is measured up from $-180°$ and negative margin is measured down from the same mark. A system is stable if both the gain and phase margins are positive. That is equivalent to the dominant, closed-loop poles being in the left half of the real-imaginary plane. The phase margin γ is the phase angle ϕ of $G(j\omega)$ at the gain crossover frequency plus $180°$,

$$\gamma = \phi + 180°. \tag{9.29}$$

Systems that have neither poles nor zeros in the right-half of the complex plane are referred to as *minimum-phase systems*, while those that do are *nonminimum-phase systems*. For minimum phase systems, a gain margin of greater than 6 dB and phase margin in the range of $30°$ to $60°$ will tend to result in a stable, closed-loop response with satisfactory performance (Ogata 2002). Hence, these can serve as frequency-based performance guidelines for designing compensators:

$$20\log K_g > 6 \text{ dB} \tag{9.30}$$
$$30° \leq \gamma \leq 60° \tag{9.31}$$

9.10.2 Resonance and the Damping Ratio

Recall in Chapter 7, particularly Section 7.5, we discussed how the transient characteristics of a higher-order system can be dictated by the presence of a dominant pole pair. If we continue on that premise, we can consider the system depicted in Figure 9.42 with the feedforward transfer function

$$G(s) = \frac{\omega_n^2}{s(s+2\zeta\omega_n)},$$

and closed-loop transfer function

$$G_{cl}(s) = \frac{y(s)}{u(s)} = \frac{\omega_n^2}{s^2 + 2\zeta\omega_n s + \omega_n^2} \tag{9.32}$$

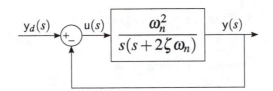

Figure 9.42: *Standard, closed-loop, second-order system.*

where we have previously defined ζ and ω_n as the damping ratio and un-damped natural frequency, respectively.

The sinusoidal transfer function is

$$
\begin{aligned}
G_{cl}(j\omega) &= \frac{\omega_n^2}{(j\omega)^2 + 2\zeta\omega_n(j\omega) + \omega_n^2} \\
&= \frac{\omega_n^2}{(\omega_n^2 - \omega^2) + j2\zeta\omega_n\omega} \\
&= \frac{1}{\left(1 - \dfrac{\omega^2}{\omega_n^2}\right) + j2\zeta\dfrac{\omega}{\omega_n}}.
\end{aligned}
$$

Recalling Euler's theorem (Section 5.2.2), the sinusoidal transfer function can be written as

$$
G_{cl}(j\omega) = M e^{j\alpha}
$$

where

$$
M = \frac{1}{\sqrt{\left(1 - \dfrac{\omega^2}{\omega_n^2}\right)^2 + \left(2\zeta\dfrac{\omega}{\omega_n}\right)^2}} \quad \text{and} \quad \alpha = -\tan^{-1}\frac{2\zeta\dfrac{\omega}{\omega_n}}{1 - \dfrac{\omega^2}{\omega_n^2}}.
$$

Maximum magnitude will occur at the resonant frequency, $\omega_r = \omega_n\sqrt{1 - 2\zeta^2}$

(recall Equation 8.5). Thus the peak magnitude at resonance is

$$M_r = \cfrac{1}{\sqrt{\left(1 - \cfrac{\omega_r^2}{\omega_n^2}\right)^2 + \left(2\zeta\cfrac{\omega_r}{\omega_n}\right)^2}}$$

$$= \cfrac{1}{\sqrt{\left(1 - \cfrac{\omega_n^2(1-2\zeta^2)}{\omega_n^2}\right)^2 + \left(2\zeta\cfrac{\omega_n\sqrt{1-2\zeta^2}}{\omega_n}\right)^2}}$$

$$= \frac{1}{\sqrt{4\zeta^4 + 4\zeta^2(1-2\zeta^2)}} = \frac{1}{\sqrt{4\zeta^2(1-\zeta^2)}}$$

$$= \frac{1}{2\zeta\sqrt{1-\zeta^2}} \tag{9.33}$$

where M_r is referred to as the *resonant peak magnitude* (Ogata 2002). Thus, the resonant peak magnitude, much like maximum overshoot in the time domain, is purely a function of the damping ratio. Figure 9.43 shows how the resonant peak magnitude varies with the damping ratio up to about $\zeta = 0.707$ because ω_r is real-valued only up to that damping ratio.

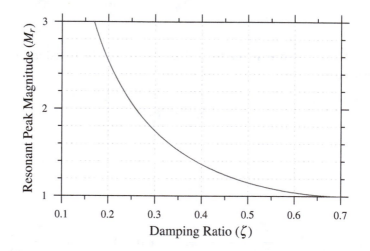

Figure 9.43: *Variation of resonant peak magnitude with damping ratio.*

Recall that for Bode plots the bandwidth is range of frequencies over which the magnitude exceeds 3 dB less than the zero-frequency value. Thus, we can solve for the cutoff frequency where $20\log M = -3$ dB or $M = 1/\sqrt{2}$:

$$\frac{1}{\sqrt{2}} = \frac{1}{\sqrt{\left(1 - \frac{\omega_c^2}{\omega_n^2}\right)^2 + \left(2\zeta\frac{\omega_c}{\omega_n}\right)^2}}$$

$$2 = \left(1 - \frac{\omega_c^2}{\omega_n^2}\right)^2 + \left(2\zeta\frac{\omega_c}{\omega_n}\right)^2$$

$$= 1 - 2\frac{\omega_c^2}{\omega_n^2} + \frac{\omega_c^4}{\omega_n^4} + 4\zeta^2 + \frac{\omega_c^2}{\omega_n^2}$$

$$0 = \frac{\omega_c^4}{\omega_n^4} + 2(2\zeta^2 - 1)\frac{\omega_c^2}{\omega_n^2} - 1.$$

By applying the quadratic equation, the resultant solution that yields a positive frequency is

$$\omega_c = \omega_n\sqrt{(1 - 2\zeta^2) + \sqrt{4\zeta^4 - 4\zeta^2 + 2}}. \tag{9.34}$$

Recollect, however, that the settling time from Equation 7.27 is $t_s = 4/(\zeta\omega_n)$, and remember that the bandwidth is the frequency range below the cutoff frequency. Ergo, the bandwidth can be specified in terms of the desired settling time and damping ratio instead:

$$\Delta\omega = \frac{4}{\zeta t_s}\sqrt{(1 - 2\zeta^2) + \sqrt{4\zeta^4 - 4\zeta^2 + 2}}. \tag{9.35}$$

9.10.3 Phase Margin and the Damping Ratio

Recall from root locus analysis that all loci must satisfy both the angle and magnitude conditions (Equations 9.20 and 9.21). Furthermore, the points on the imaginary axis where the system is marginally stable are determined by letting $s = j\omega$ as is done in the sinusoidal transfer function. Thus, given the sinusoidal form of the open-loop transfer function,

$$G(j\omega) = \frac{\omega_n^2}{j\omega(j\omega + 2\zeta\omega_n)} = \frac{\omega_n^2}{-\omega^2 + j2\zeta\omega_n\omega} = \frac{1}{-\frac{\omega^2}{\omega_n^2} + j2\zeta\frac{\omega}{\omega_n}},$$

the magnitude condition is satisfied when $|G(j\omega)| = 1$ or

$$1 = \left| -\frac{\omega^2}{\omega_n^2} + j2\zeta\frac{\omega}{\omega_n} \right|$$

$$= \sqrt{\left(\frac{\omega^2}{\omega_n^2}\right)^2 + \left(2\zeta\frac{\omega}{\omega_n}\right)^2}.$$

We rewrite the above equation as

$$1 = \frac{\omega^4}{\omega_n^4} + 4\zeta^2\frac{\omega^2}{\omega_n^2}$$

$$0 = \frac{\omega^4}{\omega_n^4} + 4\zeta^2\frac{\omega^2}{\omega_n^2} - 1$$

and then use the quadratic equation to determine ω^2. The only solution that results in a positive frequency is

$$\omega = \omega_n\sqrt{-2\zeta^2 + \sqrt{1 + 4\zeta^4}}.$$

Given this frequency, the phase angle

$$\underline{/G(j\omega)} = -\underline{/-\frac{\omega^2}{\omega_n^2} + j2\zeta\frac{\omega}{\omega_n}}$$

becomes

$$\underline{/G(j\omega)} = -\tan^{-1}\frac{\left(2\zeta\frac{\omega}{\omega_n}\right)}{\left(-\frac{\omega^2}{\omega_n^2}\right)}$$

$$= -\tan^{-1}\frac{2\zeta\omega_n}{-\omega}$$

$$= -\tan^{-1}\frac{2\zeta}{\sqrt{-2\zeta^2 + \sqrt{1 + 4\zeta^4}}},$$

making the phase margin

$$\gamma = \underline{/G(j\omega)} + 180°$$

$$= \tan^{-1}\frac{2\zeta}{\sqrt{-2\zeta^2 + \sqrt{1 + 4\zeta^4}}}. \qquad (9.36)$$

Thus, the phase margin, like the resonant peak magnitude, is a function of the damping ratio. Figure 9.44 illustrates how the phase margin varies with damping ratio. Notice that when the system is lightly damped (i.e., $\zeta < 0.6$), the phase margin can be approximated as

$$\gamma \approx 100\zeta .$$

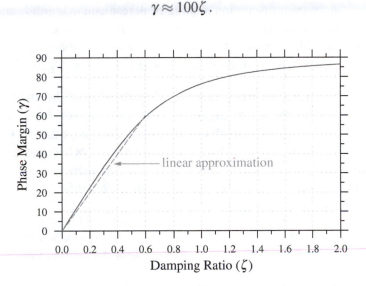

Figure 9.44: *Variation of phase margin with damping ratio.*

Thus far we have discussed how the gain margin, phase margin, and peak resonant magnitude can be used as performance specifications. Additionally, the cutoff frequency and, consequently, the bandwidth can be used to specify performance criteria as discussed in the following section.

9.10.4 Cutoff Frequency and Bandwidth

The cutoff frequency is the frequency at which the magnitude drops 3 decibels below its value at zero frequency. Remember that we normalized the first- and second-order terms discussed in Section 8.7 so that the zero-frequency magnitude would be unity or zero decibels. Generally, however, the cutoff frequency is where

$$20\log|G(j\omega)| < 20\log|G(j0)| - 3 \text{ dB} .$$

The system attenuates frequencies above the cutoff and transmits those below. The higher the cutoff frequency, the larger the bandwidth over which the system tends to transmit; in other words, the larger the frequency range over which the system tends to immediately respond to and transfer. Hence, high cutoff frequency or bandwidth correlates to rise time or roughly the swiftness of the transient response. However, a large bandwidth also tends to make the system susceptible to noise.

Based on the prior discussion relating peak resonant magnitude and phase margin to the damping ratio, we can make the following observations regarding the relationship between the time and frequency domain characteristics of higher-order systems with dominant closed-loop poles (Ogata 2002):

1. The peak resonant magnitude, M_r, is indicative of the stability. The desired peak resonant magnitude is in the range of $1.0 < M_r < 1.4$ (0 dB $< M_r < 3$ dB) which correlates to a damping ratio of $0.4 < \zeta < 0.7$. As the M_r exceeds 1.5 the maximum overshoot becomes excessive. The output is amplified near the resonant frequency, and should the input be noisy, the noise can be substantially amplified in the output.

2. The resonant frequency ω_r is representative of the quickness of the transient response. Basically, the higher the resonant frequency, the faster the transient response, and the shorter the settling time.

3. The bandwidth and cutoff frequency are indicative of the closed-loop system's capacity to replicate the input signal. Accordingly, they are linked to the transient response. The greater the cutoff frequency, the faster the transient tends to occur.

Using these guidelines, one can utilize the frequency response to design compensators. In the section that follows, we shall explore the use of the Bode plot to design lead-lag compensators.

9.11   Compensator Design Using Bode Plot Analysis

Thus far we have discussed PID compensators. In this section, we introduce an alternative type of cascaded compensation – *lead-lag compensation*.

Lead compensation is used to improve the stability margin and transient response characteristics. It increases bandwidth but can result in increased noise at high frequency due to the gains at those frequencies (Ogata 2004). Conversely, *lag compensation* is utilized to improve the steady-state accuracy. It reduces the system bandwidth and slows the system response (Ogata 2004).

It also attenuates high frequency noise. *Lead-lag* compensators are cascaded combinations of the two. They are used to simultaneously improve both the transient response and the steady-state accuracy.

We shall investigate the frequency response characteristics of each and effects each has when utilized in a closed-loop system.

9.11.1 Design of Lead Compensators

A lead compensator has a transfer function of the form

$$G_c(s) = K_c \alpha \frac{Ts+1}{\alpha Ts+1} = K_c \frac{s + \dfrac{1}{T}}{s + \dfrac{1}{\alpha T}} \tag{9.37}$$

where α is referred to as the *attention factor* and is defined such that $0 < \alpha < 1$. Notice that the lead compensator has first-order terms in the numerator and denominator. It has a zero at $s = -1/T$ and a pole at $s = -1/(\alpha T)$. The compensator is referred to as a lead compensator because the zero "leads" the pole. That is, the zero is closer to the origin. If one imagines the pole and zero to be traveling to the right, it appears as though the zero "leads" the pole as delineated in Figure 9.45 (a).

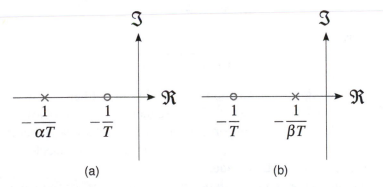

(a)   (b)

Figure 9.45: *Pole-zero positions for (a) lead compensator and (b) lag compensator.*

Recall the Bode diagrams for the first-order terms $s+a$ and $1/(s+a)$ plotted in Figures 8.21 and 8.22. Also recall that the bode plots of the individual terms can be graphically summed to derive the overall plot. Given that $0 < \alpha < 1$, $1/T$ is less than $1/(\alpha T)$. Thus, the corner frequency for the numerator (i.e., $\omega = 1/T$) will occur at a lower frequency than that for the denominator ($\omega = 1/(\alpha T)$). When the terms are summed the resulting Bode diagram looks like the plot in Figure 9.46 where α was set to 0.1. Notice the maximum phase angle contribution, ϕ_m, occurs at the geometric mean of the two corner frequencies, or

$$\log \omega_m = \frac{1}{2}\left(\log\frac{1}{T} + \log\frac{1}{\alpha T}\right)$$

where ω_m is the frequency at the maximum phase angle. Using logarithmic identities, it can be shown that

$$\log \omega_m = \frac{1}{2}\left(\log\frac{1}{T} + \log\frac{1}{\alpha T}\right)$$
$$= \frac{1}{2}\log\frac{1}{\alpha T^2}$$
$$= \log\frac{1}{\sqrt{\alpha}\, T}$$
$$\omega_m = \frac{1}{\sqrt{\alpha}\, T}. \tag{9.38}$$

It can also be shown using a Nyquist plot*, which is not discussed herein, that the relation between the *maximum phase-lead angle* and the attenuation factor is

$$\sin\phi_m = \frac{1-\alpha}{1+\alpha} \tag{9.39}$$

or alternatively

$$\alpha = \frac{1-\sin\phi_m}{1+\sin\phi_m}. \tag{9.40}$$

The lead compensator functions like a high-pass filter which attenuates the low frequencies and passes the high frequencies. It is employed to contribute phase lead at ω_m to compensate for too much phase lag inherent to the open-loop system. Note that the corner frequencies are connected by a line that rises at a rate of 20 dB/decade.

The compensator can be designed using either time- or frequency-based performance criteria. The process for deriving a lead compensator using Bode plots and the gain and phase margin guidelines (Equations 9.28 and 9.29) is as follows (Ogata 2002):

1. *Find the overall gain.* Assume a lead compensator of the form

$$G_c(s) = K\frac{Ts+1}{\alpha Ts+1}$$

---

*. Refer to (Ogata 2004) Chapter 9 for further details regarding Nyquist plots.

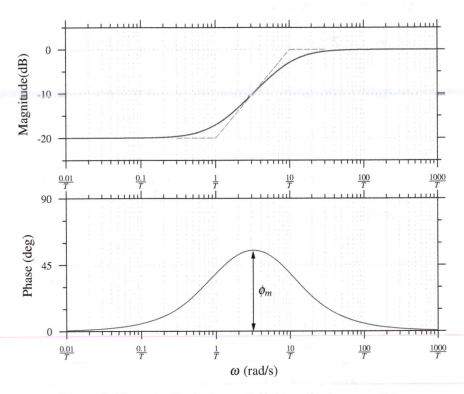

Figure 9.46: *Bode plot for* $(Ts+1)/(\alpha Ts+1)$ *where* $\alpha = 0.1$.

where $K = K_c \alpha$ and $0 < \alpha < 1$. The feedforward or open-loop transfer function is then

$$G_c(s)G(s) = K\frac{Ts+1}{\alpha Ts+1}G(s) = \frac{Ts+1}{\alpha Ts+1}KG(s) = \frac{Ts+1}{\alpha Ts+1}G_1(s)$$

where $G_1(s) = KG(s)$. Find the gain K necessary to satisfy the specified steady-state criteria typically provided as a static error constant requirement.

2. *Plot the Bode diagram.* Once K has been determined, plot the Bode diagram for $G_1(j\omega)$, and assess the phase margin.

3. *Determine the phase-lead angle.* Ascertain the additional phase angle contribution ϕ_m necessary to meet the design criteria, and be sure to compensate for the fact that the lead compensator shifts the gain margin crossover frequency by adding 5° to 12°.

4. *Calculate the attenuation factor.* Using Equation 9.40, calculate the attenuation factor. Find the frequency where the magnitude of G_1 in decibels is $-20\log(1/\sqrt{\alpha})$. This frequency is ω_m where the maximum phase shift contribution, ϕ_m, occurs. Knowing the frequency and attenuation factor, you can calculate T using Equation 9.38.

5. *Determine the corner frequencies.* Compute the corner frequencies recalling that the numerator is $s + 1/T$ and the denominator is $s + 1/(\alpha T)$.

6. *Compute the compensator gain.* Having calculated the attenuation factor and determined the gain K in the first step, one can now determine the compensator gain,

$$K_c = \frac{K}{\alpha}.$$

7. *Test the design.* Confirm that the design criteria have been met. If not, repeat the prior procedure.

Alternatively, the lead compensator can be designed using time-based criteria. The procedure detailed in (Nise 2011) is basically the same except that the requisite phase margin γ and corresponding lead-compensator contribution ϕ_m are determined by the following process:

1. Utilize the specified maximum overshoot design criteria to determine the recommended damping ratio. Recollect that the maximum overshoot is plotted as a function of the damping ratio in Figure 9.10. Given a desired overshoot, one can use the figure to determine the necessary damping ratio.

2. Again, using the damping constant found in the first step, determine the phase margin, γ. The damping ratio can be used to find the desired phase margin using Figure 9.44. Remember to add $5°$ to $12°$ to account for the fact that compensator will shift the position of the crossover frequency.

In addition, once the compensator design is derived, the closed-loop response must be checked to ensure that the settling time criterion is met. Given the specified settling time and damping ratio from the first step, the desired bandwidth is calculated using Equation 9.35. To determine if the settling time criterion is met, we check that the compensated, closed-loop frequency response has a bandwidth greater than or equal to the desired bandwidth.

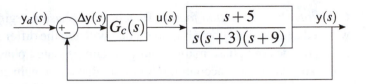

Figure 9.47: *Closed-loop system with lead compensation.*

Example 9.10

Figure 9.47 shows a closed-loop system with a lead compensator. Design a lead compensator that limits the overshoot to less than 20% and the settling time to less than 0.3 seconds. The desired static velocity error constant is $K_v = 200 \text{ sec}^{-1}$.

Solution. The plant transfer function is

$$G(s) = \frac{s+5}{s(s+3)(s+9)}.$$

First we must determine the necessary phase margin to meet the overshoot criterion. Following the time-based procedure previously provided, we first determine the desired damping ratio and then relate that to the required phase margin. From Figure 9.10 a damping ratio 0.6 should limit the overshoot to under 20%. According to the linear approximation $\gamma \approx 100\zeta$, the required phase margin is 60° which correlates well with the plot in Figure 9.44. Now that we know the desired phase margin, we can proceed with the frequency-based design process:

1. *Find the overall gain.* We use the static velocity error specification to determine the gain K:

$$K_v = \lim_{s \to 0} s\, G_c(s) G(s)$$

$$= \lim_{s \to 0} s\, K \left(\frac{Ts+1}{\alpha Ts+1} \right) \frac{s+5}{s(s+3)(s+9)} = \frac{5K}{27}$$

$$K = \frac{27}{5} K_v = \frac{27}{5}(200) = 1080.$$

2. *Plot the Bode diagram.* We can now plot the Bode diagram of $G_1(s) = KG(s)$. The Bode plot is provided in Figure 9.48. The gain compensated open-loop transfer function has an infinite gain margin and a phase margin of approximately 12°.

3. *Calculate the attenuation factor.* To achieve a phase margin of 60°
   we need a phase-lead contribution ϕ_m that will add the difference between the desired phase margin and gain-compensated phase angle
   while simultaneously accounting for offset due to the gain contribution. Thus, we will determine the difference and add 5° to 12° as
   previously explained:

$$\phi_m = 60° - 12° + 5° = 53°.$$

Given ϕ_m, we can use Equation 9.40 to calculate the attenuation
factor,

$$\alpha = \frac{1 - \sin 53°}{1 + \sin 53°} \approx 0.112.$$

Using the attenuation factor and the bode plot of G_1, we can determine the frequency ω_m where the magnitude is $-20\log(1/\sqrt{\alpha})$,
which is where the maximum phase-lead contribution occurs:

$$-20\log \frac{1}{\sqrt{\alpha}} = -20\log \frac{1}{\sqrt{0.112}} \approx -9.50 \text{ dB}.$$

According to the plot of G_1, the magnitude is -9.50 dB at about
55 rad/s. Hence, this will be our ω_m. Using Equation 9.38, we can
compute T:

$$T = \frac{1}{\sqrt{\alpha}\omega_m} = \frac{1}{\sqrt{0.112}\,(55 \text{ rad/s})} \approx 0.0545 \text{ s}.$$

4. *Determine the corner frequencies.* We now have all the pertinent
   information to calculate the corner frequencies or the zero and pole
   of the lead compensator. The corner frequencies are

$$\omega = \frac{1}{T} \approx 18.4$$

and

$$\omega = \frac{1}{\alpha T} \approx 54.8.$$

5. *Compute the compensator gain.* The compensator gain is

$$K_c = \frac{K}{\alpha} = \frac{1080}{0.112} \approx 9.62 \times 10^3.$$

6. *Test the design.* The final design for the commentator is

$$G_c(s) = K_c \frac{s + \dfrac{1}{T}}{s + \dfrac{1}{\alpha T}} = (9.62 \times 10^3) \frac{s + 18.4}{s + 54.8}.$$

The frequency response of $G_c G$ is also plotted in Figure 9.48. The result ant response has an infinite gain margin and a phase margin of approximately 60°. However, we have still yet to determine if the settling time criterion have been met. To do so, we must check the bandwidth and plot the closed-loop step response. According to Equation 9.35, the recommended bandwidth is

$$\Delta\omega = \frac{4}{\zeta t_s} \sqrt{(1 - 2\zeta^2) + \sqrt{4\zeta^4 - 4\zeta^2 + 2}}$$

$$= \frac{4}{0.6 t_s} \sqrt{[1 - 2(0.6^2)] + \sqrt{4(0.6^4) - 4(0.6^2) + 2}}$$

$$= 9.19 \text{ rad/s}.$$

The resultant bandwidth of the compensated system is 77.5 rad/s which exceeds the recommended bandwidth. The closed-loop step response plotted in Figure 9.49 shows that the design does indeed meet the transient performance criteria.

9.11.2 Design of Lag Compensators

The lag compensator is termed such because the zero "lags behind" the pole as depicted in Figure 9.45 (b). Though the lag compensator is very similar in mathematical form to the lead compensator,

$$G_c(s) = K_c \beta \frac{Ts + 1}{\beta Ts + 1} = K_c \frac{s + \dfrac{1}{T}}{s + \dfrac{1}{\beta T}}, \tag{9.41}$$

the attenuation factor β is chosen such that $\beta > 1$ rather than between 0 and 1. The zero is located at $s = 1/T$ and the pole at $s = -1/(\beta T)$. Since $\beta > 1$, the pole will be located to the right of the zero. Moreover, the left corner frequency will be located at $\omega = -1/(\beta T)$ and the right at $\omega = 1/T$. Note

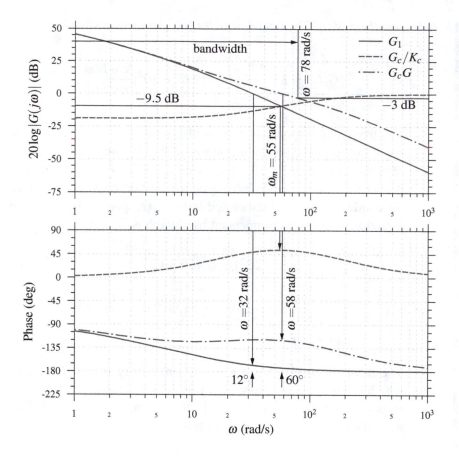

Figure 9.48: *Bode plots for G_1, G_c/K_c, and G_cG for Example 9.10.*

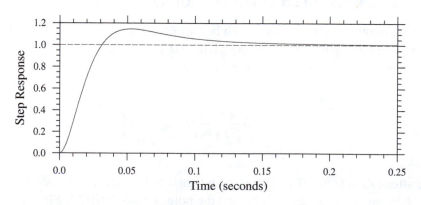

Figure 9.49: *Closed-loop step response of lead compensator example.*

that this will result in a pole and zero in close proximity to each other along the real axis between -1 and 0. Ideally, this will have minimal impact on the transient characteristics of the existing system.

The bode plot for $(Ts+1)/(\beta Ts+1)$ where $\beta = 10$ is provided in Figure 9.50. As Figure 9.50 demonstrates, the magnitude goes from 20 dB at low frequency to 0 dB at high frequency. The compensator contributes a negative phase angle that reaches its maximum absolute value halfway between the corner frequencies. It passes low frequencies and attenuates high frequencies much like a low-pass filter. The corner frequencies are connected by a line that drops at a rate of 20 dB/decade.

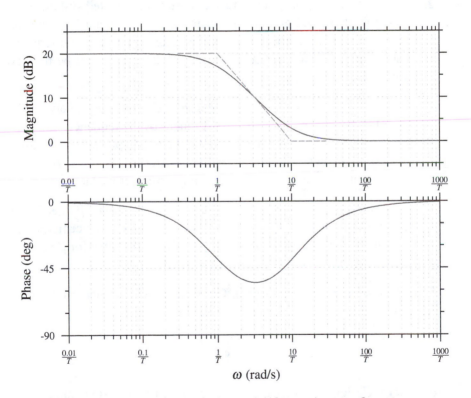

Figure 9.50: *Bode plot for $(Ts+1)/(\beta Ts+1)$ where $\beta = 10$.*

To design a lead compensator using frequency-based, performance criteria, use the following process (Ogata 2002):

1. *Find the overall gain.* Assume a lag compensator of the form

$$G_c(s) = K\frac{Ts+1}{\beta Ts+1}$$

where $K = K_c\beta$ and $\beta > 1$. The feedforward or open-loop transfer function is then

$$G_c(s)G(s) = K\frac{Ts+1}{\beta Ts+1}G(s) = \frac{Ts+1}{\beta Ts+1}KG(s) = \frac{Ts+1}{\beta Ts+1}G_1(s)$$

where $G_1(s) = KG(s)$. Find the gain K necessary to satisfy the specified steady-state criteria.

2. *Plot the Bode diagram.* Check to see if the gain-adjusted system $G_1(s)$ meets the phase and gain margin criteria. If it does not, find the frequency at which the phase angle is equal is $-180°$ plus the requisite phase margin (do not forget to add $5°$ to $12°$). This will be the new gain crossover frequency.

3. *Place the zero.* So that the lag compensator does not negatively impact phase lag, place the pole and zero at much lower frequencies than the new gain crossover frequency. This can be accomplished by choosing the corner frequency $\omega = 1/T$, which corresponds to the zero, one octave to one decade below the new gain crossover frequency.

4. *Determine the attenuation.* Ascertain the attenuation, β, needed to slide down the magnitude to 0 dB at the new gain crossover frequency. In decibels, the requisite attenuation is $-20\log\beta$. Solve for β and calculate the other corner frequency $\omega = 1/(\beta T)$.

5. *Calculate the compensator gain.* Given the previously determined gain, K, and the attenuation factor, β, solve for the compensator gain,

$$K_c = \frac{K}{\beta}.$$

6. *Test the design.* Test the compensated, closed-loop system and confirm that the design criteria were met.

The lag compensator is primarily utilized to improve stead-state characteristics. Thus, it is not typically employed to meet transient performance criteria that are normally specified in the time domain.

Example 9.11

Design a lag compensator for the system in Figure 9.47. The design criteria are that the minimum phase margin be 45° with a static velocity error constant of 100.

Solution. To derive the compensator design, we follow the process formerly described.

1. *Find the overall gain.* The static velocity error constant is used to determine the overall gain K,

$$K_v = \lim_{s \to 0} s G_c(s) G(s)$$

$$= \lim_{s \to 0} s K \left(\frac{Ts+1}{\beta Ts+1} \right) \frac{s+5}{s(s+3)(s+9)} = \frac{5K}{27}$$

$$K = \frac{27}{5} K_v = \frac{27}{5} (100) = 540.$$

2. *Plot the Bode diagram.* The Bode plot for $G_1(s)$ is provided in Figure 9.51. The gain-compensated, open-loop transfer function has a phase margin of approximately 17°, and therefore does not meet the phase margin criterion. To achieve the requisite 45° the compensator must contribute 45° plus 5° to 12°. Thus we will design a lag compensator with a maximum phase contribution of 52° (i.e., 45° + 7°). The gain-compensated, open-loop system has a phase angle of 52° at approximately 4.2 rad/sec. This will be the new gain crossover frequency.

3. *Place the zero.* To minimize the impact on the phase lag, we will place the zero one decade below the new gain crossover frequency. Thus, the compensator zero will be located at $\omega = 0.42$ rad/s which means that

$$T = \frac{1}{\omega} = \frac{1}{4.2 \text{ rad/s}} = 2.4 \text{ s}.$$

4. *Determine the attenuation.* To find the other corner frequency and determine the lag compensator pole position, we must determine the attenuation. The attenuation β is the amount needed to pull the magnitude plot down to 0 dB at the new gain crossover frequency. The magnitude at the new gain crossover frequency is approximately 24 dB. Hence, the compensator must contribute

$$-24 \text{ dB} = -20 \log \frac{1}{\beta}$$

where results in $\beta \approx 16.5$. We can now calculate the pole location,

$$\omega = \frac{1}{\beta T} = \frac{1}{(16.5)(2.4 \text{ s})} \approx 0.0252 \text{ rad/s}.$$

5. *Calculate the compensator gain.* The compensator gain is

$$K_c = \frac{K}{\beta} = \frac{540}{16.5} \approx 32.7.$$

6. *Test the design.* The final design is

$$G_c(s) = K_c \frac{s + \dfrac{1}{T}}{s + \dfrac{1}{\beta T}} = (32.7) \frac{s + 0.42}{s + 0.025}.$$

As Figure 9.51 illustrates, the compensated, open-loop frequency response indeed has approximately a 45° phase margin. Moreover, as the closed-loop, ramp responses in Figure 9.52 indicate, the lag-compensated system has a much improved steady-state response with much smaller error.

9.11.3 Design of Lead-Lag Compensators

As previously mentioned, the lead-lag compensator is a cascaded network of both lead and lag compensators. The compensator is of the form

$$G_c(s) = K_c \left(\frac{s + \dfrac{1}{T_1}}{s + \dfrac{\beta_1}{T_1}} \right) \left(\frac{s + \dfrac{1}{T_2}}{s + \dfrac{1}{\beta_2 T_2}} \right) \tag{9.42}$$

where β_1 and β_2 are both greater than 1. The first term in parentheses after K_c is the lead network and the second term in parentheses the lag network. Herein we shall discuss the special case where $\beta_1 = \beta_2$.

The Bode diagram for a lead-lag compensator is plotted in Figure 9.53. For the case plotted, $K_c = 1$, $\beta_1 = \beta_2 = 10$, and $T_2 = 10T_1$. Note at center of the plot, the phase angle crosses back through 0°. For the special case where $\beta_1 = \beta_2$, the frequency where the phase angle crosses 0° is (Ogata 2002)

$$\omega_1 = \frac{1}{\sqrt{T_1 T_2}}.$$

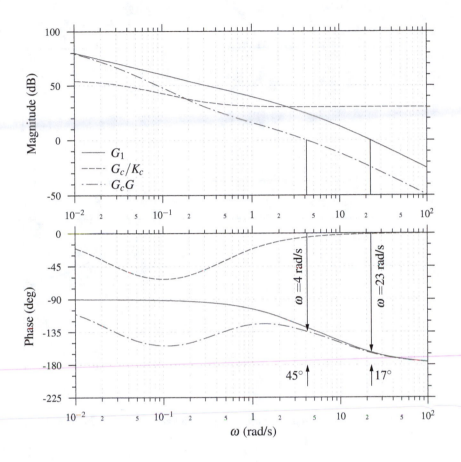

Figure 9.51: *Bode plots for G_1, G_c/K_c, and G_cG for Example 9.11.*

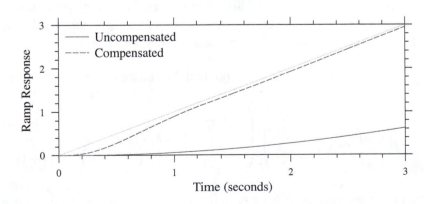

Figure 9.52: *Ramp responses for lag compensator example.*

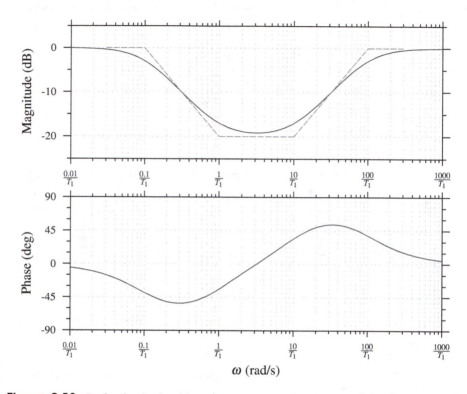

Figure 9.53: *Bode plot for lead-lag compensator where* $K_c = 1$, $\beta_1 = \beta_2 = 10$, *and* $T_2 = 10T_1$.

The magnitude is 0 dB at low and high frequencies. The design procedure for a lead-lag network is just a combination of the processes previously detailed for the lead and lag networks. The process follows Nise (2011):

1. *Find the compensator gain.* Note that the compensator is of the form

$$
G_c(s) = K_c \left(\frac{s + \dfrac{1}{T_1}}{s + \dfrac{\beta}{T_1}} \right) \left(\frac{s + \dfrac{1}{T_2}}{s + \dfrac{1}{\beta T_2}} \right) = K_c \frac{(T_1 s + 1)(T_2 s + 1)}{\left(\dfrac{T_1}{\beta} s + 1 \right)(\beta T_2 s + 1)} .
$$

(The subscripts on β have been dropped for convenience since we are assuming $\beta_1 = \beta_2$.) Calculate the compensator gain needed to meet the steady-state performance criteria.

2. *Choose a new gain crossover frequency.* Generate the Bode plot and determine the phase margin of the gain-compensated, open-loop transfer function $G_1(s) = K_c G(s)$. If overshoot and settling time are specified, then Figure 9.10 can be used to select the damping ratio and Equation 9.35 to estimate the bandwidth. The new gain crossover frequency can be chosen to meet or exceed the bandwidth. Figure 9.44 can be used to select the required phase margin based on the chosen damping ratio. Otherwise, the crossover frequency of the gain-compensated system can be used. Determine the additional phase angle contribution, ϕ_m, necessary to meet the design criteria. Remember to account for the shifting that occurs to the gain contribution by adding 5° to 12°.

3. *Determine the corner frequencies of the lag network.* Once the new gain crossover frequency is chosen, the zero for the lag portion of the compensator can be placed one octave to one decade below. The corresponding corner frequency will be $\omega = 1/T_2$. To determine the pole placement for the lag network, we must determine β. This can be accomplished using the phase margin requirement. β can be calculated from Equation 9.39 where $\alpha = 1/\beta$:

$$\sin \phi_m = \frac{1 - 1/\beta}{1 + 1/\beta} = \frac{\beta - 1}{\beta + 1}$$

$$\beta = \frac{1 + \sin \phi_m}{1 - \sin \phi_m}. \tag{9.43}$$

Knowing the upper corner frequency of the lag network zero, $\omega = 1/T_2$, and the value of β, we can calculate the lower corner frequency for the lag network, $\omega = 1/(\beta T_2)$.

4. *Determine the corner frequencies of the lead network.* To find the corner frequencies for the lead portion of the compensator, we must determine T_1. We can calculate T_1 using Equation 9.38 and assuming $\alpha = 1/\beta$:

$$\omega_m = \frac{1}{\sqrt{1/\beta}\, T_1} \quad \Rightarrow \quad T_1 = \frac{\sqrt{\beta}}{\omega_m}, \tag{9.44}$$

where ω_m is equal to our new gain crossover frequency. We can then calculate the lower and upper corner frequencies (i.e., $\omega = 1/T_1$ and $\omega = \beta/T_1$).

5. *Test the design.* As always, use the frequency response and/or step and ramp responses to assure that the design criteria have been met.

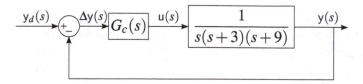

Figure 9.54: *Block diagram for a lead-lag design example.*

Example 9.12

Design a lead-lag compensator for the system depicted in Figure 9.54. The performance criteria are a static velocity error constant of 15 s^{-1}, a phase 4of 50°, and a gain margin greater than 10 dB.

   Solution. The design process is just a combination of the lead and lag procedures. The compensator is assumed to be of the form

$$G_c(s) = K_c \left(\frac{s + \dfrac{1}{T_1}}{s + \dfrac{\beta}{T_1}} \right) \left(\frac{s + \dfrac{1}{T_2}}{s + \dfrac{1}{\beta T_2}} \right).$$

Thus, we will begin by designing the lag network and then proceed to the lead portion.

1. *Find the compensator gain.* We first determine the gain K necessary to meet the specified static error constant,

$$K_v = \lim_{s \to 0} sK_c \frac{(T_1 s + 1)(T_2 s + 1)}{\left(\dfrac{T_1}{\beta} s + 1 \right)(\beta T_2 s + 1)} \frac{1}{s(s+3)(s+9)} = \frac{K_c}{27}.$$

   Ergo, the compensator gain necessary to meet the steady-state criterion is

$$K_c = 27K_v = 27(15) = 405.$$

   The resulting gain-compensated, open-loop transfer function is

$$G_1(s) = K_c G(s) = \frac{270}{s(s+3)(s+9)}.$$

2. *Choose the new gain crossover frequency.* The Bode plot of the gain-compensated system is provided in Figure 9.55. The gain-compensated system has a phase margin of approximately $-5.4°$

at 5.8 rad/s and a gain margin of -1.9 dB at 5.2 rad/s. Hence, the system is unstable, and the compensator would need to contribute approximately

$$\phi_m = 50° + 5.4° + 5° \approx 60.4°$$

to meet the phase margin performance criterion and account for any shifting due to the magnitude contribution from the lead network. We can choose to use the existing gain crossover frequency as our design point and develop a lag network a decade below and a lead network here to contribute the missing phase margin.

3. *Choose the corner frequencies of the lag network.* To design the lag network, we need to specify a new gain crossover frequency and then place the zero so that the associated upper corner frequency is an octave to a decade lower. In the previous step, we chose to use the existing gain crossover frequency 5.8 rad/s. Therefore, the upper corner frequency of the lag network is then $1/T_2 = 0.58$ rad/s if placed one decade below, making $T_2 \approx 1.73$ seconds. We can calculate β from Equation 9.43:

$$\beta = \frac{1 + \sin 60.4°}{1 - \sin 60.4°} \approx 14.3.$$

The lower corner frequency is thus

$$\omega = \frac{1}{\beta T_2} = \frac{1}{(14.3)(1.73 \text{ s})} \approx 0.0404 \text{ rad/s}.$$

4. *Choose the corner frequencies of the lead network.* Given the chosen gain crossover frequency and the value of β, we can calculate T_1 using Equation 9.44 and determine the corner frequencies of the lead network,

$$T_1 = \frac{\sqrt{14.3}}{5.8 \text{ rad/s}} \approx 0.653 \text{ seconds}.$$

Therefore, the lower and upper corner frequencies are

$$\frac{1}{T_1} \approx 1.53 \text{ and } \frac{\beta}{T_1} \approx 22.0.$$

5. *Test the design.* The final compensator design is

$$G_c = 405 \left(\frac{s+1.53}{s+22.0} \right) \left(\frac{0.580}{0.0404} \right).$$

As the lead-lag, compensated, frequency response plotted in Figure 9.55 indicates, the system exceeds both the gain and phase margin specifications. Moreover, the closed-loop, step and ramp responses in Figure 9.56 also support the fact that the system meets the performance criteria.

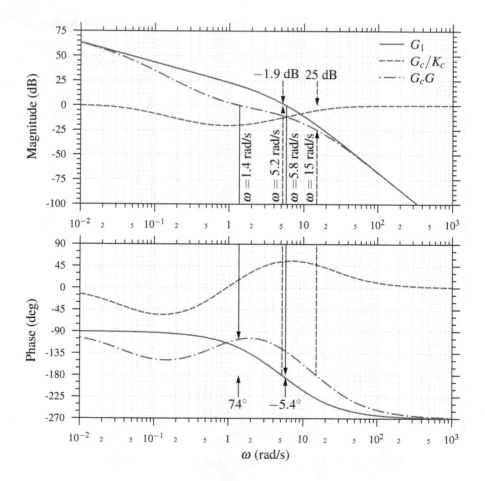

Figure 9.55: *Bode plots for G_1, G_c/K_c, and G_cG for Example 9.12.*

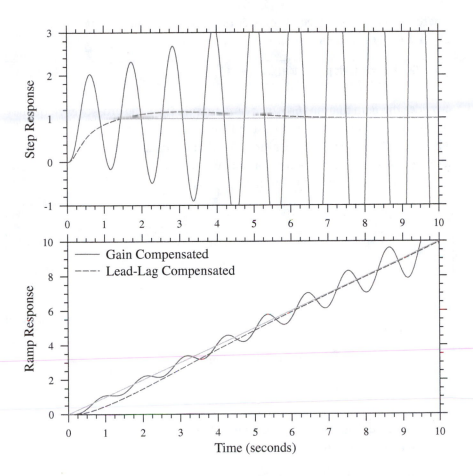

Figure 9.56: *Gain and lead-lag compensated responses for Example 9.12.*

9.11.4   Gain and Phase Margins Using MATLAB

Though the phase and gain margins can be ascertained manually from the Bode plot, MATLAB includes several tools to facilitate this process. One tool in particular – the `margin` function – plots and calculates the gain and phase margins and their respective crossover frequencies. To generate the plot the following syntax is used:

```
>> margin(sys)
```

where `sys` is a state-space or transfer function object defined using either `ss`, `tf`, or `zpk`. The alternate syntax

```
>> [Gm,Pm,Wcg,Wcp] = margin(sys)
```

provides the gain margin (`Gm`), the phase margin (`Pm`), the crossover frequency at which the gain margin is measured (`Wcg`), and the crossover frequency at which the phase margin is measured (`Wcp`). For instance, to generate the plot with margins for the gain-compensated transfer function in the last example, we could use the following commands:

```
>> G = zpk([],[0 -3  -9],1);
>> Kc = 405;
>> G1 = Kc*G

G1 =

        405
  -------------
  s (s+3) (s+9)

Continuous-time zero/pole/gain model.

>> margin(G1)
```

which produces Figure 9.57. To calculate the margins and respective frequencies, we can use the following:

```
>> [Gm,Pm,Wcg,Wcp] = margin(G1)
Warning: The closed-loop system is unstable.
> In warning at 26
  In DynamicSystem.margin at 63

Gm =

    0.8000


Pm =

   -5.4170


Wcg =

    5.1962
```

Wcp =

   5.7962

which displays the results and warns that the system is unstable.

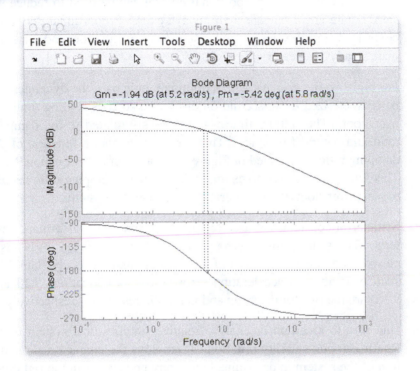

Figure 9.57: *MATLAB plot of gain and phase margins for Example 9.12.*

9.12 Summary

▷ Bond graphs can be used to derive the transfer functions of plants within a control system.

▷ Block diagram algebra is utilized to determine the overall input/output transfer function.

▷ Proportional, integral, and derivative control actions can be used to improve transient and steady-state performance of a closed-loop system.

The proportional term adds a compensation that is proportional to the error. The integral and derivative terms add control actions that are proportional to the integral and derivative of the error. A proportional-integral-derivative (PID) compensator can generally use all the control actions.

▷ For a first-order plant, proportional control can be used to reduce the steady-state error, but one will always exist. Proportional-derivative control converts the closed-loop system into second order and can eliminate steady-state error.

▷ For higher-order systems with a dominant pole pair, the dynamics can be characterized using second-order criteria such as settling time and overshoot. The settling time was previously related to the damping ratio and natural frequency. The overshoot is also a function of the damping ratio as depicted in Figure 9.10 (also refer to Equation 9.10). This figure can be used to estimate the requisite damping ratio needed to meet overshoot design criteria for higher-order systems.

▷ Static error constants are used to specify steady-state performance criteria. The static position error constant (Equation 9.11) measures a closed-loop system's ability to follow a step input at steady-state. The static velocity and acceleration error constants (Equations 9.13 and 9.15) are metrics for the ramp and parabolic responses, respectively.

▷ One way to ascertain or analyze relative stability of a system is the Routh-Hurwitz stability criterion. It employs the characteristic equation of the system to determine how many poles exist in the right half of the complex plane.

▷ One of the most utilized classical methods for compensator design is the root locus method. A root locus is a plot of all loci that satisfy the angle condition of the characteristic equation. The angle condition requires that the phase angle of open-loop transfer function be an odd integer multiple of 180°.

▷ The dominant, closed-loop poles are those points along the root locus that also satisfy the magnitude condition. The magnitude condition requires that magnitude of the open-loop transfer function be unity.

▷ The root locus can be used to design a compensator. Compensators add poles and zeros which change the shape of the root locus. A set of desired, closed-loop poles can be identified based on transient performance criteria. The compensator poles and zeros can be placed to force the root locus through the desired, closed-loop poles and then the compensator gain adjusted or selected to match the specified poles.

▷ The MATLAB rlocus function can be used to plot the root locus and aid in compensator design.

▷ In addition to the root locus and other methods, PID compensators are commonly optimized using tuning rules. A number of tuning rules have been established. Amongst them, the most commonly used are the Ziegler-Nichols rules.

▷ The frequency response can also be utilized to design compensators. However, if the performance criteria are specified in the time domain, they must be translated to frequency response characteristics. For higher-order systems with a dominant pole pair, the resonant peak magnitude can be estimated as a function of the damping ratio as shown in Figure 9.43. Furthermore, the combination of settling time and damping ratio can be translated into a cutoff frequency or bandwidth as provided in Equations 9.34 and 9.35. The phase margin is also a function of the damping ratio (Equation 9.36 and Figure 9.36).

▷ In the frequency domain, transient performance criteria are specified in terms of gain and phase margins. The gain margin is the additional gain required at the phase crossover frequency to make the system marginally stable. The phase crossover frequency is the frequency at which the phase is equal to $-180°$. Positive gain margin is measured down from 0 dB. The phase margin is the additional phase angle required at the gain crossover frequency to make the system marginally stable. The gain crossover frequency is where the magnitude is equal to 0 dB. Positive phase margin is measured up from $-180°$. The margins are measured with reference to 0 dB and $-180°$ because these correspond to the magnitude and angle conditions, respectively.

▷ A lead compensator is typically employed to improve the transient response of a closed-loop system. The compensator includes a gain, pole, and zero. The design process, using a Bode plot, involves placing the pole and zero such that the compensator contributes the needed additional phase angle at the gain crossover frequency to meet the phase margin criterion.

▷ A lag compensator is often used to improve steady-state performance. The compensator includes a gain, pole, and zero. So as to minimize impact on the transient characteristics, the pole and zero are placed close to each other between −1 and 0 along the real axis. This is accomplished by setting the upper corner frequency of the lag compensator one octave to one decade below the new gain crossover frequency.

▷ A lead-lag compensator is a combination of both a lead and a lag network. Hence, the compensator has a gain, two poles, and two zeros. The design process is a combination of the procedures for both the lead and lag compensators. This compensator is used to improve both the transient and steady-state performance.

▷ The MATLAB `margin` function can be used to plot a Bode diagram with the gain and phase margins labeled or to calculate the gain and phase margins and their corresponding crossover frequencies.

9.13   Review

R9-1 What is the closed-loop transfer function of a system with compensation and unity feedback?

R9-2 Is it possible to eliminate the steady-state error for the step response of a proportionally compensated, first-order system? Why or why not?

R9-3 For underdamped, second-order systems, how does the overshoot vary with damping ratio?

R9-4 In your own words, describe each of the static error constants and provide a mathematical formulation to accompany your description.

R9-5 Using the Routh-Hurwitz stability criterion, how can you determine if a system is stable? How can you determine how many poles, if any, are in the right-half-plane?

R9-6 Describe in your own words what the root locus is and how it can be used to aid in compensator design. Be sure to explain the magnitude and angle conditions.

R9-7 How do you plot the root locus using MATLAB?

R9-8 Describe the Ziegler-Nichols tuning rules for PID compensators.

R9-9 What is the gain margin and how is it measured?

R9-10 What is the phase margin and how is it measured?

R9-11 What are requirements for stability in the frequency domain?

R9-12 Provide a description of the frequency criteria that can be related to time domain specifications.

R9-13 Why is a lead compensator referred to as such?

R9-14 What are each of the lead-lag type compensators typically used for?

R9-15 How can you determine gain and phase margins using MATLAB?

9.14 Problems

P9-1 *Stability of Characteristic Equation Using Routh-Hurwitz Criterion.* Given the following characteristic equation

$$s^4 + 3s^3 + (2+K)s^2 + 16s + 25 = 0,$$

use the Routh-Hurwitz stability criterion to determine the range of K for the system to be stable.

P9-2 *Closed-Loop Stability Using Routh-Hurwitz Criterion.* Consider a unity-feedback, closed-loop system with the feedforward transfer function

$$G(s) = \frac{5}{s(s-2)(s+3)}.$$

Use the Routh-Hurwitz stability criterion to determine if the system is stable.

P9-3 *Static Error Constants and Steady-State Response.* Given a unit-feedback system with the feedforward transfer function

$$G(s) = \frac{50(s+8)(s+16)}{s(s+24)(s^2+2s+28)},$$

calculate the static error constants and the steady-state errors.

P9-4 *Static Error Constants and Steady-State Response of System with Inner Loop.* Consider the feedback system depicted in Figure 9.58. Find the static error constants and determine the steady-state error for unit step, ramp, and parabolic inputs.

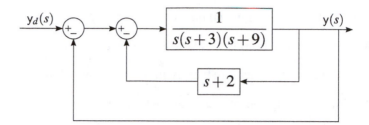

Figure 9.58: *Closed-loop system with inner loop.*

P9-5 *Sketching Root Locus Plots Given Pole-Zero Map.* Derive and sketch the root locus for each of the open-loop pole zero plots depicted in Figure 9.59.

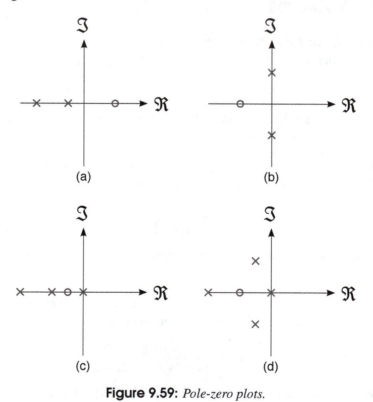

Figure 9.59: *Pole-zero plots.*

P9-6 *Sketching Root Locus Plot Given Feedforward Transfer Function.* Derive and sketch the root locus for the unity feedback systems with the following open-loop transfer functions:

(a) $G(s) = \dfrac{K(s+4)(s+6)}{s^2+3s+13}$

(b) $G(s) = \dfrac{K(s^2+16)}{s^2+4}$

(c) $G(s) = \dfrac{K(s^2+4)}{s^2}$

P9-7 *Design of Proportional Compensator Using Root Locus.* Consider a unity-feedback system with a feedforward transfer function of

$$G(s) = \frac{K}{(s^2+4s+5)(s+9)}.$$

Use the root locus method to select K such that $M_p \le 20\%$ and $t_s \le 2.5$ seconds.

P9-8 *Design of PID Compensator Using Root Locus.* Using the root locus method, design a PID compensator for a unity feedback system with the feedforward transfer function

$$G(s) = \frac{1}{s(s+3)(s+9)}.$$

The desired overshoot and settling time should be less than 40% and 2 seconds, respectively. Note that the PID controller adds a pole at the origin and two zeros along the real axis in the left-half-plane.

P9-9 *Design of Lead Compensator using a Bode Plot.* Design a lead compensator for a unity-feedback system with plant transfer function

$$G(s) = \frac{10}{s(s+10)(s+2)}.$$

The desired phase and gain margins are 45° and 12 dB and the desired static velocity error constant is $K_v = 3 \text{ s}^{-1}$.

P9-10 *Design of Lag Compensator Using Bode Plot.* Design a lag compensator for unity-feedback system whose plant is

$$G(s) = \frac{1}{s(s+3)}.$$

The desired static velocity error constant is 30 s^{-1} with a phase margin of no less than 50° and a gain margin of greater than 6 dB.

P9-11 *Design of Lead-Lag Compensator Using Bode Plot.* Design a lead-lag compensator for a unity feedback system with a plant model

$$G(s) = \frac{1}{s(s+1)(s+5)}$$

to achieve an overshoot of under 20%, a settling time of less than 15 seconds, and static velocity error constant of $K_v = 11 \text{ s}^{-1}$.

9.15   Challenges

C9-1 *A Toy Electric Car.* Some cruise controls are smart enough to track a varying speed and not just maintain constant velocity. Design a smart cruise control for the toy car. Specify your design criteria. Use a method of your choosing, and be sure to test the compensated response using an appropriate input. Note that this is a tracking and not a set point problem.

C9-2 *A Quarter-Car Suspension Model.* Some vehicles have active suspension systems that employ shock absorbers with adaptive damping. Some use an electrorheological fluid whose viscosity can be varied by applying an electric field. Assume that the shock absorber in your quarter-car suspension model is replaced by a force that can be controlled. Design a compensator to improve the performance of your suspension. You can use a method of your choosing, but remember that suspensions are often tested using a shaker and inputs of varying frequency. Specify your design criteria and use a variety of roadway inputs to test your design.

Chapter 10

Modern Control Systems

The previous chapter relied heavily on analysis methods for the frequency- and s-domains. The models that were used were exclusively transfer functions. Moreover, the control systems were strictly SISO systems. SISO systems are ones in which the objective is to control the dynamic response of a single output by varying a single input. Controllers for MIMO systems are more commonly designed in the state space. The methods from Chapter 9 focus on placing the dominant closed-loop poles such that they have the desired overall overshoot and settling time. This presumes that there exists a set of poles that are sufficiently dominant to dictate the overall response. That is not always the case. A more robust method would allow one to specify placement of all closed-loop poles. As we saw in the previous chapter, PID and lead-lag compensation rely on adjustment of a gain and the placement of at most two zeros and poles to modify the closed-loop dynamic response. For higher-order systems, these are not sufficient terms to control the placement of all closed-loop poles. State-space methods overcome this shortcoming by introducing additional adjustable parameters and methods for deriving those parameter values (Nise 2011). They utilize state feedback to generate control inputs that are generally a function of all the system states. Given your knowledge of the previous chapter, ask yourself the following:

▷ How do you specify the placement of all closed-loop poles to meet specified design criteria?

▷ What state-space methods exist for determining controller gains?

▷ How can a state-space controller be optimized based on design criteria and system limitations?

▷ If some state variables are not available for feedback, can you still implement a state-feedback control?

10.1 Introduction

Unlike the classical methods from the previous chapter, which rely primarily on frequency or s-domain analysis, *modern control* methods are implemented in the time domain and utilize state-space as opposed to transfer-function representations. State-space control methods rely primarily on Linear Algebra. Stability is determined using eigenvalue analysis. The control inputs are generally a function of all the states and not simply some function of the output and its integral and derivative. State-space methods provide more flexibility to modify the position of all the closed-loop poles and to control more than one output. As such, it is not limited to the placement of the dominant, closed-loop poles. Methods exist for multiple pole-pair placement and optimization based on state error, control effort, and response time. As with the last chapter, this chapter is only a brief overview of some of the more commonly used state-space approaches.

In state-feedback compensation, the controlled input is a function of the state vector. To implement state-feedback control, a system must be first and foremost controllable. Moreover, state-space methods require that all the dynamic state trajectories be known by way of modeling or measurement. The capability to measure or observe system states is referred to as *observability*. When states cannot be readily measured, they can be predicted in terms of the remaining states and the control inputs using a *state observer*. The ability to do so is predicated on the existence of a valid model. There are advanced methods that are designed to overcome unmodeled disturbances and uncertainties. Among those methods is *robust control theory*. This is, however, beyond the scope of this text and will not be covered herein. In this chapter, we shall explore a few state-space methods including general *pole placement* and the *linear quadratic regulator*. In addition, we will investigate state-space methods for determining stability, controllability, and observability.

The chapter objectives and outcomes are:

▷ *Objectives:*

1. To understand how to design controllers in the state-space,

2. To ascertain whether a system can be controlled,

3. To understand how to devise state observers,

▷ *Outcomes:* Upon completion, you should be able to

1. determine whether a system can be controlled via state-feedback,
2. derive state-feedback gains through pole placement,
3. synthesize state observers, and
4. optimize state-feedback gains through design of a linear quadratic regulator.

10.2 State Feedback Control

As has been seen in prior chapters, an nth-order system will have n poles and an nth-order characteristic equation (Nise 2011),

$$a_0 s^n + a_1 s^{n-1} + \cdots + a_{n-1} s + a_n = 0.$$

In order to place the n poles, n adjustable parameters are necessary. The classical methods covered in Chapter 9 are limited to a gain adjustment and the addition of at most two poles and zeros. Though sufficient for low-order systems, or systems where a single set of closed-loop poles significantly dominate, such compensation does not have enough parameters to generally address the complexity of higher-order systems. Instead, state-space compensation like that depicted by the block diagram in Figure 10.1 is necessary.

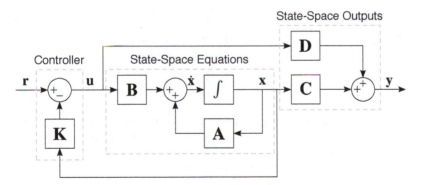

Figure 10.1: *Block diagram of state-feedback compensation.*

The classical compensation methods discussed in the previous chapter were limited to control inputs that were functions of the output. That is, the output, $y(s)$, was fed back to compare with the reference or desired output $y_d(s)$. The error, $\Delta y(s)$, was then used to calculate the control input, $u(s)$.

The compensators in the previous chapter were limited to a few adjustable parameters to achieve the desired response. The PID compensators, for instance, had three adjustable parameters – K_P, T_I (or K_I), and T_D (or K_D). What if, instead, the state vector, $\mathbf{x}(t)$, was fed back to determine the control? In state-feedback control, the mth-order control input can generally be a function of all the system states and reference an input,

$$\mathbf{u} = -\mathbf{Kx} + \mathbf{r}, \tag{10.1}$$

where $\mathbf{K}$ is an $m \times n$ matrix,

$$\mathbf{K} = \begin{bmatrix} k_{11} & k_{12} & \cdots & k_{1n} \\ k_{21} & k_{22} & \cdots & k_{2n} \\ \vdots & \vdots & \ddots & \vdots \\ k_{m1} & k_{m2} & \cdots & k_{mn} \end{bmatrix},$$

and $\mathbf{u}$ is an $m \times 1$ vector,

$$\mathbf{u} = \begin{bmatrix} u_1 \\ u_2 \\ \vdots \\ u_m \end{bmatrix}.$$

As illustrated in Figure 10.1, the state vector is fed back to calculate the control input.

The form of the reference, r, depends on the control scenario. For a SISO system, if the intent is to return a system to equilibrium at steady-state after it has been disturbed (i.e., $\mathbf{x}_{ss} = \mathbf{0} = \mathbf{x}_d$), the reference is zero and the input is simply

$$u = -\mathbf{Kx}. \tag{10.2}$$

Alternatively, for set point problems where it is desired that the outputs reach specified values at steady-state, the control should be (Franklin, Powell, and Emami-Naeini 1994)

$$\begin{aligned} u &= -\mathbf{Kx} + r \\ &= -\mathbf{Kx} + (\mathbf{Kx}_{ss} + u_{ss}) \\ &= u_{ss} - \mathbf{K}(\mathbf{x} - \mathbf{x}_{ss}), \end{aligned} \tag{10.3}$$

where u_{ss} and $\mathbf{x}_{ss}$ are the steady-state input and state vectors. Where as $u = -\mathbf{Kx}$ returns the system to equilibrium, the set point is control equal to u_{ss} and drives the steady-state error to zero so that $\mathbf{x} = \mathbf{x}_{ss}$. Hence, u_{ss} is the input value that maintains zero error. At steady-state, the output should equal the reference (or the desired final condition),

$$0 = y_{ss} - y_d \;\rightarrow\; y_{ss} = y_d \,,$$

and the error should be zero. Recall from Equations 7.55 and 7.57 that

$$\mathbf{0} = \mathbf{A}\mathbf{x}_{ss} + \mathbf{B}u_{ss} \text{ and}$$
$$y_{ss} = \mathbf{C}\mathbf{x}_{ss} + Du_{ss}\,.$$

To calculate the control input, we desire to find the values $\mathbf{x}_{ss}$ and u_{ss} that satisfy these equations. To derive the solution in terms of the reference, we let $\mathbf{x} = \mathbf{N_x}\mathbf{y}$ and $u = N_u y$ (Franklin, Powell, and Emami-Naeini 1994) so that at steady-state

$$\begin{aligned}
\mathbf{0} &= \mathbf{A}\mathbf{x}_{ss} + \mathbf{B}u_{ss} \\
&= \mathbf{A}\mathbf{N_x}y_{ss} + \mathbf{B}N_u y_{ss} \\
&= \mathbf{A}\mathbf{N_x} + \mathbf{B}N_u
\end{aligned} \tag{10.4}$$

and

$$\begin{aligned}
y_{ss} &= \mathbf{C}\mathbf{x}_{ss} + Du_{ss} \\
y_{ss} &= \mathbf{A}\mathbf{N_x}y_{ss} + \mathbf{B}N_u y_{ss} \\
1 &= \mathbf{A}\mathbf{N_x} + \mathbf{B}N_u\,.
\end{aligned} \tag{10.5}$$

In matrix form, Equations 10.4 and 10.5 become

$$\begin{bmatrix} \mathbf{0} \\ 1 \end{bmatrix} = \begin{bmatrix} \mathbf{A} & \mathbf{B} \\ \mathbf{C} & D \end{bmatrix} \begin{bmatrix} \mathbf{N_x} \\ N_u \end{bmatrix},$$

which has the solution

$$\begin{bmatrix} \mathbf{N_x} \\ N_u \end{bmatrix} = \begin{bmatrix} \mathbf{A} & \mathbf{B} \\ \mathbf{C} & D \end{bmatrix}^{-1} \begin{bmatrix} \mathbf{0} \\ 1 \end{bmatrix}. \tag{10.6}$$

The solutions for $\mathbf{N_x}$ and N_u are substituted into Equation 10.3 and we recall that for a set point problem we desire that the output reach the desired value at steady-state (i.e., $\mathbf{y}_{ss} = \mathbf{y}_d$),

$$\begin{aligned}
\mathbf{u} &= N_u y_d - \mathbf{K}(\mathbf{x} - \mathbf{N_x}y_d) \\
&= -\mathbf{Kx} + (\mathbf{N_u} + \mathbf{KN_x})y_d \\
&= -\mathbf{Kx} + \overline{\mathbf{N}}y_d\,,
\end{aligned} \tag{10.7}$$

so that $r = \overline{\mathbf{N}} y_d$.

Because the control input vector is a function of all the system states, each individual control input (i.e., u_1, u_2, ..., u_m) is inherently composed of n adjustable parameters (i.e., k_{i1}, ..., k_{in} where $i = 1, 2, \ldots, m$):

$$u_1 = k_{11}x_1 + k_{12}x_2 + \cdots + k_{1n}$$
$$u_2 = k_{21}x_1 + k_{22}x_2 + \cdots + k_{2n}$$
$$\vdots$$
$$u_m = k_{m1}x_1 + k_{m2}x_2 + \cdots + k_{mn}$$

The pole placement problem is now one of selecting the $m \times n$ gains to place the n poles. Before a state-feedback control can be designed, it must first be determined whether the system can be controlled given the system and its inputs.

10.3   Control System Analysis in the State Space

In order to implement arbitrary pole placement, a system must be state controllable (Ogata 2002). A system is said to be controllable if within a finite amount of time an unconstrained control input can transition the system from some initial condition to some final state (Ogata 2002). Kalman in 1960 (Kalman 1960) introduced the concepts of *controllability* and *observability*, which are used to determine the existence of a control solution. Herein we shall review the conditions for state and output controllability.

Systems are said to be *state controllable* if a control can be devised to transition all their states from an initial to final condition in a finite time span. State controllability is not a function of the system outputs. To facilitate deriving the controllability condition, let us assume a single input system. The only portion of the state-space model that is considered in ascertaining state controllability is

$$\dot{\mathbf{x}} = \mathbf{A}\mathbf{x} + \mathbf{B}u.$$

It can be shown that the solution to the differential equation is

$$\mathbf{x}(t) = e^{\mathbf{A}t}\mathbf{x}(0) + \int_0^t e^{\mathbf{A}(t-\tau)}\mathbf{B}u(\tau)\,d\tau. \tag{10.8}$$

Without loss of generality, we can assume that the desired final condition is

$$\mathbf{x}(t_f) = \begin{bmatrix} 0 \\ 0 \\ \vdots \\ 0 \end{bmatrix} = \mathbf{0}.$$

As detailed in Ogata (2002), the controllability condition requires that

$$\mathbf{x}(t_f) = \mathbf{0} = e^{\mathbf{A}t_f}\mathbf{x}(0) + \int_0^{t_f} e^{\mathbf{A}(t_f-\tau)}\mathbf{B}u(\tau)\,d\tau$$

$$= e^{\mathbf{A}t_f}\mathbf{x}(0) + \int_0^{t_f} e^{\mathbf{A}t_f}e^{-\mathbf{A}\tau}\mathbf{B}u(\tau)\,d\tau$$

$$= e^{\mathbf{A}t_f}\left[\mathbf{x}(0) + \int_0^{t_f} e^{-\mathbf{A}\tau}\mathbf{B}u(\tau)\,d\tau\right]$$

which necessitates that

$$\mathbf{x}(0) = -\int_0^{t_f} e^{-\mathbf{A}\tau}\mathbf{B}u(\tau)\,d\tau. \tag{10.9}$$

The exponential $e^{-\mathbf{A}\tau}$ can be calculated a number of ways as shown in Ogata (2002). Therein it is shown that the exponential can be written in the form

$$e^{-\mathbf{A}\tau} = \sum_{k=0}^{n-1} \alpha_k(\tau)\mathbf{A}^k, \tag{10.10}$$

which when substituted into Equation 10.11 yields

$$x(0) = -\sum_{k=0}^{n-1} \mathbf{A}^k\mathbf{B}\int_0^{t_f} \alpha_k(\tau)u(\tau)\,d\tau. \tag{10.11}$$

If we let

$$\beta_k = \int_0^{t_f} \alpha_k(\tau)u(\tau)\,d\tau$$

Equation 10.11 then becomes

$$\mathbf{x}(0) = -\sum_{k=0}^{n-1} \mathbf{A}^k \mathbf{B}\beta_k$$
$$= -\left(\mathbf{B}\beta_0 + \mathbf{A}\mathbf{B}\beta_1 + \mathbf{A}^2\mathbf{B}\beta_2 + \cdots + \mathbf{A}^{n-1}\mathbf{B}\beta_{n-1}\right)$$
$$= -\begin{bmatrix} \mathbf{B} & \mathbf{A}\mathbf{B} & \mathbf{A}^2\mathbf{B} & \cdots & \mathbf{A}^{n-1}\mathbf{B} \end{bmatrix} \begin{bmatrix} \beta_0 \\ \beta_1 \\ \beta_2 \\ \vdots \\ \beta_{n-1} \end{bmatrix}. \tag{10.12}$$

For the system to be state controllable, Equation 10.12 must be satisfied, which requires that the matrix

$$\mathscr{C} = \begin{bmatrix} \mathbf{B} & \mathbf{A}\mathbf{B} & \mathbf{A}^2\mathbf{B} & \cdots & \mathbf{A}^{n-1}\mathbf{B} \end{bmatrix} \tag{10.13}$$

be full rank. Equation 10.13 is referred to as the *state controllability matrix*. For the system to be state controllable, we require that the state controllability matrix be full rank.[*]

Typically, when we design a control system, we desire to control the output and not necessarily the states of the system. The question then rises: "For a given system, can we control the outputs using the provided inputs?" The state controllability matrix was dependent on only the state-space matrices **A** and **B**. When judging output controllability, one must consider the output equation

$$\mathbf{y} = \mathbf{C}\mathbf{x} + \mathbf{D}\mathbf{u}$$

where **y** includes r outputs (recall from Chapter 4 that $\mathbf{y}(t) \in \mathbb{R}^r$). A system is said to be *output controllable* if the outputs, rather than the states, can be transitioned from an initial to a final condition within a finite time span. It seems intuitive that the output controllability should also be dependent on the matrices **C** and **D**. Using a similar process to that previously used to derive the state controllability matrix, it can be shown that the condition of output controllability is that the matrix

$$\overline{\mathscr{C}} = \begin{bmatrix} \mathbf{C}\mathbf{B} & \mathbf{C}\mathbf{A}\mathbf{B} & \mathbf{C}\mathbf{A}^2\mathbf{B} & \cdots & \mathbf{C}\mathbf{A}^{n-1}\mathbf{B} & \mathbf{D} \end{bmatrix} \tag{10.14}$$

be of rank r. This is referred to as the *output controllability matrix*.

---

[*]. For further details regarding the above derivation, refer to pp.779-781 in Ogata (2002).

Example 10.1

Recall the single-input, two-output mass-spring-damper system from Examples 3.1, 3.11, 4.2, 6.2, and 7.4. For this and the following examples, however, let us assume that the only output of interest is the displacement of the second mass. The output equation therefore becomes

$$y = \begin{bmatrix} 1 & 0 & 1 & 0 \end{bmatrix} \begin{bmatrix} x_1 \\ p_1 \\ \delta_2 \\ p_2 \end{bmatrix}.$$

Determine whether the system is state and output controllable.

Solution. From Equation 4.7 the state-space differential equation is

$$\begin{bmatrix} \dot{x}_1 \\ \dot{p}_1 \\ \dot{\delta}_2 \\ \dot{p}_2 \end{bmatrix} = \begin{bmatrix} 0 & 1/m & 0 & 0 \\ -k & -b/m & k & 0 \\ 0 & -1/m & 0 & 1/m \\ 0 & 0 & -k & 0 \end{bmatrix} \begin{bmatrix} x_1 \\ p_1 \\ \delta_2 \\ p_2 \end{bmatrix} + \begin{bmatrix} 0 \\ 0 \\ 0 \\ 1 \end{bmatrix} F(t).$$

This is a fourth-order system. Hence, the state controllability matrix is

$$\mathscr{C} = \begin{bmatrix} \mathbf{B} & \mathbf{AB} & \mathbf{A^2B} & \mathbf{A^3B} \end{bmatrix}$$

$$= \begin{bmatrix} 0 & 0 & 0 & k/m^2 \\ 0 & 0 & k/m & -bk/m^2 \\ 0 & 1/m & 0 & -2k/m^2 \\ 1 & 0 & -k/m & 0 \end{bmatrix}. \qquad (10.15)$$

The matrix is full rank, thus the system is state controllable. The matrix and rank can be readily found using MATLAB.

```
>> syms m b k
>> A = [0 1/m 0 0; -k -b/m k 0; 0 -1/m 0 1/m; 0 0 -k 0];
>> B = [0 0 0 1]';
>> CTRL = [B A*B A^2*B A^3*B]

CTRL =

[ 0,    0,     0,       k/m^2]
[ 0,    0,   k/m,  -(b*k)/m^2]
[ 0,  1/m,     0,  -(2*k)/m^2]
[ 1,    0, -k/m,           0]

>> rank(CTR)
```

```
ans =

4
```

The output controllability matrix is

$$\overline{\mathscr{C}} = \begin{bmatrix} \mathbf{CB} & \mathbf{CAB} & \mathbf{CA^2B} & \mathbf{CA^3B} & \mathbf{D} \end{bmatrix}$$
$$= \begin{bmatrix} 0 & 1/m & 0 & -k/m^2 & 0 \end{bmatrix},$$

which has a rank of one and matches the number of outputs. Ergo, the system is also output controllable. The rank of output controllability matrix can also be determined in MATLAB.

```
>> C = [1 0 1 0];
>> D = 0;
>> CTRLo = [C*B C*A*B C*A^2*B C*A^3*B D]

CTRLo =

[ 0, 1/m, 0, -k/m^2, 0]

>> rank(CTRLo)

ans =

1
```

A system is observable if every state $\mathbf{x}(t)$ can be determined from the measurement or observation of the outputs $\mathbf{y}(t)$. For reasons discussed in Ogata (2002),[†] to derive the condition for observability, it suffices to consider the state-space system

$$\dot{\mathbf{x}} = \mathbf{Ax} \text{ and}$$
$$\mathbf{y} = \mathbf{Cx}.$$

The solution to the differential equation is

$$\mathbf{x}(t) = e^{\mathbf{A}t}\mathbf{x}(0)$$

making the output

$$\mathbf{y}(t) = \mathbf{C}e^{\mathbf{A}t}\mathbf{x}(0).$$

---

†. Refer to pp.786-787 in Ogata (2002) for a more detailed derivation of the observability matrix.

Substituting Equation 10.10 into the above garners

$$\mathbf{y}(t) = \sum_{k=0}^{n-1} \alpha_k(t)\mathbf{C}\mathbf{A}^k\mathbf{x}(0)$$

$$= \alpha_0\mathbf{C}\mathbf{x}(0) + \alpha_1\mathbf{C}\mathbf{A}\mathbf{x}(0) + \alpha_2\mathbf{C}\mathbf{A}^2\mathbf{x}(0) + \cdots + \alpha_{n-1}\mathbf{C}\mathbf{A}^{n-1}\mathbf{x}(0)$$

$$= \begin{bmatrix} \alpha_0 & \alpha_1 & \alpha_2 & \cdots & \alpha_{n-1} \end{bmatrix} \begin{bmatrix} \mathbf{C} \\ \mathbf{C}\mathbf{A} \\ \mathbf{C}\mathbf{A}^2 \\ \vdots \\ \mathbf{C}\mathbf{A}^{n-1} \end{bmatrix} \mathbf{x}(0). \tag{10.16}$$

To satisfy Equation 10.16 the matrix

$$\mathcal{O} = \begin{bmatrix} \mathbf{C} \\ \mathbf{C}\mathbf{A} \\ \mathbf{C}\mathbf{A}^2 \\ \vdots \\ \mathbf{C}\mathbf{A}^{n-1} \end{bmatrix} \tag{10.17}$$

must have a row rank equal to the number of states, n. This matrix is referred to as the *observability matrix*.

Example 10.2

For the previous example (Example 10.1) determine whether, given the outputs, the states are observable.

Solution. The system is fourth-order, so the row rank of the observability matrix must be four. The observability matrix for this problem is

$$\mathcal{O} = \begin{bmatrix} \mathbf{C} \\ \mathbf{C}\mathbf{A} \\ \mathbf{C}\mathbf{A}^2 \\ \mathbf{C}\mathbf{A}^3 \end{bmatrix}$$

$$= \begin{bmatrix} 1 & 0 & 1 & 0 \\ 0 & 0 & 0 & 1/m \\ 0 & 0 & -k/m & 0 \\ 0 & k/m^2 & 0 & -k/m^2 \end{bmatrix},$$

which has four linearly independent rows. It is relatively obvious that the four rows are linearly independent. This can be confirmed with MATLAB.

```
>> OBSRV = [C;C*A;C*A^2;C*A^3]

OBSRV =

[ 1,      0,      1,        0]
[ 0,      0,      0,      1/m]
[ 0,      0,   -k/m,        0]
[ 0,  k/m^2,      0,  -k/m^2]

>> rank(OBSRV)

ans =

4
```

Therefore, the system is observable. That is, the states $\mathbf{x}(t)$ can be surmised provided the outputs $\mathbf{y}(t)$.

10.4 Control Design Using Pole Placement

The pole placement problem via state feedback involves specifying through compensation the position of all the closed-loop eigenvalues so that each has associated desirable dynamics. Recall the state-space representation

$$\dot{\mathbf{x}} = \mathbf{Ax} + \mathbf{Bu}$$
$$\mathbf{y} = \mathbf{Cx} + \mathbf{Du}.$$

Substituting the control input from Equation 10.1, the state-space model becomes

$$\dot{\mathbf{x}} = \mathbf{Ax} + \mathbf{B}(-\mathbf{Kx} + \mathbf{r})$$
$$= (\mathbf{A} - \mathbf{BK})\mathbf{x} + \mathbf{Br}$$
$$= \overline{\mathbf{A}}\mathbf{x} + \mathbf{Br} \tag{10.18}$$
$$\mathbf{y} = \mathbf{Cx} + \mathbf{D}(-\mathbf{Kx} + \mathbf{r})$$
$$= (\mathbf{C} - \mathbf{DK})\mathbf{x} + \mathbf{Dr}$$
$$= \overline{\mathbf{C}}\mathbf{x} + \mathbf{Dr}. \tag{10.19}$$

Because the reference is bounded, the stability of the system depends on the eigenvalues of $\overline{\mathbf{A}} = \mathbf{A} - \mathbf{BK}$. Thus, the gain matrix can be chosen such that $\overline{\mathbf{A}} = \mathbf{A} - \mathbf{BK}$ has the desired closed-loop eigenvalues (or poles) $\lambda_1, \lambda_2, \ldots, \lambda_n$.

Should the desired eigenvalues include a complex pole, its complex conjugate should also be a desired eigenvalue. Remember complex poles always exist as complex conjugate pairs.

The steps for designing a state-feedback controller for arbitrary pole placement are as follows (Ogata 2002; Nise 2011):

1. *Check the system controllability.* Full state controllability is the necessary and sufficient condition for arbitrary pole placement. Thus, in order to to use a state-feedback control to implement pole placement, the first thing is to determine that the system is controllable.

2. *Determine the coefficients of the characteristic polynomial.* The system characteristic polynomial (Equation 7.51) is derived from the eigenvalue problem,

$$\det(s\mathbf{I} - \mathbf{A}) = s^n + a_1 s^{n-1} + \cdots + a_{n-1}s + a_n = 0, \qquad (10.20)$$

and its roots are the poles of the system. To calculate the gain matrix, we must determine the coefficients $a_1, \cdots, a_n$.

3. *Determine the transformation matrix.* As discussed in Ogata (2002), pole placement is most readily accomplished when the system is in *controllable canonical form*; that is, when the matrices $\mathbf{A}$ and $\mathbf{B}$ are of the form

$$\tilde{\mathbf{A}} = \begin{bmatrix} 0 & 1 & 0 & \cdots & 0 \\ 0 & 0 & 1 & \cdots & 0 \\ \vdots & \vdots & \vdots & \ddots & \vdots \\ 0 & 0 & 0 & \cdots & 1 \\ -a_n & -a_{n-1} & -a_{n-2} & \cdots & -a_1 \end{bmatrix} \quad \text{and} \quad \tilde{\mathbf{B}} = \begin{bmatrix} 0 \\ 0 \\ 0 \\ \vdots \\ 0 \\ 1 \end{bmatrix}.$$

However, most system, state-space representations are not in controllable canonical form. If the state controllability matrix is full rank, there exists a transformation matrix of the form

$$\mathbf{T} = \mathscr{C}\mathbf{W} \qquad (10.21)$$

where

$$W = \begin{bmatrix} a_{n-1} & a_{n-2} & a_{n-3} & \cdots & a_1 & 1 \\ a_{n-2} & a_{n-3} & a_{n-4} & \cdots & 1 & 0 \\ a_{n-3} & a_{n-4} & a_{n-5} & \cdots & 0 & 0 \\ \vdots & \vdots & \vdots & \ddots & \vdots & \vdots \\ a_1 & 1 & 0 & \cdots & 0 & 0 \\ 1 & 0 & 0 & \cdots & 0 & 0 \end{bmatrix}. \tag{10.22}$$

The matrix $\mathbf{T}$ can be used to transform the system into an equivalent system in controllable canonical form with state vector $\widetilde{\mathbf{x}}$ such that

$$\mathbf{x} = \mathbf{T}\widetilde{\mathbf{x}}. \tag{10.23}$$

Applying the transformation to the single-input, state-space differential equation yields

$$\dot{\mathbf{x}} = \mathbf{A}\mathbf{x} + \mathbf{B}u$$
$$\mathbf{T}\dot{\widetilde{\mathbf{x}}} = \mathbf{A}\mathbf{T}\widetilde{\mathbf{x}} + \mathbf{B}u$$
$$\dot{\widetilde{\mathbf{x}}} = \mathbf{T}^{-1}\mathbf{A}\mathbf{T}\widetilde{\mathbf{x}} + \mathbf{T}^{-1}\mathbf{B}u$$
$$= \widetilde{\mathbf{A}}\widetilde{\mathbf{x}} + \widetilde{\mathbf{B}}u, \tag{10.24}$$

which is in controllable canonical form. Though it is not required to transform the system, it is necessary to know the transformation in order to calculate the gain matrix.

4. *Specify the desired characteristic polynomial.* Define the desired characteristic polynomial using the desired eigenvalues,

$$(s - \lambda_1)(s - \lambda_2) \cdots (s - \lambda_n) = s^n + \alpha_1 s^{n-1} + \cdots + \alpha_{n-1} s + \alpha_n, \tag{10.25}$$

to determine the coefficients $\alpha_1, \alpha_2, \ldots, \alpha_n$.

5. *Calculate the state-feedback gain matrix* $\mathbf{K}$. The resultant gain matrix is

$$\mathbf{K} = \begin{bmatrix} (\alpha_n - a_n) & (\alpha_{n-1} - a_{n-1}) & \cdots & (\alpha_1 - a_1) \end{bmatrix} \mathbf{T}^{-1}. \tag{10.26}$$

Example 10.3
Recall from Example 7.4 that the system discussed in Examples 10.1 and 10.2 has as poles

$$p_{1,2} = -0.7106 \pm 3.7722j \quad \text{and}$$
$$p_{3,4} = -0.2894 \pm 1.5361j.$$

and as zeros
$$z_{1,2} = -1.0000 \pm 3.3166j.$$

Also recall that the response is dominated by $p_{3,4}$. Design a state-feedback controller so that the output responses settle within 10 seconds with no more than 20% overshoot. Remember that the mass, damping constant, and stiffness are 10 kg, 20 N-s/m, and 60 N/m, respectively.

   Solution. The pole placement design procedure is as follows:

1. We already determined in Example 10.1 that the system is controllable.

2. The characteristic polynomial is also the denominator of the system transfer functions which were previously derived and provided in Equations 6.19 and 6.20. Thus the characteristic equation is

$$
\begin{aligned}
0 &= m^2 s^4 + mbs^3 + 3mks^2 + bks + k^2 \\
&= (10)^2 s^4 + (10)(20)s^3 + 3(10)(60)s^2 + (20)(60)s + (60)^2 \\
&= 100s^4 + 200s^3 + 1800s^2 + 1200s + 3600 \\
&= 1000(s^4 + 0.2s^3 + 1.8s^2 + 1.2s + 3.6) \\
&= s^4 + 2s^3 + 18s^2 + 12s + 36 \\
&= s^4 + a_1 s^3 + a_2 s^2 + a_3 s + a_4.
\end{aligned}
$$

3. To calculate the transformation matrix we need $\mathscr{C}$ and $\mathbf{W}$. From Equation 10.15 we know that

$$
\mathscr{C} = \begin{bmatrix}
0 & 0 & 0 & k/m^2 \\
0 & 0 & k/m & -bk/m^2 \\
0 & 1/m & 0 & -2k/m^2 \\
1 & 0 & -k/m & 0
\end{bmatrix}
$$

$$
= \begin{bmatrix}
0 & 0 & 0 & 0.6 \\
0 & 0 & 6 & -12 \\
0 & 0.1 & 0 & -1.2 \\
1 & 0 & -6 & 0
\end{bmatrix} \tag{10.27}
$$

and from Equation 10.22

$$\mathbf{W} = \begin{bmatrix} a_3 & a_2 & a_1 & 1 \\ a_2 & a_1 & 1 & 0 \\ a_1 & 1 & 0 & 0 \\ 1 & 0 & 0 & 0 \end{bmatrix}$$

$$= \begin{bmatrix} 12 & 18 & 2 & 1 \\ 18 & 2 & 1 & 0 \\ 2 & 1 & 0 & 0 \\ 1 & 0 & 0 & 0 \end{bmatrix}.$$

Hence the transformation matrix is

$$\mathbf{T} = \mathscr{C}\mathbf{W}$$

$$= \begin{bmatrix} 0 & 0 & 0 & 0.6 \\ 0 & 0 & 6 & -12 \\ 0 & 0.1 & 0 & -1.2 \\ 1 & 0 & -6 & 0 \end{bmatrix} \begin{bmatrix} 1.2 & 1.8 & 0.2 & 1 \\ 1.8 & 0.2 & 1 & 0 \\ 0.2 & 1 & 0 & 0 \\ 1 & 0 & 0 & 0 \end{bmatrix}$$

$$= \begin{bmatrix} 0.6 & 0 & 0 & 0 \\ -10.8 & 6 & 0 & 0 \\ -1.02 & 0.02 & 0.1 & 0 \\ 0 & -4.2 & 0.2 & 1 \end{bmatrix}$$

This can be accomplished with the aid of MATLAB.

```
>> m = 10; b = 20; k = 60;
>> W = ...
[1.2 1.8  0.2 1; ...
1.8 0.2 1 0; ...
0.2 1 0 0; ...
1 0 0 0]

W =

    1.2000    1.8000    0.2000    1.0000
    1.8000    0.2000    1.0000         0
    0.2000    1.0000         0         0
    1.0000         0         0         0

>> T = eval(CTRL)*W

T =

    0.6000         0         0         0
```

$$
\begin{vmatrix}
-10.8000 & 6.0000 & 0 & 0 \\
-1.0200 & 0.0200 & 0.1000 & 0 \\
-0.0000 & -4.2000 & 0.2000 & 1.0000
\end{vmatrix}
$$

The eval command is used to evaluate the previously defined symbolic matrix CTRL.

4. From Example 7.4 we know that the dominant poles have an associated damping ratio and natural frequency of 0.1851 and 1.5631 rad/s. According Figure 9.10, the existing dominant poles have a maximum overshoot of over 50%. The corresponding settling time is

$$
t_s = \frac{4}{\zeta \omega_n} = \frac{4}{(0.1851)(1.5631 \text{ rad/s})} \approx 13.2 \text{ s}.
$$

To meet the specified settling time a damping ratio of approximately 0.5 should suffice. Given the settling time requirement ($t_s \leq 10$ s), the desired natural frequency is

$$
\omega_n = \frac{4}{\zeta t_s} = \frac{4}{(0.5)(10 \text{ s})} = 0.8 \text{ rad/s}.
$$

The system is fourth-order and has four closed-loop poles. To achieve the desired dynamics we can choose a complex pole pair with the desired characteristics

$$
\lambda_{3,4} = -\zeta \omega_n \pm j \omega_n \sqrt{1 - \zeta^2}
$$
$$
= -(0.5)(0.8 \pm j(0.8)\sqrt{1 - (0.5)^2}
$$
$$
= -0.4 \pm j0.6928.
$$

This provides us a starting point for choosing two of the closed-loop poles. The poles need not be exact to meet the design criteria. The two other poles can be chosen such that they are five times or more farther from the imaginary axis so that their associated dynamics will attenuate relatively quickly. Therefore we can choose the desired closed loop poles to include two poles at $\lambda_{3,4} = -0.4 \pm j0.7$ and two more at $\lambda_{1,2} = -2 \pm j4$. The pair $\lambda_{1,2}$ is chosen such that the real part is five times more than the real part of the other pair $\lambda_{3,4}$. The imaginary parts of $\lambda_{1,2}$ are chosen so that they are similar to those of the original pair $p_{1,2}$. This is done in an attempt to minimize the gain needed to realize the state feedback. *Generally, the more you move the poles, the more the gain required.* The desired characteristic polynomial is thus

$$
\begin{aligned}
(s+0.4+j0.7)&(s+0.4-j0.7)(s+2+j4)(s+2-j4)\\
&= (s^2+0.8s+0.65)(s^2+4s+20)\\
&= s^4+4.8s^3+23.85s^2+18.6s+13 \qquad (10.28)\\
&= s^4+\alpha_1 s^3+\alpha_2 s^2+\alpha_3 s+\alpha_4.
\end{aligned}
$$

5. We can now calculate the gain matrix,

$$
\begin{aligned}
\mathbf{K} &= \left[(\alpha_4-a_4)\quad(\alpha_3-a_3)\quad(\alpha_2-a_2)\quad(\alpha_1-a_1)\right]\mathbf{T}^{-1}\\
&= \left[(13-36)\quad(18.6-12)\quad(23.85-18)\quad(4.8-2)\right]\dots\\
&\qquad\begin{bmatrix}0.6 & 0 & 0 & 0\\ 0 & 6 & 0 & 0\\ 0.6 & 0.2 & 0.1 & 0\\ 0 & 12 & 2 & 1\end{bmatrix}\\
&\approx \left[-40.8\quad -4.58\quad 2.50\quad 2.80\right].
\end{aligned}
$$

Now, let us examine the resultant responses to check if indeed the pole-placement design achieved the desired responses.

The question now remains, "How does the system perform using this state-feedback control?" To evaluate the control design, we can examine the step response. For the purpose of this example, let us assume that the only output we wish to control is the displacement of the second mass so that

$$
y = x_1 + \delta_2 = \begin{bmatrix}1 & 0 & 1 & 0\end{bmatrix}\begin{bmatrix}x_1\\ p_1\\ \delta_2\\ p_2\end{bmatrix} = \mathbf{Cx} + Du.
$$

When using the step response, in effect what we are doing is assessing how well the system responds when the reference is specified as a desired output. Hence, to calculate the appropriate reference input, we must use Equations 10.6 and 10.7 and recall that for a set point problem $r = \overline{N}y_d$. First we calculate the matrices $\mathbf{N_x}$ and N_u (note that N_u is scalar because there is a single input),

$$
\begin{bmatrix} \mathbf{N_x} \\ N_u \end{bmatrix} = \begin{bmatrix} \mathbf{A} & \mathbf{B} \\ \mathbf{C} & D \end{bmatrix}^{-1} \begin{bmatrix} \mathbf{0} \\ 1 \end{bmatrix}
$$

$$
= \begin{bmatrix}
0 & 1/m & 0 & 0 & 0 \\
-k & -b/m & k & 0 & 0 \\
0 & -1/m & 0 & 1/m & 0 \\
0 & 0 & -k & 0 & 1 \\
\hline
1 & 0 & 1 & 0 & 0
\end{bmatrix}^{-1}
\begin{bmatrix} 0 \\ 0 \\ 0 \\ 0 \\ 1 \end{bmatrix}
$$

$$
= \begin{bmatrix}
0 & 0.1 & 0 & 0 & 0 \\
-60 & -2 & 60 & 0 & 0 \\
0 & -0.1 & 0 & 0.1 & 0 \\
0 & 0 & -60 & 0 & 1 \\
\hline
1 & 0 & 1 & 0 & 0
\end{bmatrix}^{-1}
\begin{bmatrix} 0 \\ 0 \\ 0 \\ 0 \\ 1 \end{bmatrix}
= \begin{bmatrix} 0.5 \\ 0 \\ 0.5 \\ 0 \\ 30 \end{bmatrix}.
$$

Thus $\mathbf{N_x} = \begin{bmatrix} 0.5 & 0 & 0.5 & 0 \end{bmatrix}^T$ and $N_u = 30$. The matrices in addition to $\overline{N}$ can be readily calculated with the aid of MATLAB.

```
>> A = eval(A); B = eval(B);
>> N = inv([A B; C D])*[zeros(4,1);1]

N =

    0.5000
         0
    0.5000
         0
   30.0000

>> Nx = N(1:4)

Nx =

    0.5000
         0
    0.5000
         0

>> Nu = N(5)

Nu =

    30
```

```
>> K = [-40.8,-4.58,2.50,2.80];
>> Nbar = Nu+K*Nx

Nbar =

   10.8500
```

Knowing the gain matrix and the reference in terms of the desired output, we can write an alternate set of state-space equations that incorporate the state feedback and set point scenario. From Equations 10.18 and 10.19 we know that

$$\dot{\mathbf{x}} = \overline{\mathbf{A}}\mathbf{x} + \mathbf{B}r$$
$$y = \overline{\mathbf{C}}\mathbf{x} + Dr$$

where $\overline{\mathbf{A}} = \mathbf{A} - \mathbf{B}\mathbf{K}$ and $\overline{\mathbf{C}} = \mathbf{C} - \mathbf{K}D$. If we substitute $r = \overline{N}y_d$ the equations become

$$\dot{\mathbf{x}} = \overline{\mathbf{A}}\mathbf{x} + \overline{\mathbf{B}}y_d$$
$$y = \overline{\mathbf{C}}\mathbf{x} + \overline{D}y_d$$

where $\overline{\mathbf{B}} = \mathbf{B}\overline{N}$ and $\overline{D} = D\overline{N}$. The resultant equations include the state feedback and are in a set point conducive form. The resultant unit step response is plotted with MATLAB using commands similar to those that follow.

```
>> Abar = A-B*K;
>> Bbar = B*Nbar;
>> Cbar = C-D*K;
>> Dbar = D*Nbar;
>> sys_cl = ss(Abar,Bbar,Cbar,Dbar);
>> step(sys_cl)
```

As Figure 10.2 demonstrates, the step response has less than 20% overshoot and a settling time of about 10 seconds. If we compare to the open-loop impulse response plotted in Figure 7.19, this state feedback controlled system seems to exhibit less oscillation and an improved settling time.

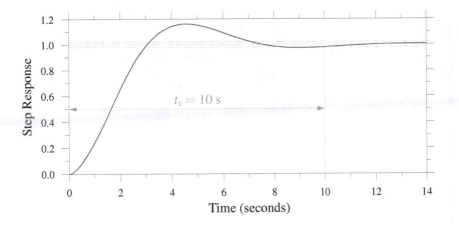

Figure 10.2: *Step response for mass-spring-damper system with state feedback designed through pole placement.*

10.5 Ackermann's Formula

As an alternative to deriving the transformation, *Ackermann's formula* provides an abbreviated and compact means of finding a suitable feedback gain matrix (Franklin, Powell, and Emami-Naeini 1994). It circumvents the need for deriving the transformation. Ackermann's formula is a method used for determining the state-feedback gain matrix for a SISO system (Dorf and Bishop 2001). According to Ackermann's formula, the state-feedback gain matrix is given by (Franklin, Powell, and Emami-Naeini 1994)

$$\mathbf{K} = \begin{bmatrix} 0 & 0 & \cdots & 1 \end{bmatrix} \mathscr{C}^{-1} \alpha_c(\mathbf{A}) \tag{10.29}$$

where $\mathscr{C}$ is the controllability matrix (Equation 10.13) and where, given the desired characteristic equation (Equation 10.25) and its coefficients,

$$\alpha_c = \mathbf{A}^n + \alpha_1 \mathbf{A}^{n-1} + \alpha_2 \mathbf{A}^{n-2} + \cdots + \alpha_n \mathbf{I}. \tag{10.30}$$

Example 10.4

Use Ackermann's formula to find a state-feedback gain matrix for the previous example (Example 10.3).

Solution. Given that the controllability matrix was previously calculated in Equation 10.27 and the desired characteristic equation was specified in Equation 10.28, the gain matrix, according to Ackermann's formula is

$$\mathbf{K} = \begin{bmatrix} 0 & 0 & 0 & 1 \end{bmatrix} \mathscr{C}^{-1} \begin{bmatrix} \mathbf{A}^4 + \alpha_1 \mathbf{A}^3 + \alpha_2 \mathbf{A}^2 + \alpha_3 \mathbf{A} + \alpha_4 \mathbf{I} \end{bmatrix}$$

$$= \begin{bmatrix} 0 & 0 & 0 & 1 \end{bmatrix} \begin{bmatrix} 0 & 0 & 0 & 0.6 \\ 0 & 0 & 6 & -12 \\ 0 & 0.1 & 0 & -1.2 \\ 1 & 0 & -6 & 0 \end{bmatrix}^{-1} \cdots$$

$$\begin{bmatrix} \mathbf{A}^4 + 4.8\mathbf{A}^3 + 23.85\mathbf{A}^2 + 18.6\mathbf{A} + 13\mathbf{I} \end{bmatrix}$$

$$\approx \begin{bmatrix} -40.8 & -4.58 & 2.50 & 2.80 \end{bmatrix}$$

where

$$\mathbf{A} = \begin{bmatrix} 0 & 1/m & 0 & 0 \\ -k & -b/m & k & 0 \\ 0 & -1/m & 0 & 1/m \\ 0 & 0 & -k & 0 \end{bmatrix}$$

$$= \begin{bmatrix} 0 & 0.1 & 0 & 0 \\ -60 & 0 & 60 & 0 \\ 0 & -0.1 & 0 & 0.1 \\ 0 & 0 & -60 & 0 \end{bmatrix}.$$

Note that Ackermann's formula resulted in the same gain matrix derived using the pole placement procedure without the need for determining the transformation.

10.6 Optimal Control and the Linear Quadratic Regulator

For high-order systems, it may become a bit intractable to design a compensator via pole placement. Furthermore, the poles in the previous examples were chosen to meet specified design criteria. More than one solution would readily satisfy the design criteria. No effort was exerted to optimize the solution.

Optimization involves maximizing performance within the design constraints. This can be accomplished by minimizing the output error, the control effort, and/or the response time. The most commonly used technique for state-feedback optimization is the *linear quadratic regulator* (LQR).

Take a state feedback system of the form

$$\dot{x} = \mathbf{A}x + \mathbf{B}u$$

where

$$\mathbf{u}(t) = -\mathbf{K}\mathbf{x}(t).$$

To determine the optimal gain matrix $\mathbf{K}$, we define a quadratic *performance index*

$$J = \int_0^\infty (\mathbf{x}^T \mathbf{Q} \mathbf{x} + \mathbf{u}^T \mathbf{R} \mathbf{u}) \, dt \qquad (10.31)$$

where $\mathbf{Q}$ and $\mathbf{R}$ are generally *positive-definite* Hermitian or real symmetric matrices. The matrix $\mathbf{Q}$ can be *positive semidefinite*.[‡]

The matrices are square and are sized to match the dimensions of the state and input vector, respectively (i.e., $\mathbf{Q}$ is $n \times n$ and $\mathbf{R}$ is $m \times m$). The matrix $\mathbf{Q}$ determines the relative importance of minimizing the state vector error at steady-state, and $\mathbf{R}$ minimizes the respective control effort. The simplest way to ensure that the matrices are positive definite is to use diagonal matrices with positive elements along the diagonal,

$$\mathbf{Q} = \begin{bmatrix} q_1 & 0 & \cdots & 0 \\ 0 & q_2 & \cdots & 0 \\ \vdots & \vdots & \ddots & \vdots \\ 0 & 0 & \cdots & q_n \end{bmatrix}$$

and

$$\mathbf{R} = \begin{bmatrix} r_1 & 0 & \cdots & 0 \\ 0 & r_2 & \cdots & 0 \\ \vdots & \vdots & \ddots & \vdots \\ 0 & 0 & \cdots & r_m \end{bmatrix}.$$

To emphasize minimization of a select error or input effort, we increase the value of the respective diagonal element (i.e., q_2 correspond to the minimization of the error associated with x_2 and r_3 impacts minimization of u_3). As described in Ogata (2002), the gain matrix that minimizes the performance index is

$$\mathbf{K} = \mathbf{R}^{-1} \mathbf{B}^T \mathbf{P} \qquad (10.32)$$

where $\mathbf{P}$ is a positive definite and must satisfy the *algebraic Riccati equation*

$$\mathbf{A}^T \mathbf{P} + \mathbf{P} \mathbf{A} - \mathbf{P} \mathbf{B} \mathbf{R}^{-1} \mathbf{B}^T \mathbf{P} + \mathbf{Q} = \mathbf{0}. \qquad (10.33)$$

[‡]. A matrix $\mathbf{M}$ is positive definite if $\mathbf{z}^T \mathbf{M} \mathbf{z} > 0$ and positive semidefinite if $\mathbf{z}^T \mathbf{M} \mathbf{z} \geq 0$ for any non-zero vector $\mathbf{z}$.

Equation 10.31 is used for optimization of an *infinite horizon* LQR. An infinite horizon LQR problem is one in which there is no time constraint or term in the performance index that minimizes the time expended in transitioning the system from initial to final condition. A term can be added to the performance index to minimize time. The performance index for a *finite horizon* LQR is

$$\int_{t_0}^{t_1} (\mathbf{x}^T \mathbf{Q} \mathbf{x} + \mathbf{u}^T \mathbf{R} \mathbf{u})\, dt + \frac{1}{2} \mathbf{x}^T (t_1) \mathbf{S} \mathbf{x}(t_1). \qquad (10.34)$$

This, however, is beyond the scope of this text.[§]

Example 10.5

Design an optimal control for the mass-spring-damper system from Examples 10.3 and 10.4.

Solution. The problem statement is rather open-ended. By examining the problem we can make some conclusions that will enable us to choose appropriate values for the matrices $\mathbf{Q}$ and $\mathbf{R}$. Based on the fact that we wish to control the output $y = x_2 = x_1 + \delta_2$ which is the displacement of the second mass, it seems tantamount that we minimize the errors on the first and third states, x_1 and δ_2. We are not especially concerned with the second and fourth states, p_1 and p_2. Moreover, the problem statement specifies no criteria for the control effort. We can initially assume that the control effort is unconstrained. Thus, to begin we can use

$$\mathbf{Q} = \begin{bmatrix} 10 & 0 & 0 & 0 \\ 0 & 1 & 0 & 0 \\ 0 & 0 & 10 & 0 \\ 0 & 0 & 0 & 1 \end{bmatrix} \quad \text{and} \quad R = 1.$$

Note that the first and third diagonal elements in $\mathbf{Q}$ are ten times more than the second and fourth. This is to increase the relative importance of minimizing the errors associated with the first and third states, x_1 and δ_2.

Given the system state-space model and the chosen performance index matrices $\mathbf{Q}$ and R, we can proceed to solve the algebraic Riccati equation for $\mathbf{P}$. Though this can be calculated analytically through matrix factorization or iteration, typically the algebraic Riccati equation is solved numerically. The solution provided below was garnered using the MATLAB

---

§. For further details, reference the texts *Linear Optimal Control Systems* (Kwakernaak and Sivan 1972).

care function which will detailed later in Section 10.8. The solution to the algebraic Riccati equation is

$$
\mathbf{P} = \begin{bmatrix} 322.8731 & 0.0098 & 86.2499 & 2.9715 \\ 0.0098 & 0.3886 & -2.8228 & 0.1109 \\ 86.2499 & -2.8228 & 312.3883 & -2.8051 \\ 2.9715 & 0.1109 & -2.8051 & 0.6626 \end{bmatrix}
$$

making the gain matrix

$$
\mathbf{K} = \left(1^{-1}\right) \begin{bmatrix} 0 & 0 & 0 & 1 \end{bmatrix} \begin{bmatrix} 322.8731 & 0.0098 & 86.2499 & 2.9715 \\ 0.0098 & 0.3886 & -2.8228 & 0.1109 \\ 86.2499 & -2.8228 & 312.3883 & -2.8051 \\ 2.9715 & 0.1109 & -2.8051 & 0.6626 \end{bmatrix}
$$

$$
= \begin{bmatrix} 2.9715 & 0.1109 & -2.8051 & 0.6626 \end{bmatrix}.
$$

The corresponding step and control input responses are plotted in Figure 10.3 with a solid line. Notice that the overshoot is greater than 20% and the settling time is beyond 7 seconds.

By reducing R we effectively loosen the constraint on the control input allowing more range of control effort to be used to minimize the output error. Take, for example, $R = 0.5$. The solution of the algebraic Riccati equation now yields

$$
\mathbf{P} = \begin{bmatrix} 273.2461 & 0.0108 & 62.9928 & 2.0865 \\ 0.0108 & 0.3454 & -1.9422 & 0.0664 \\ 62.9928 & -1.9422 & 271.8199 & -1.9203 \\ 2.0865 & 0.0664 & -1.9203 & 0.5549 \end{bmatrix},
$$

which results in a gain matrix of

$$
\mathbf{K} = \left(0.5^{-1}\right) \begin{bmatrix} 0 & 0 & 0 & 1 \end{bmatrix} \begin{bmatrix} 273.2461 & 0.0108 & 62.9928 & 2.0865 \\ 0.0108 & 0.3454 & -1.9422 & 0.0664 \\ 62.9928 & -1.9422 & 271.8199 & -1.9203 \\ 2.0865 & 0.0664 & -1.9203 & 0.5549 \end{bmatrix}
$$

$$
= \begin{bmatrix} 4.1731 & 0.1328 & -3.8407 & 1.1099 \end{bmatrix}.
$$

This design allows for more control effort and, as a consequence, reduced overshoot and settling time. As illustrated in Figure 10.3, the new step response exhibits just under 20% overshoot and a settling time of less than 6 seconds. Notice that the control input varies between about 20 and 32 N as opposed to just over 23 and 32 N. Both designs result in settling times less than those attained using pole placement or Ackermann's formula (refer to Figure 10.2).

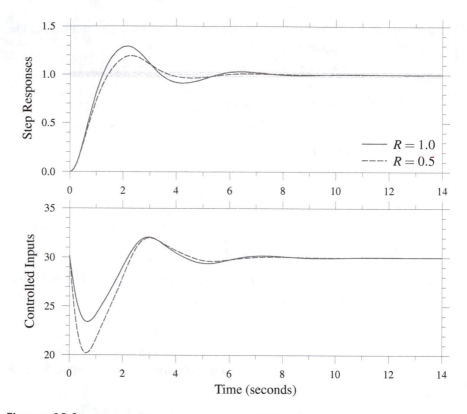

Figure 10.3: *Step and input responses for LQR compensation of mass-spring-damper system.*

10.7   State Observers

The state-feedback controllers discussed thus far assume that all the states can be readily fed back for measurement or observation. For real systems, this may not always be possible. States must be estimated when not available due to cost or physical constraints. An *observer* or *estimator* is employed to approximate state variables that cannot be measured.

Given a reasonably accurate model of the system, it seems intuitive to use the model to estimate the unobservable states. Take the state-space representation

$$\dot{\mathbf{x}} = \mathbf{A}\mathbf{x} + \mathbf{B}\mathbf{u} \qquad (10.35a)$$

$$\mathbf{y} = \mathbf{C}\mathbf{x} \qquad (10.35b)$$

and the corresponding estimator

$$\dot{\hat{\mathbf{x}}} = \mathbf{A}\hat{\mathbf{x}} + \mathbf{B}\mathbf{u} \qquad (10.36a)$$
$$\hat{\mathbf{y}} = \mathbf{C}\hat{\mathbf{x}} \qquad (10.36b)$$

where $\hat{\mathbf{x}}$ is the estimate of the state vector $\mathbf{x}$ and $\hat{\mathbf{y}}$ the estimate of the output vector $\mathbf{y}$. If we subtract the observer (Equation 10.36) from the state-space model (Equation 10.35) we obtain

$$\dot{\mathbf{x}} - \dot{\hat{\mathbf{x}}} = \mathbf{A}(\mathbf{x} - \hat{\mathbf{x}})$$
$$\dot{\mathbf{e}}_\mathbf{x} = \mathbf{A}\mathbf{e}_\mathbf{x}$$
$$\mathbf{y} - \hat{\mathbf{y}} = \mathbf{C}(\mathbf{y} - \hat{\mathbf{y}})$$
$$\mathbf{e}_\mathbf{y} = \mathbf{C}\mathbf{e}_\mathbf{x}$$

where $\mathbf{e}_\mathbf{x}$ and $\mathbf{e}_\mathbf{y}$ are the state and output estimate errors. The resultant differential equation has the same eigenvalues as the original state-space model,

$$\det(\lambda\mathbf{I} - \mathbf{A}) = 0.$$

The estimator error will be zero at steady-state. Figure 10.4 (a) illustrates this observer architecture. The shortcoming of this method, however, is that the estimator will have the same transient characteristics as the system. The rate of convergence between the actual and estimated state vectors is the same as the transient response of the system (Nise 2011). This form of state estimation introduces undesirable delay.

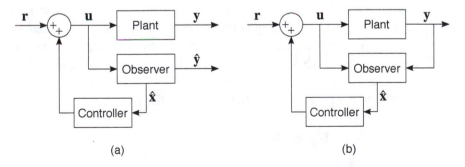

Figure 10.4: *Block diagrams for (a) open-loop observer and (b) closed-loop observer.*

Instead, we seek an observer with comparatively quick dynamics along with negligible, steady-state error. Much like a sensor, the observer should appear to respond near instantaneously. Typically, we ignore the dynamics of a sensor because its response tends to be orders of magnitude faster than the dynamics it measures. To accelerate the rate of convergence, we use instead a closed-loop observer architecture like that illustrated in Figure 10.4 (b). The observer is now a function of the input and output vectors. The observer model is

$$\dot{\hat{x}} = A\hat{x} + Bu + L(y - \hat{y}) \tag{10.37a}$$

$$\hat{y} = C\hat{x}. \tag{10.37b}$$

Accordingly, the difference between the state-space (Equation 10.35) and observer models is

$$\dot{x} - \dot{\hat{x}} = A(x - \hat{x}) - L(y - \hat{y})$$

$$\dot{e}_x = Ae_x - Le_y \tag{10.38}$$

$$y - \hat{y} = C(y - \hat{y})$$

$$e_y = Ce_x. \tag{10.39}$$

By substituting Equation 10.39 into Equation 10.38 we derive

$$\dot{e}_x = Ae_x - LCe_x$$

$$= (A - LC)e_x. \tag{10.40}$$

The poles of the closed-loop model of the observer will be the eigenvalues of $A - LC$. Computing L is the same as computing the feedback gain matrix K, except that the desired closed-loop poles of the observer should be chosen so that it seemingly responds instantaneously relative to the controlled system. Furthermore, the observability, not controllability, is the critical issue for the observer. Ackermann's formula for observer design is transposed and uses the observability matrix in place of the controllability matrix,

$$L = \varphi_o(A)\mathscr{O}^{-1}\begin{bmatrix} 0 & 0 & \cdots & 1 \end{bmatrix}^T \tag{10.41}$$

where

$$\varphi_o = A^n + \varphi_1 A^{n-1} + \varphi_2 A^{n-2} + \cdots + \varphi_n I \tag{10.42}$$

and where φ_1, φ_2, ..., φ_n are the coefficients of the desired characteristic equation for the observer. Remember that Ackermann's is applicable to SISO systems.

If we assume a SISO set point problem (i.e., $u = -\mathbf{Kx} + \overline{N}y_d$) and use the observer (Equation 10.37), the closed-loop differential equation is

$$\dot{\mathbf{x}} = \mathbf{Ax} + \mathbf{Bu}$$
$$= \mathbf{Ax} + \mathbf{B}(-\mathbf{K}\hat{\mathbf{x}} + \overline{N}y_d)$$
$$= \mathbf{Ax} - \mathbf{BK}\hat{\mathbf{x}} + \mathbf{B}\overline{N}y_d. \tag{10.43}$$

Similarly, if we substitute the output in the observer differential equation (Equation 10.37) it becomes

$$\dot{\hat{\mathbf{x}}} = \mathbf{A}\hat{\mathbf{x}} + \mathbf{Bu} + \mathbf{L}(\mathbf{y} - \hat{\mathbf{y}})$$
$$= \mathbf{A}\hat{\mathbf{x}} + \mathbf{B}(-\mathbf{Kx} + \overline{N}y_d) + \mathbf{L}(\mathbf{Cx} - \mathbf{C}\hat{\mathbf{x}})$$
$$= \mathbf{A}\hat{\mathbf{x}} - \mathbf{BK}\hat{\mathbf{x}} + \mathbf{B}\overline{N}y_d + \mathbf{LCx} - \mathbf{LC}\hat{\mathbf{x}}$$
$$= \mathbf{LCx} + (\mathbf{A} - \mathbf{BK} - \mathbf{LC})\hat{\mathbf{x}} + \mathbf{B}\overline{N}y_d. \tag{10.44}$$

Equations 10.43 and 10.44 for the closed-loop, state-space system with observer, can be combined to form the representation

$$\begin{bmatrix} \dot{\mathbf{x}} \\ \dot{\hat{\mathbf{x}}} \end{bmatrix} = \begin{bmatrix} \mathbf{A} & -\mathbf{BK} \\ \mathbf{LC} & \mathbf{A} - \mathbf{BK} - \mathbf{LC} \end{bmatrix} \begin{bmatrix} \mathbf{x} \\ \hat{\mathbf{x}} \end{bmatrix} + \begin{bmatrix} \mathbf{B}\overline{N} \\ \mathbf{B}\overline{N} \end{bmatrix} y_d. \tag{10.45}$$

The output can be written in terms of both the state vector and its estimate

$$\mathbf{y} = \begin{bmatrix} \mathbf{C} & \mathbf{0} \end{bmatrix} \begin{bmatrix} \mathbf{x} \\ \hat{\mathbf{x}} \end{bmatrix}. \tag{10.46}$$

Example 10.6

The mass-spring-damper from the previous examples has a single output, $x_2(t)$. Let us assume that the displacement $x_2(t)$ is the the only measurable output and that the states $\mathbf{x} = \begin{bmatrix} x_1 & p_1 & \delta_2 & p_2 \end{bmatrix}^T$ must be estimated to implement the state feedback. Design a state observer that is at least an order of magnitude faster than the dominant closed-loop system.

Solution. If we use the pole placement solution from Example 10.3, the chosen closed-loop poles were $\lambda_{1,2} = -2 \pm j4$ and $\lambda_{3,4} = -0.4 \pm j0.7$. The eigenvalues $\lambda_{3,4}$ are closest to the imaginary axis and will have associated dynamics that take the longest to die out. In contrast, we choose the other set of poles, $\lambda_{1,2}$, so they would die out five times faster. Let us choose the closed-loop poles of our observer so that they have corresponding dynamics that are at least ten times faster than the dominant poles, $\lambda_{3,4}$. Recall that the settling time for a second-order system is $4/(\zeta\omega_n)$ and that for a pair of underdamped complex poles,

$$s = -\zeta\omega_n \pm j\omega_d = -\zeta\omega_n \pm j\omega_n\sqrt{1 - \zeta^2},$$

the natural frequency is (from Equation 7.29)

$$\omega_n = \sqrt{(\zeta\omega_n)^2 + \omega_d^2} = \sqrt{(\zeta\omega_n^2) + \omega_n^2(1 - \zeta^2)}.$$

If we multiply the closed-loop poles by ten, the resultant natural frequency will be ten times more. We can, therefore, use

$$p_{1,2} = 10\lambda_{1,2} = -20 \pm j40$$
$$p_{3,4} = 10\lambda_{3,4} = -4 \pm j7$$

as our desired closed-loop poles for the observer design. The desired characteristic equation is

$$(s+20+j40)(s+20-j40)(s+4+j7)(s+4-j7)$$
$$= (s^2 + 40s + 2000)(s^2 + 8s + 65)$$
$$= s^4 + 48s^3 + 2,385s^2 + 18,600s + 130,000$$
$$= s^4 + \varphi_1 s^3 + \varphi_2 s^2 + \varphi_3 s + \varphi_4.$$

The observability is

$$\mathcal{O} = \begin{bmatrix} \mathbf{C} \\ \mathbf{CA} \\ \mathbf{CA}^2 \\ \mathbf{CA}^3 \end{bmatrix}$$

$$= \begin{bmatrix} 1 & 0 & 1 & 0 \\ 0 & 0 & 0 & 0.1 \\ 0 & 0 & -6 & 0 \\ 0 & 0.6 & 0 & -0.6 \end{bmatrix}.$$

Thus the observer gain matrix is

$$\mathbf{L} = \begin{bmatrix} \mathbf{A}^4 + 48\mathbf{A}^3 + 2,385\mathbf{A}^2 + 18,600\mathbf{A} + 130,000\mathbf{I} \end{bmatrix} \dots$$

$$\begin{bmatrix} 1 & 0 & 1 & 0 \\ 0 & 0 & 0 & 0.1 \\ 0 & 0 & -6 & 0 \\ 0 & 0.6 & 0 & -0.6 \end{bmatrix}^{-1} \begin{bmatrix} 0 \\ 0 \\ 0 \\ 1 \end{bmatrix}$$

$$= \begin{bmatrix} 2,250 \\ 126,150 \\ -2,200 \\ 22,750 \end{bmatrix}$$

where **A** has been previously defined. The matrix can be readily calculated with the aid of MATLAB.

```
>> m = 10; b = 20; k = 60;
>> A = [0 1/m 0 0; -k -b/m k 0; 0 -1/m 0 1/m; 0 0 -k 0];
>> C = [1 0 1 0];
>> OBSRV = [C;C*A;C*A^2;C*A^3];
>> L = (A^4+48*A^3+2385*A^2+18600*A+130000*eye(4))...
*inv(OBSRV)*[0 0 0 1]'

L =

   1.0e+05 *

   0.0225
   1.2615
  -0.0220
   0.2275
```

10.8 State-Space Control Design Using MATLAB

Thus far, we have utilized basic MATLAB commands to solve the examples in this chapter. However, the MATLAB Control Systems Toolbox includes a series of commands that facilitate state-feedback controller design. Table 10.1 provides a list of some of the more commonly used functions in the toolbox.

The controllability matrix can be computed using the ctrb function, and then the rank can be readily checked with rank as was seen in previous examples to determine system controllability.

```
>> CTRL = ctrb(A,B);
>> rank(CTRL)
```

Note that this is state and not output controllability. Similarly, the observability matrix can be computed and checked using obsv and rank.

```
>> OBSRV = obsv(A,C);
>> rank(OBSRV)
```

It should be noted that the MATLAB functions presented herein are not compatible with symbolic matrices. The matrices must be passed as arrays with numbers and not parameters.

There are also several commands for computing controller gains through pole placement. The standard pole placement can be numerically calculated with the aid of place.

```
>> K = place(A,B,E)
```

The command returns the gain matrix when provided the arrays A and B, which represent the state matrices **A** and **B**, and a polynomial array E, which contains the desired closed-loop poles λ_1, λ_w, ..., λ_n (recall Equation 10.25). To compute the gain matrix using Ackermann's formula instead, use the acker command instead; remember, however, that this is restricted to SISO systems.

```
>> K = acker(A,B,E)
```

The MATLAB Control Systems Toolbox includes commands that can aid in the computation of a LQR design. Amongst those are care and lqr. The command care computes the solution to the continous-time algebraic Riccati equation. It also returns the gain matrix and a vector of closed-loop eigenvalues.

```
>> [P,E,K] = care(A,B,Q,R)
```

The inputs include arrays that represent **A**, **B**, **Q**, and **R**. The outputs are an array of the solution to the algebraic Riccati equation (P), an array with the closed-loop eigenvalues (E), and the gain matrix in array form (K). The lqr command functions similarly except that a state-space system object is passed as an input, and the outputs are returned in a different order than the care command.

```
>> [K,P,E] = lqr(sys,Q,R)
```

Both commands have additional, optional inputs to solve more generalized cases including, for example, the finite horizon case.

Example 10.7

Using the MATLAB functions introduced in this section, check the state controllability and observability. Also confirm the gain matrices derived using pole placement, Ackermann's formula, and the LQR method. Check the observer solution and simulate the closed-loop response with the observer.

Solution. First we specify all the necessary arrays for the state-space representation.

Table 10.1: *MATLAB command for state-feedback control design.*

Function	Description
ctrb(A,B)	computes the controllability matrix
obsv(A,C)	computes the observability matrix
place(A,B,E)	computes closed-loop pole placement
acker(A,B,E)	computes pole placement using Ackermann's formula
care(A,B,Q,R)	computes solution to algebraic Riccati equation
lqr(sys,Q,R)	computes linear-quadratic regulator

```
>> m = 10; b = 20; k = 60;
>> A = [0 1/m 0 0; -k -b/m k 0; 0 -1/m 0 1/m; 0 0 -k 0];
>> B = [0 0 0 1]';
>> C = [1 0 1 0];
>> D = 0;
```

Now we calculate the controllability and observability matrices and check their ranks.

```
>> CTRL = ctrb(A,B)

CTRL =

         0         0         0    0.6000
         0         0    6.0000  -12.0000
         0    0.1000         0   -1.2000
    1.0000         0   -6.0000         0

>> rank(CTRL)

ans =

     4

>> OBSRV = obsv(A,C)

OBSRV =

    1.0000         0    1.0000         0
         0         0         0    0.1000
         0         0   -6.0000         0
         0    0.6000         0   -0.6000
```

```
>> rank(OBSRV)

ans =

     4
```

The system is both state controllable and observable. We must define an array with the desired closed-loop poles to compute the gain matrix using both pole placement and Ackermann's formula.

```
>> E = [-0.4+0.7*i -0.4-0.7*i -2+4*i -2-4*i];
```

The solutions garnered using place and acker functions are

```
>> K = place(A,B,L)

K =

  -40.8333    -4.5833     2.5000     2.8000

>> K = acker(A,B,E)

K =

  -40.8333    -4.5833     2.5000     2.8000
```

which match the solutions detailed in Examples 10.3 and 10.4. To compute the LQR solution, we specify **Q** and R as arrays and define a state-space object with the system model.

```
>> sys = ss(A,B,C,D);
>> Q = [10 0 0 0; 0 1 0 0; 0 0 10 0; 0 0 0 1];
>> R = 1.0;
>> [K,P,E] = lqr(sys,Q,R)

K =

    2.9715     0.1109    -2.8051     0.6626


P =

  322.8731     0.0098    86.2499     2.9715
    0.0098     0.3886    -2.8228     0.1109
   86.2499    -2.8228   312.3883    -2.8051
    2.9715     0.1109    -2.8051     0.6626
```

```
E =

  -0.7864 + 3.7803i
  -0.7864 - 3.7803i
  -0.5449 + 1.4575i
  -0.5449 - 1.4575i
```

Notice that the resultant gain matrix matches the solution described in Example 10.5. The observer can be computed using the place or acker functions. However, according to MATLAB documentation, to use these functions to calculate the observer gain matrix the appropriate syntax is L = place(A',C',Eo)'. Notice that the **A** and **C** matrices are transposed as is the result.

```
>> L = place(A',C',Eo)'

L =

   1.0e+05 *

    0.0225
    1.2615
   -0.0220
    0.2275
```

This is the result reported in Example 10.6. We utilize Equations 10.45 and 10.46 to define the closed-loop system with the observer.

```
>> Ahat = [A -B*K; L*C A-B*K-L*C];
>> Bhat = [B*Nbar; B*Nbar];
>> Chat = [C zeros(1,4)];
>> Dhat = 0;
>> sys_cl_w_obsv = ss(Ahat,Bhat,Chat,Dhat);
```

If we wish to plot the response, $x_2(t)$, and controlled input, $u(t)$, we have to recreate the control input array recalling that $u(t) = -\mathbf{K}x + \overline{N}y_d$ where y_d is the desired position of the second mass. For the unit step response, the desired position would be one meter to the right. First we generate the response using the step command.

```
>> t = (0:0.01:14)';
>> n = length(t);
>> [Y,T,X] = step(sys_cl_w_obsv,t);
```

The time array was specified to facilitate generating the controlled input array. The outputs from the step command are column arrays where each element in each column corresponds to a specific time in the array T. To calculate the input, we must multiply the gain matrix by the state vector at each time step. To do so, we must transpose the state-vector array X. This, however, will generate a row array. After we multiply the negative of the gain matrix array -K by each column of the transposed row array X' we need to transpose the result to convert it back to a column array where the elements represent $-\mathbf{Kx}$ at each time step. To each of these elements we add $\overline{N}y_d$ where y_d is simply 1 for a unit step response.

```
>> U = -(K*X(:,1:4)')'+Nbar*1;
>> [AX,H1,H2] = plotyy(T,Y,T,U);
>> set(get(AX(1),'YLabel'),'String','x_2 (cm)')
>> set(get(AX(2),'YLabel'),'String','F (N)')
>> xlabel('Time (seconds)')
```

The response and controlled input are plotted on a two-axis plot with independent scales using the plotyy command. The individual vertical-axis titles are specified using the set command. The resultant step response plotted in Figure 10.5 is indistinguishable from the response in Figure 10.2. For all intents and purposes, the response of the observer is instantaneous and the loss of performance is negligible.

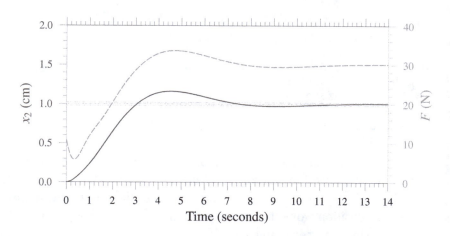

Figure 10.5: *Step and control responses for mass-spring-damper system using pole placement and observer.*

10.9 MIMO Control

Though the problem examined thus far is SISO, the methods presented are generally applicable to MIMO systems. The equations and processes have, for the most part, been generalized to include MIMO control systems. The addition of control inputs does not change A. Therefore, the eigenvalues and pole placement do not change. Additional control inputs increase the flexibility with which we can place the poles as desired.

Imagine that instead of a single force, the system has two forces, one driving each mass as depicted in Figure 10.6. The intent is to control the position of each mass independently. Therefore, the two displacements are outputs. The following example will illustrate.

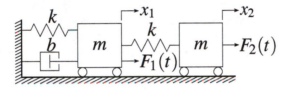

Figure 10.6: *MIMO mass-spring-damper systems.*

Example 10.8

Regarding the mass-spring-damper system in Figure 10.6, design a feedback controller to position the first mass 10 cm left of equilibrium and the second mass 20 cm right of equilibrium. Limit the overshoot to 20% and the settling time to under 10 seconds.

Solution. It can be shown that the state-space representation for MIMO system is

$$\begin{bmatrix} \dot{x}_1 \\ \dot{p}_1 \\ \dot{\delta}_2 \\ \dot{p}_2 \end{bmatrix} = \begin{bmatrix} 0 & 1/m & 0 & 0 \\ -k & -b/m & k & 0 \\ 0 & -1/m & 0 & 1/m \\ 0 & 0 & -k & 0 \end{bmatrix} \begin{bmatrix} x_1 \\ p_1 \\ \delta_2 \\ p_2 \end{bmatrix} + \begin{bmatrix} 0 & 0 \\ 1 & 0 \\ 0 & 0 \\ 0 & 1 \end{bmatrix} \begin{bmatrix} F_1(t) \\ F_2(t) \end{bmatrix}$$

$$\begin{bmatrix} x_1 \\ x_2 \end{bmatrix} = \begin{bmatrix} 1 & 0 & 0 & 0 \\ 1 & 0 & 1 & 0 \end{bmatrix} \begin{bmatrix} x_1 \\ p_1 \\ \delta_2 \\ p_2 \end{bmatrix} + \begin{bmatrix} 0 & 0 \\ 0 & 0 \end{bmatrix} \begin{bmatrix} F_1(t) \\ F_2(t) \end{bmatrix}.$$

Though the system now has two inputs, the desired close-loop poles remain the same because A has not changed. The difference is that now there are two inputs,

$$\mathbf{u} = \begin{bmatrix} F_1 \\ F_2 \end{bmatrix} = -\mathbf{Kx} + \overline{\mathbf{N}}\mathbf{y}_d \,,$$

that can be used to place the poles. We specify the same poles as before, but now when we compute the gain matrix it will be 2×4.

```
>> A = [0 1/m 0 0; -k -b/m k 0; 0 -1/m 0 1/m; 0 0 -k 0];
>> B = [0 0; 1 0; 0 0; 0 1];
>> E = [-0.4+0.7*i -0.4-0.7*i -2+4*i -2-4*i];
>> K = place(A,B,E)

K =

  -29.0838     0.3322    58.2351    -3.2412
   41.8022     3.3507   -20.3372     2.4678
```

To simulate the closed-loop response we need to compute $\overline{\mathbf{N}} = \mathbf{N}_u + \mathbf{K}\mathbf{N}_x$:

$$\begin{bmatrix} \mathbf{N_x} \\ \mathbf{N_u} \end{bmatrix} = \begin{bmatrix} \mathbf{A} & \mathbf{B} \\ \mathbf{C} & \mathbf{D} \end{bmatrix}^{-1} \begin{bmatrix} \mathbf{0} \\ \mathbf{1} \end{bmatrix}$$

$$= \begin{bmatrix} 0 & 1/m & 0 & 0 & 0 & 0 \\ -k & -b/m & k & 0 & 1 & 0 \\ 0 & -1/m & 0 & 1/m & 0 & 0 \\ 0 & 0 & -k & 0 & 0 & 1 \\ \hline 1 & 0 & 0 & 0 & 0 & 0 \\ 1 & 0 & 1 & 0 & 0 & 0 \end{bmatrix}^{-1} \begin{bmatrix} 0 & 0 \\ 0 & 0 \\ 0 & 0 \\ 0 & 0 \\ \hline 1 & 0 \\ 0 & 1 \end{bmatrix}$$

$$= \begin{bmatrix} 0 & 0.1 & 0 & 0 & 0 & 0 \\ -60 & -2 & 60 & 0 & 1 & 0 \\ 0 & -0.1 & 0 & 0.1 & 0 & 0 \\ 0 & 0 & -60 & 0 & 0 & 1 \\ \hline 1 & 0 & 0 & 0 & 0 & 0 \\ 1 & 0 & 1 & 0 & 0 & 0 \end{bmatrix}^{-1} \begin{bmatrix} 0 & 0 \\ 0 & 0 \\ 0 & 0 \\ 0 & 0 \\ \hline 1 & 0 \\ 0 & 1 \end{bmatrix} = \begin{bmatrix} 1 & 0 \\ 0 & 0 \\ -1 & 1 \\ 0 & 0 \\ \hline 120 & -60 \\ -60 & 60 \end{bmatrix}$$

$$\overline{\mathbf{N}} = \begin{bmatrix} 120 & -60 \\ -60 & 60 \end{bmatrix} + \begin{bmatrix} -29.0838 & 0.3322 & 58.2351 & -3.2412 \\ 41.8022 & 3.3507 & -20.3372 & 2.4678 \end{bmatrix} \begin{bmatrix} 0 & 0 \\ 0 & 0 \\ 0 & 0 \\ 0 & 0 \end{bmatrix}$$

$$= \begin{bmatrix} 32.6811 & -1.7649 \\ 2.1395 & 39.6628 \end{bmatrix} \,.$$

```
>> N = inv([A B; C D])*[zeros(4,2);eye(2)];
>> Nx = N(1:4,:);
>> Nu = N(5:6,:);
>> Nbar = Nu+K*Nx

Nbar =

   32.6811    -1.7649
    2.1395    39.6628
```

By substituting the control into the state-space representation, we can derive the closed-loop system in terms of the desired position vector

$$\dot{\mathbf{x}} = \mathbf{Ax} + \mathbf{Bu}$$
$$= \mathbf{Ax} + \mathbf{B}(-\mathbf{Kx} + \overline{\mathbf{N}}\mathbf{y}_d)$$
$$= (\mathbf{A} - \mathbf{BK})\mathbf{x} + \mathbf{B}\overline{\mathbf{N}}\mathbf{y}_d$$
$$= \overline{\mathbf{A}}\mathbf{x} + \overline{\mathbf{B}}\mathbf{y}_d$$
$$\mathbf{y} = \mathbf{Cx} + \mathbf{Du}$$
$$= \mathbf{Cx} + \mathbf{D}(-\mathbf{Kx} + \overline{\mathbf{N}}\mathbf{y}_d)$$
$$= (\mathbf{C} - \mathbf{DK})\mathbf{x} + \mathbf{D}\overline{\mathbf{N}}\mathbf{y}_d$$
$$= \overline{\mathbf{C}}\mathbf{x} + \overline{\mathbf{D}}\mathbf{y}_d.$$

Now we can define the closed-loop system.

```
>> Abar = A-B*K;
>> Bbar = B*Nbar;
>> Cbar = C-D*K;
>> Dbar = D*Nbar;
>> sys_cl = ss(Abar,Bbar,Cbar,Dbar);
```

Because we wish to simulate the response to two distinct forces that are neither unit step nor impulse functions, we will need to use the more generalized lsim command which requires that we define the two inputs as arrays. The inputs are constant values of -10 cm for the position of the first mass and 20 cm for the position of the second mass. Thus, we will specify a time array and for each element in that array, specify a corresponding element for the arrays representing the desired positions of the two masses.

```
>> t = (0:0.01:16)';
>> n = length(t);
```

```
>> x1d = -0.1*ones(n,1);
>> x2d = 0.2*ones(n,1);
```

Note that the ones command was used to create column arrays for each of the desired mass positions. The arrays are scaled by the desired displacements. The responses can be readily generated using the lsim function.

```
>> lsim(sys_cl,[x1d x2d],t)
```

However, if we wish a more detailed view of the results, we can plot the responses and inputs on dual axis plots like was done in the previous example. The closed-loop system is defined in MATLAB.

```
>> Abar = A-B*K;
>> Bbar = B*Nbar;
>> Cbar = C-D*K;
>> Dbar = D*Nbar;
>> sys_cl = ss(Abar,Bbar,Cbar,Dbar);
```

The desired positions are specified and multiplied by an appropriated sized array of ones to generate the control inputs for the lsmim function.

```
>> x1d = -0.1;
>> x2d = 0.2;
>> t = (0:0.01:16)';
>> n = length(t);
>> [Y,T,X] = lsim(sys_cl,[x1d*ones(n,1) x2d*ones(n,1)],t);
>> x1 = Y(:,1);
>> x2 = Y(:,2);
```

We confront the same numerical issue when composing the control vectors as we did with the previous example. Because there are two control inputs, the gain matrix has two rows. Hence, we recreate the control input arrays separately using the individual rows of the gain matrix and recall that $\mathbf{u}(t) = -\mathbf{K}\mathbf{x} + \overline{\mathbf{N}}\mathbf{y}_d$.

```
>> Yd = [x1d;x2d];
>> u1 = -(K(1,:)*X')'+Nbar(1,:)*Yd;
>> u2 = -(K(2,:)*X')'+Nbar(2,:)*Yd;
```

We are now ready to plot the results.

```
>> subplot(2,1,1)
>> [AXa,H1a,H2a] = plotyy(T,100*x1,T,u1)
```

```
>> set(get(AXa(1),'YLabel'),'String','x_1 (cm)')
>> set(get(AXa(2),'YLabel'),'String','F_1 (N)')
>> subplot(2,1,2)
>> [AXb,H1b,H2b] = plotyy(T,100*x2,T,u2)
>> set(get(AXb(1),'YLabel'),'String','x_2 (cm)')
>> set(get(AXb(2),'YLabel'),'String','F_2 (N)')
>> xlabel('Time (seconds)')
```

The output and control responses are plotted in Figure 10.7. The process
we employed to implement this MIMO control system is quite similar to
that which we used for the SISO system.

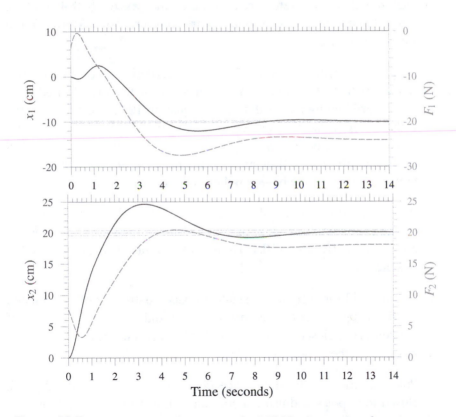

Figure 10.7: *Step and control responses for MIMO mass-spring-damper system.*

10.10 Summary

▷ Whereas classical approaches are primary implemented in the *s*-domain, modern control methods are implemented in the time domain using state-space models.

▷ State-space methods are more general and applicable to multi-input-multi-output (MIMO) systems, where the methods from the previous chapter are primarily used for single-input-single-output (SISO) systems.

▷ The control input is the difference between the reference and the feedback signal $\mathbf{Kx}$. The reference depends on the type of control problem. The reference input for set point or tracking problems is provided in Equation 10.7.

▷ A system is controllable if it is possible to derive a control that will transition the system from some initial condition to a final desired state. The controllability matrix defined in Equation 10.13 must be full rank if the control system is state controllable, meaning that, given the dynamic state vector signals, the system can be controlled. The output controllability matrix (Equation 10.14) must be full rank for the system to be controllable based on the available outputs.

▷ A system is observable if the state vector $\mathbf{x}(t)$ can be determined from measurement or observation of the available outputs $\mathbf{y}(t)$. The observability matrix (Equation 10.17) must have row rank equal to the number of states n.

▷ One method for designing a state-feedback control is pole placement. Unlike using the root locus method which only allows positioning of the dominant, closed-loop poles, pole placement in the state space can position *all* the closed-loop poles.

▷ Pole placement in the state space involves specifying all the desired, closed-loop poles and using a coordinate transformation to derive the gain matrix that will move the existing closed-loop poles to the desired positions.

▷ Several methods exist for placing the closed-loop poles in the state space including Ackermann's formula (Equations 10.29 and 10.30), which is only applicable to SISO systems.

▷ Numerous methods exist to optimize the feedback gain matrix. One such method, the linear quadratic regulator, optimizes based on a cost function or performance index that can minimize error and control effort.

▷ Sometimes it is physically impossible or impractical to measure sufficient outputs to recreate all the system states. In such cases, state observers are used to estimate the missing states from the measured outputs and the supplied inputs. The state observers are designed in much the same manner as the feedback control. However, to ensure appropriate performance, the state observer must respond significantly faster than the dynamics of the system. The closed-loop poles of the state observer are placed to achieve faster transient performance than the overall system.

▷ MATLAB includes a series of functions summarized in Table 10.1 for checking the controllability and observability of a state-space system and for designing a controller through pole placement or linear quadratic regulation.

10.11   Review

R10-1 What advantages, if any, do state-space control methods have over the traditional methods discussed in the previous chapter?

R10-2 What is the form of the control input for a set point or tracking problem?

R10-3 What is the requirement for a system to be state controllable? What does it mean to be controllable?

R10-4 What is the requirement for a system to be output controllable?

R10-5 Describe the difference between state controllable and output controllable?

R10-6 How do you determine whether a system is observable? What does it mean that the system is observable?

R10-7 Describe the pole placement process in your own words.

R10-8 When can you use Ackermann's formula?

R10-9 Describe what a linear quadratic regulator is and explain basically how it works. (You do not need to provide mathematical details.)

R10-10 What is a state observer and why is it used?

R10-11 How does the design of a state observer differ from the design of a state-feedback control?

R10-12 What MATLAB functions can be used to facilitate designing controls in the state space?

10.12 Problems

P10-1 *Pole Placement of a State-Space Model.* Consider a system with the state-space representation

$$\dot{\mathbf{x}} = \begin{bmatrix} 0 & 1 & 0 \\ 0 & 0 & 1 \\ -1 & -3 & -4 \end{bmatrix} \mathbf{x} + \begin{bmatrix} 0 \\ 1 \\ 1 \end{bmatrix} u.$$

Design a state-feedback control to place the desired closed-loop poles at $s = -1 \pm j2$ and $s = -12$. Determine the gain matrix $\mathbf{K}$ and plot the step response if $y = x_1(t)$.

P10-2 *Controllability of a State-Space Model.* Show that the system

$$\begin{bmatrix} \dot{x}_1 \\ \dot{x}_2 \end{bmatrix} = \begin{bmatrix} 0 & 2 \\ 0 & 3 \end{bmatrix} \begin{bmatrix} 1 \\ 0 \end{bmatrix} u$$

cannot be controlled regardless of the state feedback.

P10-3 *Controllability and Observability of a Mass-Spring-Damper System.* Check the state controllability, output controllability, and observability of the mass-spring-damper system depicted in Figure 1.15 (b). The input is the velocity $v(t)$ and the outputs are $x_1(t)$ and $v_2(t)$. Assume that $m = 100$ kg, $k = 50$ N/m, $b_1 = 25$ N-s/m, and $b_2 = 75$ N-s/m.

P10-4 *Controllability and Observability of a Torsion Plant.* Determine the state controllability, output controllability, and observability of the rotational system depicted in Figure 7.17. Recall that the system parameters were provided in Table 7.1. The input is the torque and the outputs are the angular displacements of the two rotational inertias.

P10-5 *Pole Placement of a Mass-Spring-Damper System.* Design a state feedback control using the pole placement method for the mass-spring-
damper system in Problem P10-3 to position the second mass. Select
your desired closed-loop poles. Design a separate control using Ackermann's formula. Compare your designs. Plot the step responses for
both. Calculate the designs and then compare with those attained using
MATLAB.

P10-6 *Pole Placement of a Torsion Plant.* Using the pole placement method,
design a state-feedback control to attenuate the vibration of the torsion
plant in Problem P10-4 to position the second rotational inertia. Select
your desired closed-loop poles. Also, design a separate control using
Ackermann's formula. Compare your designs, and plot the impulse
responses for both. Calculate the design and then compare with results
attained using MATLAB.

P10-7 *Linear Quadratic Regulation of a State-Space Model.* Consider the
system

$$\begin{bmatrix} \dot{x}_1 \\ \dot{x}_2 \end{bmatrix} = \begin{bmatrix} 0 & 2 \\ 0 & 0 \end{bmatrix} \begin{bmatrix} x_1 \\ x_2 \end{bmatrix} + \begin{bmatrix} 0 \\ 1 \end{bmatrix} u.$$

Derive a linear quadratic regulator that minimizes the performance index

$$J = \int_0^\infty (\mathbf{x}^T \mathbf{Q} \mathbf{x} + 2u^2) \, dt$$

where

$$\mathbf{Q} = \begin{bmatrix} 1 & 0 \\ 0 & 1 \end{bmatrix}.$$

The output is $x_1(t)$. Plot the step responses.

P10-8 *Linear Quadratic Regulation of a Mass-Spring-Damper System.* Synthesize a linear quadratic regulator for the mass-spring-damper system
in Problem P10-3. Assume, however, that the system parameters are
$m = 1$ kg, $k = 5$ N/m, $b_1 = 0.01$ N-s/m, and $b_2 = 0.02$ N-s/m. Make
sure to minimize the position error for the displacement of the second
mass and reasonably limit the control effort. Be sure to justify the performance index used to derive the gain matrix. Compare your results
with those attained through pole placement. Provide a plot of the step
response.

P10-9 *Linear Quadratic Regulation of a Torsion Plant.* Design a regulator to
attenuate the vibration of the torsion plant in Problem P10-4. Make

sure to minimize the position error for the angular displacement of the second rotational inertia and reasonably limit the control effort. Be sure to justify the performance index used to derive the gain matrix. Compare your results with those attained through pole placement. Provide a plot of the impulse response.

P10-10 *State Observer for a State-Space Model.* Consider the system

$$\dot{x} = \begin{bmatrix} -2 & 1 \\ 1 & -3 \end{bmatrix} x$$

$$y = \begin{bmatrix} 1 & 0 \end{bmatrix} x.$$

Design a state observer with dynamics that are sufficiently faster than that of the system.

P10-11 *State Observer for a Torsion Plant.* Assume that the only measurable output for the system in Problem P10-4 is the angular velocity of the upper rotational inertia. Design a state observer that is at least an order of magnitude faster than the ideal closed-loop system.

P10-12 *MIMO Control of a Torsion Plant.* Assume that the torsion plant from Problem P10-4 has two torque inputs instead of one – one to actuate the bottom inertia and another to actuate the top. Design a MIMO control system to control the positions of the rotational inertias. Use a method of your choosing with appropriate design specifications. Plot the angular displacements of the two inertias for specified displacements of your choosing.

10.13 Challenges

C10-1 *A Toy Electric Car.* Recall the cruise control you designed in the previous chapter. Design a smart state-feedback cruise control for the toy car. Specify your design criteria. Use a method of your choosing, and be sure to test the compensated response using an appropriate input. Compare with your design from Chapter 9.

C10-2 *A Quarter-Car Suspension Model.* Recollect the active suspension you designed in Chapter 9. Design a state-feedback control to improve the performance of your suspension. You can use a method of your choosing. Specify your design criteria and use a variety of roadway inputs to test your design. Compare with the compensator from the previous chapter.

Bibliography

Armstrong, Brian, and C. C. de Wit. 1995. The control handbook. Chap. Friction Modeling and Compensation in. Boca Raton, FL: CRC Press.

Beaman, J., and H. Paynter. 1993. Modeling of physical systems.

Bennet, Stuart. 1996. A brief history of automatic control. *IEEE Control Systems* 16 (3): 17–25.

Chua, L. O. 1971. Memristor - the missing circuit element. *IEEE Transactions on Circuit Theory* CT18 (5): 507–519.

Commons, Wikimedia. 2009. *Main page — Wikimedia commons,* [Online; accessed 9-January-2009].

Crowell, B. 2010. Light and matter. Chap. Vibrations and Waves in. Creative Commons Attribution-ShareAlike.

Dorf, R. C., and R. H. Bishop. 2001. *Modern control systems.* Englewood Cliffs, NJ: Prentice Hall, Inc.

Evans, W. R. 1948. Graphical analysis of control systems. *Transactions of the AIEE* 67 (1): 547–551.

———. 1950. Control systems synthesis by root locus method. *Transactions of the AIEE* 69 (1): 66–69.

Fox, R. W., and A. T. McDonald. 1992. *Introduction to fluid mechanics.* New York, NY: John Wiley & Sons, Inc.

Franklin, G. F., J. D. Powell, and A. Emami-Naeini. 1994. *Feedback control of dynamic systems.* Addison Wesley Publishing Company, Inc.

Harlow, J. H. 2004. *Electric power transfomer engineering.* Boca Raton, FL: CRC Press.

Kalman, R. E. 1960. Contributions to the theory of optimal control. *Boletin de la Sociedad Matematica Mexicana* 5:102–119.

Karnopp, D. C., D. L. Margolis, and R. C. Rosenberg. 2000. *System dynamics: modeling of mechatronic systems*. 3rd. New York, NY: Wiley-Interscience.

Kwakernaak, H., and R. Sivan. 1972. *Linear optimal control systems*. New York, NY: Wiley-Interscience.

Nise, N. S. 2011. *Control systems engineering*. New York, NY: John Wiley & Sons, Inc.

Ogata, K. 2002. *Modern control engineering*. Englewood Cliffs, NJ: Prentice Hall.

———. 2004. *System dynamics*. 4th. Upper Saddle River, NJ: Pearson-Prentice Hall.

Paynter, H. P. 1961. *Analysis and design of engineering systems*. Cambridge, MA: MIT Press.

Rugh, W. J., and J. S. Shamma. 2000. Research on gain scheduling. *Automatica* 36:1401–1425.

Schwarz, S. E., and W. G. Oldham. 1993. *Electrical engineering: an introduction*. Philadelphia, PA: Saunders College Publishing.

Tooley, M. H. 2006. *Electronic circuits: fundamentals and applications*. Bulrlington, MA: Newnes.

Wikipedia. 2009a. *Hero of Alexandria — Wikipedia, the free encyclopedia*. [Online; accessed 9-January-2009].

———. 2009b. *Space race — Wikipedia, the free encyclopedia*. [Online; accessed 16-June-2009].

———. 2009c. *Water clock — Wikipedia, the free encyclopedia*. [Online; accessed 9-January-2009].

Ziegler, J. G., and N. B. Nichols. 1942. Optimum settings for automatic controllers. *Transactions of the ASME* 64:759–768.

Index

CPSIA information can be obtained
at www.ICGtesting.com
Printed in the USA
LVHW051502190122
708829LV00005B/281

9 781466 560758